T0176033

Processes of Formation of Micro- and Nanodispersed Systems

Processes of Formation of Micro- and Nanodispersed Systems

A. A. Bochkarev
V. I. Polyakova

CRC Press
Taylor & Francis Group
Boca Raton London New York

CISP

CRC Press is an imprint of the
Taylor & Francis Group, an **informa** business

CRC Press
Taylor & Francis Group
6000 Broken Sound Parkway NW, Suite 300
Boca Raton, FL 33487-2742

First issued in paperback 2020

© 2016 by Taylor & Francis Group, LLC
CRC Press is an imprint of Taylor & Francis Group, an Informa business

No claim to original U.S. Government works

ISBN 13: 978-0-367-57534-2 (pbk)
ISBN 13: 978-1-4822-5145-6 (hbk)

This book contains information obtained from authentic and highly regarded sources. Reasonable efforts have been made to publish reliable data and information, but the author and publisher cannot assume responsibility for the validity of all materials or the consequences of their use. The authors and publishers have attempted to trace the copyright holders of all material reproduced in this publication and apologize to copyright holders if permission to publish in this form has not been obtained. If any copyright material has not been acknowledged please write and let us know so we may rectify in any future reprint.

Except as permitted under U.S. Copyright Law, no part of this book may be reprinted, reproduced, transmitted, or utilized in any form by any electronic, mechanical, or other means, now known or hereafter invented, including photocopying, microfilming, and recording, or in any information storage or retrieval system, without written permission from the publishers.

For permission to photocopy or use material electronically from this work, please access www.copyright.com (http://www.copyright.com/) or contact the Copyright Clearance Center, Inc. (CCC), 222 Rosewood Drive, Danvers, MA 01923, 978-750-8400. CCC is a not-for-profit organization that provides licenses and registration for a variety of users. For organizations that have been granted a photocopy license by the CCC, a separate system of payment has been arranged.

Trademark Notice: Product or corporate names may be trademarks or registered trademarks, and are used only for identification and explanation without intent to infringe.

Visit the Taylor & Francis Web site at
http://www.taylorandfrancis.com

and the CRC Press Web site at
http://www.crcpress.com

Contents

3, The formation of disperse systems on a solid substrate in heterogeneous condensation   77

4. The formation of disperse systems by homogeneous condensation with vapour jet expansion   130

4.1. Methods of calculating the start of homogeneous condensation of

5. 'Genetics' and the evolution of the structure of vacuum condensates   179

6. The formation of columnar structures in co-condensation of two components as a way of producing nanosized composites 214

Contents

8. Numerical simulation of molecular processes on the substrate   299

9. Methods for direct conversion of kinetic energy to surface energy of the disperse system   335

10. Production of metallic powders of the required composition by melting a rotating rod   367

11. Some aspects of coalescence processes in aerosol systems   393

12. Capture of microparticles from gas flows by condensation processes   416

Introduction

The disperse system is a set of individual elements of material with the characteristic dimensions considerably smaller than the dimensions of the system as a whole. The disperse medium is the medium consisting of the individual elements of the material of one phase, separated from each other by the material of the other phase. The disperse system or medium is microdisperse if the total surface area of its disperse elements influences the macroscopic properties of the material, forming the system or medium. The microdisperse system or medium is nanodisperse if the elements forming it are so small that their molecular properties greatly influence the properties of the system or medium as a whole and even lead to the formation of new properties which the homogeneous matter does not have. The new properties of the disperse systems make it possible to develop and create new technologies which are referred to as nanotechnologies. The process as a result of which the material is transformed from the homogeneous to the disperse state is referred to as dispersion. As a result of dispersion the individual elements of the matter of the produced disperse system are separated by the interlayers of a foreign medium. In the process of refining the matter the increase of the number N of the individual elements with the size r in the closed system the total area of the boundaries of the matter S with the foreign medium increase in inverse proportion to the size of the elements $S \sim r^{-1}$ and in proportion to the cubic root of the number of the elements $S \sim N^{1/3}$. The dependence of the area of the interface of the phases of the disperse system has a strong effect on its properties and, secondly, determines the processes of dispersion in the area in which the total surface energy of the interfaces becomes high on the background of the energy of the system as a whole. The main physical content of the dispersion processes, leading to the formation of microdisperse systems, is the development of the specific interface of the phases and supplying the surface energy to the system. The absence of the direct path of complete transition of some types of energy to surface energy results in the occurrence of a large number of phenomena accompanying the processes of formation of microdisperse systems.

The variability of the paths leading to the formation of the microdisperse systems and the existence of a large number of accompanying processes and phenomena are discussed in the first chapter of the book. The processes of dispersion extend outside the framework of a specific science and all the problems cannot be sufficiently clarified in a single book. Consequently, it is necessary to choose a specific direction of the investigations. In the present book, the selection was motivated by the tendency to use the restricted volume of the investigations to confirm the procedural recommendations for the development of new and improvement of the available technologies of producing materials in the dispersed state [1]. This approach has greatly reduced the number of scientific directions which should be analyzed, and has made it possible to formulate the specific targets and results for each of them.

The methodological analysis of the currently available methods of dispersion, presented in the first chapter, shows that the processes with the most advantageous energy parameters in different ranges of the characteristic dimensions of the disperse particles are the processes of condensation and disintegration of the films. This determined the main scientific directions of long-term work of the authors. It is clear that in the range of the dimensions of the dispersed particles 10^{-9}–10^{-7} m and greater the process with the most advantageous energy parameters is the synthesis of the dispersed system by vapour phase condensation, and in the range 10–7–10^{-3} m it is the dispersion of the melt film, formed by the Taylor bell procedure. At the Third National Coordination Conference in 1984 dealing with the problems of disperse materials, the future prospects of the formation of ultrafine systems by condensation methods were reflected in the conclusions of the conference. The procedural approach to the planning of investigations has made it possible to determine efficiently the direction of investigations.

The synthesis of the disperse phase by vapour phase condensation starts with the formation of nanosized nuclei of the condensed phase. This physical phenomenon is referred to as nucleation which in most cases is of the spontaneous and controlled nature owing to the fact that it only starts when the supercritical super saturation of the vapour is reached at which the frequency of formation of the nuclei is very high. Nucleation has been studied by many investigators. Nevertheless, there are still a number of ion source problems associated with nucleation. To solve these problems, it is necessary to carry out investigations using complicated sensitive methods.

Chapter 2 describes the preparation for experimental studies of homogeneous and heterogeneous nucleation and vacuum condensation.

Preparation includes the selection of the configuration and conditions of condensation, the construction of a specialised vacuum chamber, the development of specialised vapour sources, including the sources of metal vapours [2–4], the characterisation of the vapour flow, investigation of the role of the residual atmosphere in the vacuum chamber and, finally, the selection of the material with the vapours of this material suitable for the physical simulation of condensation processes. In addition to these attributes of the experiments, chapter 2 also describes the original method developed in the investigations: calorimetric measurements of the process of heterogeneous nucleation [5], measurement of the heat content of the vapour in the jet, specific consumption of the metal vapour in the jet, the efficiency of condensation, determination of the orientation of the structures of vacuum condensates on a gradient substrate. Tests were also carried out to determine the possibilities of using electron beam diagnostics [6] for calculating the density of metal vapours on the example of magnesium vapours. In addition to these investigations, other methods, described in the literature have also been used. These methods are not described in detail in the present book because the authors did not intend the book to be a review of these investigations.

The results of preparation and continuing modernisation of the experimental base for the investigations of homogeneous and heterogeneous condensation of metal vapours in vacuum were used in the construction of Metagus-2 equipment which in fact is a unique gas-dynamic low-density pipe in which the superheated metal vapour is used as a working gas. The application of the metal vapours instead of the gas leads to an important advantage of this equipment. To produce the flows suitable for studying gas dynamics, this structure requires a powerful pumping vacuum system because the flow of the metal vapours condenses at room temperature so that it is not necessary to use expensive cryogenic coolers. The possibilities of this gas dynamic equipment are limited only by the power supplied to the vapour source.

Chapter 3 is concerned with the examination of heterogeneous condensation in the conditions in which it results in the formation of a disperse system on the surface of the solid body – substrate. The first fundamental problem described in this chapter is the occurrence of the mysterious so-called 'collapse of condensation' of the metal vapours, observed by L.S. Palatnik in the 50s of the previous century [7]. L.S. Palatnik and his colleagues found in the experiments the ranges of temperature and vapour flow density in which the heterogeneous condensation does not take place. The authors of the present book repeated some of these experiments using magnesium vapours and confirmed that the fields of 'collapse' of condensation is influenced by the residual pressure of the gas in the vacuum chamber [8]. It is now clear that the 'collapses' are caused by the harmful effect of the impurities on the condensation process. Therefore, the mathematical simulation of heterogeneous condensation was carried out [9]. Attention was given to three characteristic stages of heterogeneous condensation: adsorption, nucleation and the growth of nuclei. The simulation of the initial stage – adsorption – was carried out on the basis of the modified Langmuir linear single-layer model. This simple model was sufficient for qualitative understanding of the critical phenomena taking place during heterogeneous condensation. The atoms and/or molecules of the impurities occupied part of the adsorption vacancies and, consequently, reduce the concentration of the adsorbed atoms of the condensed metal vapour and inhibit nucleation. The superposition of the binding energies of the metal, impurities and their temperature dependences create a complicated pattern from the areas of possible condensation in the field with the temperature–density of the metal vapour flow coordinates. Some of the areas in this field permit condensation with the formation of disperse condensates with filling of the gaps between the particles of the impurity condensate in the capillary condensation mode. As a result of the interest in the formation of the disperse systems, the authors paid special attention to these areas.

The successful application of the Langmuir method led to its further modification. Models were constructed for describing the adsorption of the two-component gas mixture with a binary chemical reaction on the surface and without considering the reactions, with the transfer of the adsorbate to the centres of nucleation of the condensed phase. The models were used to describe the heterogeneous nucleation on the growth of nuclei in the presence of the impurities resulting in the classification of the chain mechanism of nucleation [10, 11] often observed in the experiments. It was found that this phenomenon is based on the fact that the formation of metal nuclei is accompanied by improvement of the conditions for the nucleation of capillary nuclei of the impurities which in turn creates more suitable conditions for the formation of the next metal nucleus. This consecutive nucleation of the nuclei of different components at the periphery of each other is the start of formation of the base of the columnar structures of the condensates. The metallic columns in these condensates are separated by the interlayers of the condensate of the impurities, forming a compact disperse system [12]. These disperse condensates can be easily transformed by milling to the ultrafine powder.

Further development of the sorption–desorption model resulted in the successful application of the model for describing surface ionisation phenomena [13–18].

Chapter 3 is also concerned with the investigation of the effect of the gas dynamic parameters of the vapour flows on the formation of disperse structures. It was noted in the experiments that the individual crystals of the condensates have a characteristic parachute form. It was found that the characteristic size of these crystals increases in proportion to the condensation time. The similarity of their form is retained and this means [19] that the shape of these crystals should be determined by the vapour flow parameters. Therefore, investigations were carried out to develop a theory of the normal growth of the surface of the crystals as a result of the local mass exchange determined by the structure of the vapour flow [20]. The theory led to the fundamental result: the shape of any single crystal of the vacuum condensate can be described by a universal formula in which the main parameter is the Mach number of the vapour flow. The geometrical interpretation of the shape consists of the surface of the moving and expanding sphere [21, 23].

The success of this approach and the 'gas dynamic' theory of the growth of crystals have been extended to describing the reasons for the orientation of columnar structures of the vacuum condensate with inclined incidence of the vapour flow on the substrate. The lateral interaction of the adjacent columns creates suitable conditions for their growth only in the direction characterised by the maximum growth rate predicted by the theory, taking into account the angle of incidence of the vapour flow on the substrate. This method is confirmed by comparison with the experimental data, whereas other theories of the orientation of the structure of the vacuum condensate have not been confirmed by experiments [23].

The 'gas dynamic theory' has also made it possible to describe the faceting of the tips of the columns and the appearance of stresses in the condensate with the columnar structure. These were fundamental problems in the area of heterogeneous vacuum condensation. The 'gas dynamic' theory of growth of the individual crystals explains the phenomenon of the 'starry sky' which results in rejection in the production of thin films for electronics applications. Any cluster, falling on the condensation front, develops in accordance with the law described by the 'gas dynamic' theory. This resulted in critical defects of the film which could not be repaired.

Chapter 4 describes the results of experimental studies of the homogeneous condensation processes, leading to the formation of aerosol disperse systems. The principal difference of these investigations from the studies of the gas phase processes, for example, studies by I.V. Frinberg, is that attention is given to the condensation of the pure vapour in the absence of gas-cooler. The supersaturation and nucleation of the clusters take place as a result of the expansion of the vapour in the low-pressure range. Difficulties in experimental studies of these processes have been indicated in the experimental procedures described in [24–26]. Chapters 3 and 4 the book are closely linked by the common features of the experimental methods and the methods of processing and analysis of the experimental data.

There is a large number of literature data obtained in studies of homogeneous nucleation. A review of the studies show that the majority of the principal problems have already been solved. However, all the results presented in the literature relate to homogeneous condensation in the vapour flows from open sources. This means that the researchers studied saturated metal vapours. In these conditions, the vapour flow always contains unstable aggregates of the molecules or atoms formed as a result of the fluctuations of inter-particle collisions. Cooling of these vapours results in condensation on their unstable aggregates and this removes the supersaturation and prevents nucleation in the classic mode. The authors of the book used vapour sources with superheating of the vapour which brings the investigations closer to the classic concept: supercooling as a result of the isoentropic expansion up to critical supersaturation when the formation of the nuclei of the condensed phase – clusters – takes place with the dimensions of the order of the size of the atom, the growth of the clusters as a result of vapour condensation on them with removal of the heat into the surrounding vapours. Therefore, in formulating the experiments, special attention was given to calculating the described classic procedure. Calculations carried out using the methods proposed by O. Hagen [27] showed that the majority of investigators worked in the range of the parameters in which homogeneous nucleation is not possible owing the fact that the inter-particle collisions disappeared before the critical supersaturation, sufficient for homogeneous nucleation, was reached. Nevertheless, the majority of the authors of the published studies detected clusters in their systems and investigated them. In particular, the well-known studies of Japanese investigators with cluster spraying were also carried out in the range of the parameters in which the homogeneous nucleation of the clusters cannot take place. This resulted in the fundamental problem of the origin of these 'unlawful' clusters.

To solve the problem, experimental equipment was fitted with additional measurement devices not perturbed by the vapour flow, in order to avoid uncontrollable nucleation. The methods, associated with ionisation of the flow or with scanning of the flow with high-intensity electron laser beams are not suitable for this purpose because there has been direct confirmation of the activation of condensation by high energy particles and electromagnetic radiation. Therefore, the conditions of the start of homogeneous condensation in the experiments was investigated only by the methods of measuring the flow rate, enthalpy, specific consumption of the vapour in the flow and the local density of the vapour atoms using a gas spectrometer. In this case, nucleation was recorded as the appearance of a 'jump' of the parameters when reducing the pressure of deceleration of the vapour flow.

On the whole, the experimental results confirm the classic assumptions regarding the nucleation of clusters in the jet expansion of the vapours. At the same time, several previously unknown effects have been detected. Firstly, the expansion of the metal vapours in the jet is more complicated because of the generation of the latent heat of condensation which in the case of metals is considerably larger than for the previously investigated gases. Secondly, the transfer of liquid metal to the outlet of the nozzle was detected as a result of the condensation of the vapour on the internal walls of the nozzle. The open evaporation of the liquid metal from the end of the nozzle results in considerable distortion of the experimental results. Thirdly, the results of deposition on the substrates, accommodated for electron microscopic studies, show that the clustered flow contains two fractions of the microparticles, with greatly differing dimensions. Repeated measurements by the methods of ionisation and analysis of the charged particles with respect to the energy confirm this. The latter fact indicates that evidently there are some other reasons for the formation of clusters in the vapour flows not associated with homogeneous condensation.

The model of heterogeneous nucleation was developed to find additional reasons for the formation of clusters in the vapour flows [28–31]. The results of investigations using this model show that not only the single atom is but also the clusters, consisting of several atoms and formed on the solid surface, are capable of desorbing to the vapour medium. The estimates show that the probability of such a process is not small because the frequency of heterogeneous nucleation is the presence of orders of magnitude higher than the frequency of homogeneous nucleation. The results of these estimates have been confirmed in the experiments using the probe of the delay potential which made it possible to record the departure of the charged particles from the collector also in the special experiments with nucleation on the target and with the trapping of clusters leaving the target. This means that in the experiments with clustering of the expanding vapour, in addition to the homogeneous mechanism, there is an additional heterogeneous mechanism of generation of the clusters by any solids surfaces present in equipment. This explains the experimentally detected formation of clusters at the parameters at which homogeneous nucleation cannot take place.

The results have also a significant applied importance. In the technologies of production of thin films the additional mechanism of formation of the clusters is the reason for the formation of dust with the nanosized particles which are the sources of production defects.

Chapter 4 discusses the effect of non-condensed impurities on the homogeneous nucleation formation process. If the jet of the expanding vapour contains non-condensed additions, the atoms of these impurities may occupy part of the adsorption vacancies on the surface of the pre-critical aggregates of the main condensed component and, consequently, lower the rate of homogeneous condensation [32, 33]. This phenomenon is described on the basis of the experimental data obtained by Chukanov [34] for the nucleation of heavy water vapours in the presence of helium as the cooling gas. Consequently, another 'mystery of the century' has been solved, as described by the authors of [34], who detected this phenomenon in the experiments.

The results of investigations of homogeneous and heterogeneous condensation have been published in [35].

As already mentioned, under approximately identical conditions the frequency of heterogeneous nucleation in the vapour medium is tens of orders of magnitude higher than the frequency of homogeneous nucleation. This may be explained by the fact that any surface, bordering with the vapour, is the collector of the adsorbate of the atoms or molecules of the vapour. The frequency of particle collisions and, consequently, the probability of formation of the pre-critical aggregates as a result of the fluctuations in the adsorbate is considerably higher than in the vapour phase. This greatly increases the frequency of nucleation in the nucleation. In addition, in heterogeneous nucleation the removal of the latent heat of condensation is facilitated by the presence of the substrate which accelerates the growth of the nuclei in comparison with the heterogeneous nucleation where the removal of heat is ensured only by the collisions with the third particles. This circumstance indicates unambiguously the fact that the heterogeneous nucleation is more promising for developing the technologies of production of nanodispersed materials. For this reason, the authors of this book paid special attention in further studies to the heterogeneous processes. Therefore, chapter 5 is concerned with the detailed examination of the processes of nucleation of the dispersed structures of the vacuum condensate, the volume of the structure during condensation, the metamorphosis of the structures under changes of the condensation conditions [36, 37]. The model of the kinetics of sorption and condensation, described previously in chapter 3, resulted in three important conclusions regarding the heterogeneous condensation of the vapour. Firstly, it is the strong effect of the multicomponent nature of the vapour mixture. The capillary condensation of the impurities greatly changes the sorption pattern on the substrate and may cause the chain-like nucleation of the condensation centres. Secondly, the surface migration as a mechanism of the effective transfer of matter plays a certain role in nucleation. Thirdly, to ensure the unambiguous simple analysis of nucleation, it is necessary to accept the condition of the local 'circular' vapour–adsorbate–condensate–vapour equilibrium. This condition makes it possible to justify the analogue of the Kelvin–Thomson equation for the observed phase taking into account the local curvature of the condensate surface. These three conclusions have been used to form a more general view of the nucleation and evaluation of the disperse structures. Neglecting the anisotropy of the surface properties of the nuclei of the individual phases, the stability of existence of the flat, cylindrical and spherical forms can be monitored on the basis of the Kelvin–Thomson effect using the condition of their equilibrium with the adsorbed phase. This approach makes it possible to construct the homological matrix of the forms of the nuclei of the condensate on the solid substrate, and the development of these forms during condensation results in the formation of the entire set of the columnar, plate-shaped, plane cellular, and volume-cellular structures observed in the experiments [38].

The evolution of the form of the condensate nuclei takes place as a result of a number of different factors: convective and radiation heat exchange, the incident and in the upper vapour flows, surface migration of the adsorbed atoms. In vacuum condensation with the formation of the disperse condensates with the size of the crystals considerably smaller than the free path of the atoms in the vapour phase the calculations of heat and mass exchange on the surface of the crystals of the condensate are not associated with any significant problems. This requires only simple although cumbersome calculations. The appropriate preparation of the algorithm and computer programs have been developed. The calculations have yielded a number of new non-trivial results.

1. Even in the absence of capillary condensation and weak surface diffusion, the individual nuclei during condensation do not grow into

each other because of the mutual screening of the vapour flow arriving on the substrate.

2. During growth, the nuclei continuously complete with each other. The nuclei, lagging behind in growth, do not receive enough vapour flow because of the increasing screening by the neighbours and in the final analysis disappear. This explains why the experimental results showed the increase of the characteristic size of the particles of the disperse condensate. These results indicate how to construct the technology of production of the disperse materials with the required characteristic particle size.

3. There are condensation conditions in which supersaturation forms in the local areas of the surface of the nuclei and leads to the further development of these areas. In other areas, for example, screened areas, saturation is not reached and these areas evaporate. The evaporation of the screened areas may result in the complete evaporation of the nucleation base which may separate from the substrate and transfer to the vapour medium. This means that in addition to the fluctuation mechanism of dust generation with the nanosized particles, described in the third chapter, there is a mechanism based on re-evaporation [39, 40].

4. There are critical temperatures of the substrate with the condensate above which the nuclei grow into each other and the condensation mechanism changes to film condensation. This is observed even in the absence of surface diffusion by the mechanism of re-evaporation of the condensate and re-condensation of the vapour on the screened parts of the nuclei. This result explains the application of high substrate temperatures and high-temperature annealing in thin-film technologies.

The results obtained in this stage of the experiments were used to determine the parameters which can be used for controlling the growth, dimensions and form of the particles of the dispersed structures on the substrates during condensation. Using these results, in chapter 6 the authors solve more complicated practical problems, for example, application of controlled condensation of disperse structures for different components on the nanoscale. These problems arise in, for example, the production of small chemical electric current sources (ECS). The attempt to apply this concept is described in this chapter. Several condensation mechanisms are discussed. The joint and layer condensation of lithium and aluminium and lithium and an organic material is tested for feasibility in producing solid electrodes for ECS with a larger active surface. Attention was given to the evaporation and condensation of polypropylene in vacuum to produce porous condensates for the manufacture of separators for ECS. The experiments with joint condensation of Li with polypropylene showed that it is possible to produce the propylene–ultradispersed Li composites suitable for use in supercharged ECS as liquid active anodes.

To find other processes based on spontaneous nucleation, successful tests were carried out with the joint condensation of magnesium with vacuum oil. These experiments showed that it is possible to produce porous metallic components filled with oil on the microlevel. It was found that condensation is efficiently controlled by defining the ratio of the flows of the metal and oil vapours. This result may lead to the development of a new technology of self-lubricating friction materials. Successful tests of the joint condensation of magnesium with polyethylene were also carried out. Conducting nanocomposite materials were produced [41, 42]. The concept of trapping metallic nanoclusters by soft capture in a carbonic acid condensate was also verified. The expectations were not fulfilled but previously unknown cryochemical phenomena, determined by the high chemical activity of the clusters and accompanied by chemiluminescence, were observed.

This chapter also describes the results of investigations of the effect of light on the adsorption of vapour atoms on the substrate. This was caused by the fact that the external energy effect on the process of adsorption may be another tool for controlling the initial stage of condensation. It is well-known that as a result of the Lipmann effect the light reduces the binding energy of the molecules with the adsorption vacancies and this stimulates desorption and releases part of the adsorption vacancies. The released adsorption vacancies may be used for the adsorption of the vapour molecules of other components whose molecules are less sensitive to the Lipmann effect. This concept is especially efficient when using the light in the pulsed mode. The experimental results show that the pulsed mode of the light greatly intensifies the mass exchange of the layer of the adsorbate on the surface with the vapour medium [43, 44] and also increases the growth rate of plants.

The concept of the construction of micro-components by consecutive or joint condensation of the vapours of different components has been developed in a study by the authors described in chapter 7. At cryogenic substrate temperatures the calculated size of the critical nuclei of the condensed phase may decrease to the dimensions of the atoms, i.e., suitable conditions are created for the condensation of the vapour to the amorphous solid state. In this mode that is no longer any effect of the crystallographic properties of the components and it is possible to produce the amorphous condensate, consisting of the mixture of the atoms of the components. When using mutually soluble composites with the temperature increasing to room temperature this is the true solution of the components. This is of no special interest because it can also be produced without condensation. Special interest is attracted by the atomic mixture of the mutually insoluble components. Evidently, such a mixture is unstable when the temperature is increased to room temperature. With increasing temperature the atomic mixture is characterised by the formation of supersaturation of a specific component and subsequently by its nucleation. If there is a tool which can be used to arrest the nucleation at the given level of the size of the nuclei, we obtain the determined process leading to the formation of nanodisperse components. Such a tool can be, for example, the previously specified concentration of the component in the cryogenic condensate. The nucleation of the component at a restricted concentration is arrested as a result of the removal of supersaturation. Another obstacle in nucleation may be the polarity of one of the components leading to micelle formation and arrest of the process. The same role may also be played by the addition of peptisation agents with the anti-coagulation effect.

The almost 'fantastic' concept, presented here, has been partially realised by the authors on the example of the joint condensation of the vapours of zinc and butanol. Chapter 7 describes the details of these investigations [45–49]. The main results may be described as follows:

The joint condensation of the vapours of zinc and butanol on the cryogenic condensate makes it possible to produce a composite of butanol with dispersed zinc with the zinc particle size in the range 10^{-9}–10^{-5} m. After increasing temperature to room temperature the condensate consists of a viscous black liquid, metastable in a hermetically sealed glass vessel and unstable in an open vessel.

2. The dispersion of the zinc particles is regulated by the level and ratio of the consumption of the vapours of zinc and butanol. With increasing consumption and their ratio the particle size increases, the composites become grey and completely stable.

3. Measurements of the zinc concentration in the composite using especially developed so-called 'hydrogen' method gave the values of 1–80 wt.%. A small part of zinc is found in the form of organometallic substance.

4. The nanodisperse zinc of the composite is characterised by high physical activity, reflected in the capacity to stick to the metallic surfaces. The capacity of the composite to cover the metallic surfaces with zinc was also investigated in detail.

5. The composite, placed in the open vessel with the immersed dissimilar electrodes, is a source of electric current formed as a result

of the oxidation of nanodispersed zinc with atmospheric oxygen, diffusing into the composite.

6. The metastability of the composite was investigated. The results show that slow aggregation of the zinc particles, taking place in the composite, is governed by the Smoluchowsky law.

7. Pilot plant equipment was constructed for the production of composites with the productivity of up to 100 tonnes per annum.

The experimental results revealed a number of unusual phenomena complicating both the experiments to produce the composite and also the operation of the pilot plant equipment.

1. The dependence of the zinc content of the composite on the ratio of the consumption of the vapours of the components changes even during a single experiment and cannot be predicted for the subsequent experiment.

2. The combined condensation mode is extremely unstable. The operation of systems requires continuous manual control.

These problems remained a mystery to the authors until recently when it was possible to construct the mathematical models for the numerical experiments with the joint condensation of two vapour components. The results of these experiments are described in chapter 8. Assumptions, used in the modified Langmuir model, were made for the direct numerical modelling of the sorption processes in the conditions in which the situation on the surface of the adsorbent is non-stationary. The vapour molecule, falling on the surface of the crystal-substrate, is 'fixed' in one of the adsorption vacancies with the absorption energy and is instantaneously accommodated on the surface of the crystal with respect to temperature. Subsequently, as a result of thermal oscillations and fluctuations of the lattice of the vacancies, the molecule may either travel to the adjacent adsorption vacancies, resulting in surface diffusion, can be desorbed in the vapour medium. The lifetime of the molecule in a single adsorption vacancy is determined by the sum of their binding energies with the substrate and with the molecules in the adsorption vacancies. Moving on the surface, the molecules collide with adder molecules, forming unstable aggregates, or if the degree of supersaturation in the adsorbed phase is sufficiently high, the stable single layer nuclei of the condensed phase. This results in filling of the substrate and, subsequently, also of the adjacent molecule are layers, simulating condensation.

Three problems were solved [50–52] using this approach to the numerical simulation of the molecular processes on the substrate.

1. The adsorption of a binary vapour mixture of water and silver. A number of new interesting results have been obtained. At a moderate

temperature of 373 K, in the conditions in which the silver vapours are strongly supersaturated with respect to the substrate, and the water vapours are not yet saturated, the results showed stimulated adsorption of the water vapours changing the capillary condensation of water between the silver nuclei. This illustrates the transition of the molecular adsorption process to the microprocess of capillary condensation. This approach was not described previously in the literature. Another interesting fact relates to the problems of the adhesion of the silver condensate to the substrate. In filling the first molecular layer on the substrate only part of the adsorption vacancies is occupied by the silver atoms. This part also determines the adhesion of silver to the substrate. However, in filling the subsequent molecular layers the water molecules displace some of the atoms in the first molecular layer to the second layer, additionally reducing the adhesion of the silver condensate to the substrate. When investigating the effect of the angle of wetting by water on adsorption it was shown that stimulated adsorption takes place in the entire wetting angle range 0–130°. However, at a wetting angle of approximately 35.9° adsorption is accompanied by radical changes. In a narrow contact angle range the overall covering of the substrate with silver is greatly reduced and the amount of the water adsorbate rapidly increases. Detailed analysis of this result shows that at this contact wetting angle the structure of the condensate on the substrate is subjected to radical rearrangement. This is accompanied by a large change of the efficiency of both the condensation of silver and the capillary condensation of water. This indicates that the stimulation of the adsorption processes in the vapour mixture is mutual and plays a significant role in the mass exchange between the vapour medium and the substrate.

The adsorption isotherm, calculated for the water pressure range of 0.1–36 000 Pa, also indicates important phenomena taking place. At a specific pressure of the water vapours the structure of the silver condensate is greatly rearranged. The crisis phenomena in the structure of the silver condensate were also detected in the adsorption isobar, calculated in the temperature range 303–473 K.

The adsorption of the mixture of the silver and water vapours and the adsorption of these vapours separately were also calculated. The results show that the presence of the water vapours always lowers the efficiency of condensation of silver in comparison with the condensation of pure silver vapours, and the presence of the silver vapours results in the efficient capillary condensation of water. If the adsorption of pure water vapours at a supersaturation of 2.32 indicates the nucleation of the water adsorbate, then in the presence of the silver vapours the nucleation of water in the adsorbate cannot take place owing to the fact that the supersaturation of the water adsorbate is removed by capillary condensation.

2. The formation of the nanocomposites in vacuum deposition of the two-component vapour on a cryogenic condenser. Calculations using the model of combined condensation of the silver and water vapours also revealed previously unknown phenomena. The stationary mode of combined condensation of the silver and water vapours in the initial stage was not determined. The produced composite showed condensation modes with the monotonically increasing silver concentration and a monotonically decreasing concentration. Both the condensation rate of the two components and the relative number of the silver particles in the composite were unstable. This result explains why in the experiments, described in chapter 7, the condensation mode in equipment had to be continuously controlled manually. The experimental results show that the silver concentration in the composite is always higher than the concentration in the mixture supplied to the condenser. This is also confirmed by the experimental data. The results show that with increasing silver concentration of the vapour mixture the particles of the structure in the composite become elongated and their height above the substrate increases and the size of the cross-section decreases. The surface of the composite is always enriched with silver in comparison with the average silver concentration of the composite.

The above results lead to a new approach to the process of production of nanodisperse composites and the development of a more advanced production technology.

3. The conditions in which the growth of columnar and filament crystal takes place. In the investigations carried out by E.I. Givargizov the author proposed three conditions ensuring the growth of single columnar and filament crystals:

a) preferential deposition of matter at the point with the activity considerably higher than that of other areas of the crystal;

b) directional flow of matter under the effect of different factors;

c) the growth of crystals at a high anisotropy of crystallochemical bonds in the lattice.

The problem for the investigations was formulated as follows: can the columnar crystals grow if these conditions are not satisfied? Calculations of the dynamics of sorption, nucleation and condensation of the silver vapours show that in the initial stage of condensation the form of the outer surface of the condensate depends strongly on the ratio of the adsorption energies of the silver atoms to the substrate and their mutual binding energy. With decreasing adsorption energy the growth of the condensate crystals in the direction of height becomes preferred in comparison with filling the substrate surface. There is a critical adsorption energy at which only single columnar crystals grow with the rate many times greater than the growth rate of the condensate in the film mode. With a further decrease of the adsorption energy only the filament crystals can grow, but their growth rate decreases in comparison with that of the film. The results of these calculations are in qualitative agreement with a large number of experimental data and represent a suitable supplement to the studies of E.I. Givargizov.

In chapter 1 it was shown that in the particle size range of the disperse system of 10^{-7}–10^{-3} m the process with suitable energy characteristics is the formation by disintegration of the matter which had accumulated in advance a reserve of the surface energy of formation of the film. This mechanism becomes effective if the pre-formed film is sufficiently thin. The investigations of the process of formation of the relatively thin films in the liquid phase is the subject of chapter 9. The most interesting results here is the one which shows the principal needs to take into account the surface phenomena in the hydrodynamics of thin films when as a result of the bulk and surface viscosity the surface forces take part in the definition of the movement of the liquid together with the bulk forces. This circumstance can be used to formulate a more general concept according to which the formation of mechanisms with suitable energy parameters should be based on the processes in which the dynamics of matter enables the transformation of some type of energy to surface energy. Therefore, in addition to the thin-film flows, chapter 9 also investigates the formation of liquid droplets by centrifugal forces as an example of the transformation of kinetic energy to surface energy.

Two approaches to the calculation of flows are investigated for the axisymmetric thin liquid film, formed as a result of collision of two coaxial jets. The need for studying these processes was dictated by problems in taking into account the effect of surface viscosity. In the first approach when deriving the phenomenon logically equation of motion it was necessary to take into account the effect of surface forces on the free surfaces of the film only in the radial direction. In the second approach, the surface energy was included in the energy conservation equation. Experimental verification showed that the experimental points are situated between the solutions obtained in these two approaches, but closer to the calculations carried out using the energy equation.

It was also clarified that the phenomenological equation of motion for the flow of the viscous liquid in the axisymmetric thin film, obtained in the first approach, can be derived from the Navier–Stokes equation. It was concluded that in the hydrodynamics problems in which the flow takes place with the change of the free surface, it is important to take into account the tangential stresses, formed on the free surface as a result of the effect of the surface tension forces and viscosity. The specific type of the tangential component of the stress tensor at the interface depends on the geometry of the problem. When attempting to publish this conclusion, the authors were subjected to criticism because it was believed that only normal stresses exist at the free interface and the existence of surface viscosity was not widely known.

In the problem of the formation of droplets during 'dripping' of the liquid from a rotating barrier and in melting of a rotating bar the results were used to derive the equation for calculating the size distribution of the particles. This was carried out assuming that the droplet separates from the bar in the local area of melting and using the previously known assumptions regarding the processes of separation of the droplets by centrifugal forces. Comparison of the calculations of the distribution function with the experiments showed qualitative agreement.

The problem of the deformation of the melt droplet in impact on the solid surface was investigated. The problem was solved on the basis of a simple assumption of the complete transition of kinetic energy of the droplet to surface energy. The resultant equations are suitable for calculations. The estimates of the deformation and the calculation of the distribution function of the produced droplets were used in the development of a technology for the production of the powders of superpure zinc for the chemical power sources, described in chapter 10 [53].

Chapter 9 discusses the Taylor bell which forms in a collision of the liquid jet with an obstacle of a limited size. The problem was solved using the local equilibrium equation which includes the effect of surface tension, the pressure difference, inertia and gravitational forces. The conditions of spontaneous breakdown of the bell film into liquid droplets were determined. These calculations are highly promising for developing the technology of production of aerosols and powders with the economically efficient application of dispersing gases or even without using these gases. Important aspects of the technologies of production of tin and zinc powders, based on the dispersion of the Taylor bells, are described in this chapter [54, 55].

Chapter 11 investigation the problems of the interaction of the particles of the dispersed system. Collision of the liquid aerosol particles may be accompanied by their coalescence and merger into a larger droplet. A large number of these events increase the size of the aerosol particles which is a harmful phenomenon in the technology of production of powders by melt dispersion. This results in the interest in the coalescence processes.

In the conditions of the non-isothermal disperse system there are several little-known and little studied phenomena which together determine the susceptibility of the particles of the system to coalescence or coagulation. The susceptibility of the dispersed system to coagulation coalescence of the particles is the manifestation of its metastability. The investigations of the coalescence processes are complicated and, therefore, and the recently they were reduced mainly to introducing the coefficients of the efficiency of coalescence at collisions of the particles. The physical meaning of this coefficient remains unclear. Therefore, chapter 11 gives attention mainly to the principle of the phenomena influencing the efficiency of coalescence [56, 61].

In particular, the possibility of formation of the vapour interlayer at contact of two liquid droplets is investigated. To facilitate investigations, this phenomenon is simulated by the interaction of the liquids with the surface of a heated solid. In simple experiments with careful immersion of a spherical blunt solid into a liquid special attention was given to the moment and condition of adhesion of the liquid to the surface of the solid. After wetting the solid with the liquid the solid was gradually heated. The well-known boiling curve of the liquid was reproduced; in the liquid convective heat exchange takes place at the boiling point, and above the boiling point heat exchange with bubble boiling occurs. The first boiling crisis – transition to the film boiling mode – took place at strong superheating. Subsequently, the temperature of the solid was carefully reduced. In the absence of vibrations and surface roughness of the blunt solid the vapour interlayer between the solid and the liquid remained unchanged to the difference of the temperatures of the solid in the liquid of 2 K. This unique result shows that the second boiling crisis is not an independent physical phenomenon and is a random event of the coalescence of the liquid with the surface of the heated solid under the effect of vibrations and surface roughness. This conclusion is very important for the theory of boiling. The experimental results were used to calculate the minimum heat flux and the temperature difference at which the vapour interlayer of minimum thickness can still form. This result is also important for investigating the stability of the disperse systems because the aerosol particles are never isothermal. The theoretical model for the existence of the vapour interlayer between the liquid droplets was constructed on the basis of Labuntsov's theory of non-equilibrium boiling which predicts that the local pressure defect curse in the vapour layer in the vicinity of the evaporating liquid. Calculations carried out using the model showed that depending on the conditions in the surrounding space between the droplets, situated in close contact, the pressure defect may result in the formation of both of repulsive and attracting forces. This phenomenon has much in common with the well-known process of thermal transpiration in vacuum. The results can also be related to the well-known phenomenon of 'floating droplets' investigated in detail by Deryagin. In addition to the well-known phenomenon of the 'gas cushion', all these results provide a general pattern of the interactions causing preventing the coalescence of liquid droplets. The calculated criteria and data show how to ensure the stability of the aerosol particles in the developing condensation processes.

The experience accumulated in the investigation of the conditions preventing coalescence was used in proposing a concept to use the non-wetting phenomenon for the development of the processes of impact dispersion of the metal melts in order to obtain metallic powders. The estimates this technology may prove to be economically efficient even for the dispersion of the Taylor bells. On the basis of the experimental results it was concluded that this method is suitable for producing the powders of copper and its alloys, iron and its alloys, and stainless steel [62]. However, there are shortcomings in equipment which must be removed in order to make this promising method suitable for industrial application.

Chapter 12 is concerned with the investigation of the possibilities of using the condensation processes for trapping aerosol particles in a wide range of the dimensions. For the particles with the size of tens of microns there are no problems with trapping because the drift velocity of these particles in the gradient graphs flows is comparable with the flow speed. There are a large number of suitable methods. For the particles with the size of the order of units of microns the problem of trapping becomes important because their drift velocity becomes negligible. The number of suitable methods is considerably smaller in this case. There are no efficient trapping methods for particles of the sub micron size. This is the impetus for the development of new methods of trapping submicron particles. In this chapter, attention is also given to the application of selective condensation of vapours of a liquid on sub micron particles in order to increase their size to the dimensions for which the currently available trapping methods can be used [63].

Two aspects of this direction have been investigated.

1. An algorithm was developed for calculating the centrifugal separator of dust from the gas taking the compressibility of the gas into account. As a result of taking compressibility into account it was possible to construct the algorithm for calculating the decrease of the gas temperature in the vortex flow. When adding vapours of some liquid to the gas from which the submicron gas particles have been removed, it is possible to calculate the supersaturation of the vapour in the flow and the start of condensation of the vapour on the dust particles. The dust in which the particle size has been increased by this method is suitable for trapping in the vortex flow. The results of these calculations were used to construct a centrifugal separator fitted with a rotor with blades which represents an obstacle to the enlarged particles. The tests of the separator used for trapping the paint aerosol showed that the trapping efficiency equals 99.2–100% for the paint particles and 25–34% for the acetone solvent. Recommendations are given for constructing industrial systems for trapping paint aerosols. Tests of the trapping of the dispersed medium showed that the trapping efficiency is 100%, and for the fine cement fracture rates 99.89%.

2. Several types of selective condensation of vapours on micro-particles have been investigated. One of these types – dimensional selectivity of condensation – is used to develop the processes of selective trapping of the products of combustion of kerosene. The combustion products with a shortage of oxygen travel in a pipe consisting of 10 sections, with each section fitted with an injector for the water vapour. Higher supersaturation of the water vapour is produced in each subsequent section. The vapour, arriving from the injector, condenses on the microparticles and the water-cooled walls of the sections. The condensate is removed from the sections together with the trapped soot fraction. The tests of this type show that finer and finer soot fraction is trapped in each subsequent section. The final sections The ucombusted remnants of the organic material. On the whole, the tests have been successful. However, a shortcoming of the method has also been found – high energy consumption for the generation of vapour. The method is suitable for use only in cases in which the complete cleaning is required, regardless of the expenditure [64].

In the conclusions of the book the authors describe their generalized results in the physics of dispersed system.

References

1. Bochkarev A.A., Poroshk. Metall., 1982, No. 4, 40–47.
2. Bochkarev A.A., et al., in: Phase transitions in pure metals and binary alloys: Novosibirsk:. ITP SB RAS, 1980, 133–145.
3. Bochkarev A.A., et al., Pribory Tekh. Eksperimenta, 1988. V. 1. S. 37.
4. Author Cert. 1566780, The evaporator for vacuum coating deposition, A.A. Bochkarev, et al., 1990.
5. Bochkarev A.A., et al., in: Boiling and condensation (hydrodynamics and heat exchange), Collection of scientific articles, Novosibirsk, ITP SB RAS, 1986, 102–110.
6. Bochkarev A.A., et al., in: Experimental methods in rarefied gas dynamics: Novosibirsk, ITP SB RAS, 1974, 98–137.
7. Palatnik L.S., Komnik Yu.F., Dokl. AN SSSR, 1959, V. 124, No. 4, 808–811.
8. Bochkarev A.A., et al., in: Thermophysics of crystallization and high-temperature processing of materials. Novosibirsk, ITP SB RAS, 1990, 98–117.
9. Bochkarev A.A., Polyakova V.I., Proc. 2nd All-Union. Conf. Modeling of crystal growth, Riga, 1987, V. 1, 90–91.
10. Bochkarev A.A., Polyakova V.I., *Ibid*, 84–86.
11. ` Bochkarev A.A., Polyakova V.I., in: Thermal processes in the crystallization of substances, Novosibirsk, ITP SB RAS, 1987, 128–135.
12. Bochkarev A.A., Polyakova V.I., Berdnikova V.V., in: Thermophysics of crystallization of substances and materials, Novosibirsk, ITP SB RAS, 1987, 115–122.
13. Bochkarev A.A., Polyakova V.I., Telegin GG Surface photoionization of atoms in a gas discharge electrode regions, Proc. XI All-Union. Conf. "Generators of low-temperature plasma." 20-23 June 1989, Novosibirsk. Novosibirsk, 1989. V. 2. S. 147-148.
14. Bochkarev A.A., Polyakova V.I., Telegin G.G., Proc. X All-Union. Conf. of Rarefied gas dynamics, Moscow, 1989, 117.
15. Bochkarev A.A., Polyakova V.I., Telegin G.G., in: Plasma jets in the development of new materials technology: Proc. Int. Workshop, 3-9 Sept. 1990, Frunze, USSR, VSP, Netherlands, 665–672.
16. Bochkarev A.A., Polyakova V.I., Telegin G.G., in: Proceedings of X All-Union. Conf. of Rarefied gas dynamics, Moscow, 1991, V. 3, 81–86.
17. Bochkarev A.A., Polyakova V.I., Telegin G.G., Russian Journal of Engineering Thermophysics, 1991, V. 1, No. 2, 183–192.
18. Bochkarev A.A., Polyakova V.I., Telegin G.G., Sib. Fiz. Tekh. Zh., 1991, V. 4. 109–112.
19. Bochkarev A.A., Polyakova V.I.,, Berdnikova V.V., in: Thermophysics of crystallization of substances and materials,Novosibirsk, ITP SB RAS, 1987, 122–132.
20. Bochkarev A.A., Polyakova V.I., in: Proceedings of IX All-Union. Conf. Rarefied gas dynamics, Sverdlovsk, 1988, V. 2, 164–169.
21. Bochkarev A.A., Polyakova V.I., in: Proc. 2nd All-Union. Conf. Modeling crystal growth " Riga, 1987. V. 1. 87-89.
22. Bochkarev A.A., Polyakova V.I., Poverkhnost. Fizika, khimiya, mekhanika,

1988, No. 3, 29–35.

23.   Bochkarev A.A., Polyakova V.I., Soviet Journal of Appl. Phys., 1989, V. 3, No. 2. 132–149).

24.   Bochkarev A.A., et al., in: Clusters in the gas phase, Novosibirsk, ITP SB RAS, 1987, 61–65.

25.   Bochkarev A.A., et al., Izv. SO AN SSSR, Ser. Tekh. Nauka, 1987, No. 18, vol. 5, 76–81.

26.   Bochkarev A.A., et al., in: Proc. reports IX All-Union. Conf. Rarefied gas dynamics, Sverdlovsk, 1987, V. 2, 19.

27.   Hagena O., J. Surface Science, 1981, V. 106, 101–116.

28.   Bochkarev A.A., et al., in: Proc. rep. XV All-Union. Conf. Actual problems of physics of aerodisperse systems, Odessa, 1989, V. 1, 38.

29.   Bochkarev A.A., Polyakova V.I., in: Proc. reports X All-Union. Conf. Rarefied gas dynamics, Moscow, 1989, 140.

30.   Bochkarev A.A., Polyakova V.I., in: Proceedings of X All-Union. Conf. of Rarefied gas dynamics, Moscow, 1991, V. 3, 138–144.

31.   Bochkarev A.A., et al., Sib. Fiz. Tekh. Zh., 1993, V. 2, 7–13.

32.   Bochkarev A.A., Polyakova V.I., in: Proc. rep. X All-Union. Conf. Dynamics of rarefied gases, Moscow, 1989, 141.

33.   Bochkarev A.A., Polyakova V.I., Teplofizika Vysokikh temperatur, 1989, V. 3, 472–474.

34.   Chukanov V.N., Kuligin A.P., *ibid*, 1987, V. 25, No. 1, 70–77.

35.   Polyakova V.I., in: M.F. Zhukov, V.E. Panin (eds), New materials and technologies. Extreme processes. Nauka, 1992, 9–37.

36.   Bochkarev A.A., et al., Book of abstr. 14th Intern. Vacuum Congress, Birmingham, UK, 1998. 337.

37.   Bochkarev A.A., Polyakova V.I., Pukhovoy M.V., J. Thin Solid Films, 1999, V. 343–344, 9–12.

38.   Bochkarev A.A., Polyakova V.I. Genesis and evolution of a vapour-deposited film structure, Preprint No. 267-93, Institute of Thermophysics of RAS, Novosibirsk, 1993.

39.   Bochkarev A.A., Pukhovoy M.V., PMTF, 1994, No. 3, 102–111.

40.   Bochkarev A.A., Polyakova V.I., Pukhovoy M.V., Vacuum, 1999, V. 53, 335-338.

41.   Author Cert. 314848, A method of producing composite materials, A.A. Bochkarev, et al., Publ. 2.1.87.

42.   Author Cert 1314848, A method for producing the conductive composite materials based on metal and polyethylene, et al., 1987.

43.   Bochkarev A.A., et al., in:Proc. rep. X All-Union. Conf. Dynamics rarefied gases, Moscow, 1989, 109.

44.   Bochkarev A.A., et al., Izv. SO AN SSSR, Ser. Tekh. Nauk, 1989, No. 3, 25–30.

45.   Bochkarev A.A., Pukhovoy M.V., Sib. Fiz. Tekh. Zh., 1993, No. 5, 1–11.

46.   Bochkarev A.A., et al., Zavod. lab., 1994, No. 4, 34–35.

47.   Bochkarev A.A., Pukhovoy M.V., Vacuum, 1997, V. 48, No. 6, 579–584.

48.   Bochkarev A.A., et al., Condensation on the gradient substrate butanol, Preprint No. 284-97, ITP SB RAS, Novosibirsk, 1997.

49.   Bochkarev A.A., et al., J. Colloid Interface Sci., 1995, V. 175, 6–11.

50.   Bochkarev A.A., Polyakova V.I., Teplofizika Aeromekhanika, 2009, V. 16, No. 1, 103–114.

51.   Bochkarev A.A., Polyakova V.I., PMTF, 2009, No. 5, 95–106.

52.   Bo)chkarev A.A., Polyakova V.I., Doklady Physics, 2009, V. 54, No. 4, 178-

181).

53. Bochkarev A.A., Gaiskii N.V., in: Heat and Mass Transfer during crystallization and condensation of metals, Novosibirsk, ITP SB RAS, 1981, 48–51.

54. Author Cert. 928723, A device for spraying a melt, A.A. Bochkarev, et al., 1980.

55. Author Cert. 176251, An apparatus for producing round particles, A.A. Bochkarev, et al., 1980.

56. Bochkarev A.A., Avksentyuk B.P. Zh. Tekh. Fiz., 1985, V. 55, No. 4, S. 797–798.

57. Bochkarev A.A., Avksentyuk B.P., Izv. SO AN SSSR, Ser. Tekh. Nauk, 1985, No. 16, V. 3, 29–33.

58. Bochkarev A.A., Avksentyuk B.P., in: Proc. rep. VII Conf. Two-phase flow in power machines and devices, Leningrad, 1985, V. 1.

59. Bochkarev A.A., Avksentyuk B.P., Izv. SO AN SSSR, Ser. Tekh. Nauk, 1986. No. 16, V. 3, 38–44.

60. Avksentyuk B.P., et al., in: Int. Workshop 'High-temperature dust-laden jets in the processes of treatment of powder materials', Sept. 6–10, 1988. Novosibirsk: Book of Abstracts, 1988, 157–162.

61. Bochkarev A.A., et al., in: High-Temperature Dust-Laden Jets in plasma technology: Proc. Int. Workshop. 6–8 Sept. 1988. Novosibirsk, USSR; Netherlands: VSP, 1989, 499–525.

62. Author Cert. 5050115/02/031550. A method of creating a powder, A.A. Bochkarev, et al., Publ. 19.03.93.

63. Bochkarev A.A., et al., Selective condensation of steam solid microparticles, Proc. rep. XV All-Union. Conf. Current issues of the physics of aerodisperse systems. Odessa, 26–29 October, 1989, Publishing House of Odessa State University, 1989, V. 1, 37.

64. Bochkarev A.A., Golubev Yu.A., Zolotarev S.N., Method and apparatus for clea ning gas flows, Appl. su91/00155, 1991.

1

Methodological evaluation of the prospects for the development of physical dispersion methods

Traditionally, the physical processes of dispersion are the subject of attention of experts in the field of powder metallurgy. This is due to the fact that in recent decades this industry has evolved as a major field in mechanical engineering. It is urgent to expand the volume of production of powder materials using the most productive methods. By now many different dispersion technologies have been developed. There is no need to overview them here because this is done in a series of monographs [1–6]. The flow of original papers on new developments, improvement of the known methods of dispersion and the study of processes used in these technologies is not decreasing. In the foreseeable future tehre will be no radical solutions to the problem of creating a method of dispersion that meets the modern demands of scientific and technological progress.

In this regard, there is a need to look at the methods of dispersion from a generalized point of view and by analyzing the received information to give a definite forecast of further promising developments.

All dispersion methods, resulting in a system of material bodies transferring to the disperse state, i.e. individual elements of the system have a defined, at least on average, size and can be divided into two major groups – dispersion by grinding and dispersing by synthesis. Obviously, the processes used in these methods may be different. Figure 1.1 shows an incomplete scheme of these processes inherent in the two mentioned fundamentally different methods. The grinding method depending on the application to a substance in the liquid or solid states may differ by activating the stages of melting and crystallization. The methods for the synthesis of disperse systems from matter in the molecular–homogeneous state can vary substantially depending on the phase–carrier (solution, gas) and on the role of chemical reactions in one or other stage of the formation of a dispersed system.

Figure 1.1 shows that the intermediate processes that lead to transfer of the substance to the dispersed state relate to at least four different branches of natural sciences: physics, chemistry, fluid dynamics, mechanics of deformable

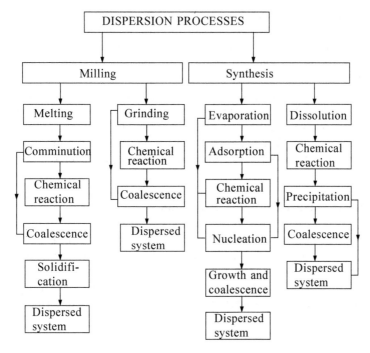

Fig. 1.1. The scheme of the processes involved in the formation of disperse systems.

media. Apparently, this is due to a variety of currently known methods of dispersion, developed by various experts. It also creates some difficulties for the selection of the criterion of the generalized evaluation of the dispersion merhod.

In assessing the rationality of the technology used to produce powders, it is usually necessary to take into account several factors, most important of which are the main parameters of the resulting powder – size, shape, microstructure of the particles, technological properties, performance in one unit, compared with the desired volume of production, the production cost of the resulting powder. The last is used as a criterion for evaluation of the technology in [5]. However, despite the universality of the production costs for the various branches of science and technology, the production cost it is not an objective stationary measure of performance of the given processes.

A more objective and fixed criterion is the value of the unit cost of energy required to achieve a certain dispersion. At the same time, the specific energy consumption is the compound part of the production cost of the resulting dispersed system. Therefore, this criterion is more convenient for evaluating the effectiveness of the chosen dispersion method. However, at present there are no distinct energy estimates of the processes used in the technology of powder production.

Simple calculation shows that the energy required for the formation of particle surfaces in the conversion of the molten metal to a powder with particles with a diameter of 20 mm is approximately equal to the energy of the droplet falling from a height of 10 m, which corresponds to a velocity of 15 m/s. But, for example, spray jet technology uses natural gas, by weight approximately equal to the melt, accelerated to a velocity of 200 m/s. This example illustrates the need to analyze the amount of energy expended in dispersing technology.

In this chapter, the calculations of energy consumption in the various processes were based on simplified models intentionally excluding features of real technologies. The peculiarities related to possible chemical reactions in the process were also ignored. For a comparative analysis of the effectiveness of various dispersion processes it is apparently necessary to develop a comparison standard.

1.1. Change in surface energy in dispersion

Since most of the technologies of powder production are based on fragmentation of the melt material, the energy expended on the formation of the surface of droplets is one of the most important indicators of the process. For ease of calculation we consider a powder consisting of spherical particles.

The surface energy of a spherical drop with radius r is described by

$$e_\sigma = 4\pi r^2 \sigma, \tag{1.1}$$

where σ is the surface tension coefficient of the melt, adopted in these calculations as independent of the droplet size. The mass of a spherical droplet is given by

$$m = \frac{4}{3}\pi r^3 \rho, \tag{1.2}$$

where ρ is the density of the melt. The surface energy of the droplet per unit mass can be obtained by dividing equation (1.1) by (1.2):

$$E_\sigma = \frac{3\sigma}{\rho r}. \tag{1.3}$$

It is convenient to operate with the surface energy, related to the latent heat of vaporization:

$$\bar{E}_\sigma = \frac{3\sigma}{\rho \lambda r}. \tag{1.4}$$

The calculation results of $\bar{E}_\sigma$ for different substances at $r = r_a$, where r_a is the atomic radius of the corresponding element, are shown in Table 1.1. The results of the calculation show that for the energy required for the dispersion

Table 1.1.

Element	Al	Cu	Pb	Sn	Ti	Zn
$\bar{E}_\sigma$	0.743	0.815	0.893	0.616	0.706	1.535

of the substance to the size of the atom, is approximately equal to the heat of evaporation.

1.2. Gas-phase method of powder production

The method of obtaining powders by evaporation–condensation is known as the gas-phase method [7] and consists of the formation of powder particles by condensation from the vapour phase. Energy costs in this process consist of the heat required to heat the material to the melting temperature, heating to the evaporation temperature and evaporation itself. The amount of energy required to evaporate 1 kg of the substance is calculated by the formula

$$\overline{E}_g = \frac{\eta_g}{\lambda}\left[(T_l - T_0)c_s + r_l + (T_g - T_l)c_l + \lambda\right], \tag{1.5}$$

where T_0, T_l, T_g are the initial temperature, melting point and evaporation temperature; c_s, c_l are the specific heats of substances in the solid and molten states; r_l is the latent heat of fusion; η_g is the efficiency of the gas-phase process.

A characteristic feature of the method of evaporation and condensation is that the energy consumption is independent of the size of the resulting powder.

1.3. The method of melting a rotating rod

Preparation of powders by melting a rotating shaft is described in [6]. Energy consumption for 1 kg of powder in the process can be written as

$$\overline{E}_c = \eta_c(\overline{E}_\sigma + \overline{E}_l + \overline{E}_k), \tag{1.6}$$

where η_c is the efficiency of the centrifugal process.

In the formula (1.6) the second term means the contribution of energy to heating and melting of the substance and can be determined by the formula

$$\overline{E}_l = \left[(T_l - T_0)c_s + r_l\right]/\lambda. \tag{1.7}$$

The third term in equation (1.6) represents the kinetic energy of the droplets emitted from the rod. To determine E_k consider a simple model of the process. The rotating rod with a radius R_0 melts from the end face. The melt in the form of a film or separate jets, rotating together with the rod, flows in the radial direction on its front surface. The separation of the droplet occurs from the edge with a radius R_0. In this case, the surface tension of the melt overcomes the centrifugal forces. The condition for the separation of droplets can be written as the equation of balance of forces

$$F_c = F_\sigma, \tag{1.8}$$

where F_c is the centrifugal force acting on the droplet; F_σ is the surface tension force. Assuming that the separated droplet is spherical, to determine the centrifugal force we can write

$$F_c = \frac{4}{3}\pi r^3 \rho \frac{v^2}{R_0}. \tag{1.9}$$

Here v is the peripheral speed at the edge of the end of the rod. The surface tension force can be approximately written as

$$F_\sigma = 2\pi r \sigma. \tag{1.10}$$

The approximate expression (1.10) lies in the assumption that the break occurs on a line equal to the circumference of the droplet.

By substituting (1.9) and (1.10) in (1.8) we obtain

$$\frac{2\rho r^2 v^2}{3R_0} = \sigma. \tag{1.11}$$

From equation (1.11) it follows

$$\frac{r}{R_0} = \sqrt{\frac{3}{2}}\, \mathrm{We}_c^{-1/2}, \tag{1.12}$$

where the Weber number $\mathrm{We}_c = \rho v^2 R_0/\sigma$ is defined by the outer radius of the rod. The physical meaning of We_c is the ratio of the centripetal force acting on unit volume of the rod material to the surface force of the melt acting on the film, radially flowing down from the edge.

The results of calculation by formula (1.12) in comparison with experimental results [8], obtained in the range of Weber numbers $9.5 \cdot 10^2$–

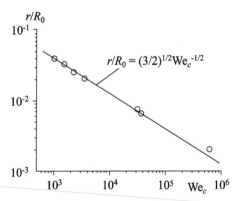

Fig. 1.2. The dependence of the relative radius of the powder particles, produced by melting a rotating rod, on the Weber number: points – experiment [8], the line - calculation by formula (1.12).

$6 \cdot 10^5$, are shown in Fig. 1.2. Good agreement between the experimental data and the calculation indicates the correctness of the process model.

The kinetic energy of a droplet after separation can be defined as

$$e_k = mv^2 / 2, \tag{1.13}$$

where v can be determined from (1.11). The kinetic energy of droplets in the sum equalling 1 kg of matter, related to the specific heat of evaporation, is equal to

$$\bar{E}_k = \frac{3\sigma R_0}{4\rho r^2 \lambda}. \tag{1.14}$$

Substitution of (1.4), (1.7), and (1.14) in (1.6) gives the formula

$$\bar{E}_c = \frac{\eta_c}{\lambda}\left[r_l + (T_l - T_0)c_s + \frac{3\sigma}{\rho r} + \frac{3\sigma R_0}{4\rho r^2} \right], \tag{1.15}$$

which can be used to define energy costs for 1 kg of powder produced by melting the rotating rod.

1.4. Dispersion by a 'focus' of jets

Details of the method for obtaining powders by dispersion of the melt with gas are contained in [5]. The energy costs during the spraying process are made up of the cost of melting and spraying:

$$\bar{E}_F = \eta_f(\bar{E}_l + \bar{E}_f). \tag{1.16}$$

The energy spent on spraying 1 kg of melt, related to the specific heat of evaporation, can be determined by the formula

$$\bar{E}_f = \chi \frac{u_g^2}{2\lambda}, \tag{1.17}$$

where u_g is the velocity of the spraying gas; $\chi = G_g / G$ is the ratio of gas flow to the flow rate of the melt. The dependence on χ and u_g on the particle radius of the produced powder can be approximately determined by considering the dispersion process. The minimum radius of droplets formed under the influence of the spraying gas can be obtained from the ratio of the dynamic pressure of the gas, flowing around the droplet, and the Laplace pressure inside the droplet. This approach is known [5] in the theory of atomization of liquids:

$$\frac{\rho_g(u_g - u)}{2} = A\frac{2\sigma}{r}. \tag{1.18}$$

In equation (1.18) ρ_g is the gas density; u is the velocity of the melt droplet, and A is a dimensionless coefficient that determines the minimum ratio of the dynamic pressure of the gas and the Laplace pressure in the droplet large enough for comminution. The magnitude of the coefficient A depends mainly on the viscosity of the melt in the droplet. For the evaluation calculations the value of A can be assumed to be unity.

A solution of (1.18) is

$$\rho_g u_g = \rho_g u \pm \sqrt{\frac{A \rho_g \sigma}{r}}. \tag{1.19}$$

To determine the value χ of a specific spraying installation, we can write an approximate formula

$$\chi = \frac{\rho_g u_g S_g}{\rho u S}, \tag{1.20}$$

where S_g and S are the cross-sectional areas of nozzles for the spraying gas and the melt. Substitution of (1.19) in (1.20) gives

$$\chi = \frac{\rho_g S_g}{\rho S} \pm \frac{2 S_g}{\rho_s u} \sqrt{\frac{A \rho_g \sigma}{r}}. \tag{1.21}$$

The values of S_g and S are the structural characteristics of the facility, usually close to each other. For most gas–melt couples the ratio is $\rho_g / \rho \approx 10^{-4}$. In this regard, the first term in (1.21) can be neglected, leading to the form

$$\chi = \frac{2 S_g}{\rho S u} \sqrt{\frac{A \rho_g \sigma}{r}}. \tag{1.22}$$

From (1.19), neglecting the first term we obtain

$$u_g = 2 \sqrt{\frac{A \sigma}{\rho_g r}}. \tag{1.23}$$

Using (1.22), (1.23) in (1.17), we obtain the expression

$$\bar{E}_f = \frac{8 S_g A^{3/2}}{\rho S u \lambda \sqrt{\rho_g}} \left(\frac{\sigma}{r} \right)^{3/2}. \tag{1.24}$$

From (1.24) it can be seen that the unit costs of energy are highly dependent on the coefficient of surface tension and the radius of the particles of the produced powder. Substituting (1.24) and (1.7) in (1.16), we have the formula

$$\overline{E}_F = \frac{\eta_f}{\lambda}\left[\eta + (T_l - T_0)c_s + \frac{8S_g A^{3/2}}{\rho S u\sqrt{\rho_g}}\left(\frac{\sigma}{r}\right)^{3/2}\right],$$ (1.25)

with which to evaluate the unit cost of energy, referred to the latent heat of vaporization. The dependence of $\overline{E}_f$ on the radius of the particles can be indirectly verified. The results of calculation by formula (1.22) reduced to the form $\chi \approx B/\sqrt{r}$, are shown in Fig. 1.3 in comparison with experimental data [6]. The coefficient B in the calculation of the curves was determined by the leftmost experimental points. There is good agreement with experiment, demonstrating the correctness of the dependence (1.22) on the particle radius.

To estimate the unit cost of energy in the technology of powder spraying, formula (1.25) should be used as

$$\overline{E}_F = \frac{\eta_f}{\lambda}\left[\eta + (T_l - T_0)c_s + D\left(\frac{\sigma}{r}\right)^{3/2}\right],$$ (1.26)

determining factor D from experimental data.

1.5. Dispersion of the melt layer

There are experimental data on the method of powder production by spraying a film formed on the nozzle walls by spinning the melt at the inlet [9]. The data for the sputtering of tin in the averaged form are shown in Fig. 1.3 in the form of curve 5. Comparison of the curves 3 and 5 shows that the amount of gas required in sputtering of the melt layer is several times smaller. Thus, the cost of energy in the process of dispersion of the melt layer can be estimated by the formula

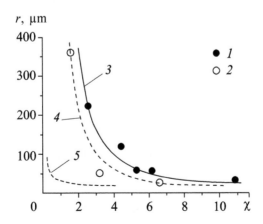

Fig. 1.3. The average radius of powder particles obtained by spraying gas. 1, 2 – experiment [6]. copper, nozzle diameter for metal 5.6 and 7.9 mm, respectively, 3, 4 – calculation by formula (1.22), 5 – average experimental data for tin [9].

$$\bar{E}_F = \frac{\eta_F}{\lambda}\left[\eta + (T_l - T_0)c_s + \beta D\left(\frac{\sigma}{r}\right)^{3/2}\right].$$
(1.27)

1.6. Milling of ductile materials

The energy costs of dispersing the materials that exhibit plastic properties consist of the surface energy of solid particles and the energy expended in overcoming the friction forces in the formation of the surface:

$$\bar{E}_M = \eta_M(\bar{E}_\sigma + \bar{E}_m).$$
(1.28)

Consider the model of a process under the following simplifying assumptions:

a) a piece of spherical material subjected to shear deformation;
b) milling results in the formation of spherical particles;
c) the surface of the particles that make up the mass of substance M, formed by a single act of shear deformation. Then

$$\bar{E}_\sigma = \frac{3\sigma_s}{\rho r\lambda},$$
(1.29)

where σ_s is the surface tension coefficient of the substance in the solid state; r is the radius of the particles obtained in the course of deformation. The mechanical energy, expended during shear deformation, in differential form is

$$dE_m = -F_\tau\, dl.$$
(1.30)

Here F_τ is the shear force; its value depends on the surface S_τ, on which the shear takes place and on the tensile strength of the material in shear σ_τ; l is displacement:

$$F_\tau = S_\tau\sigma_\tau.$$
(1.31)

The characteristic size of the shear surface l can be roughly defined as $l = \sqrt{S_\tau}$, from where

$$dl = dS_\tau / 2\sqrt{S_\tau}.$$
(1.32)

Substitution of (1.31) and (1.32) in (1.30) and integrating over S_τ within the range $S_H - S_\tau$ give

$$\bar{E}_m = \frac{\sigma_\tau}{3}(S_\tau^{3/2} - S_H^{3/2}).$$
(1.33)

In (1.33) S_H is the surface of a piece of material before milling. For the amount of substance M the values S_τ and S_H are given by

$$S_\tau = \frac{3M}{\rho r}, \; S_H = \frac{3M}{\rho r_H}. \tag{1.34}$$

Using (1.34) in (1.33), we obtain

$$E_m = \frac{\sqrt{3}\sigma_\tau M^{3/2}}{\rho^{3/2}} \left(\frac{1}{r^{3/2}} - \frac{1}{r_H^{3/2}} \right),$$

which after reduction to 1 kg of the substance and dividing by the heat of vaporization is converted to the form

$$E_m = \frac{\sqrt{3}\sigma_\tau M^{3/2}}{\rho^{3/2}} \left(\frac{1}{r^{3/2}} - \frac{1}{r_H^{3/2}} \right), \tag{1.35}$$

Substitution of (1.29) and (1.35) into (1.28) gives the formula

$$\overline{E}_M = \frac{\eta_M}{\lambda} \left[\frac{3\sigma_s}{\rho r} + \frac{\sqrt{3}\sigma_\tau}{\rho^{3/2}} \left(\frac{1}{r^{3/2}} - \frac{1}{r_H^{3/2}} \right) \right], \tag{1.36}$$

suitable for the evaluation calculations of the unit cost of energy during milling of ductile materials.

It is advisable to consider the process of milling of plastic materials by grinding. Assume that before milling the substance has the form of a cylinder with a radius and height of r_H. In the actual process milling is the result of many deformation cycles. To simplify the model, it is assumed that grinding of the material occurs in a single act of crushing a piece to the film with thickness r. The cylinder at the ends is deformed by the surfaces of the working parts of the mill. Assume that there is no friction of the material on the surface of the working parts of the mill, which is equivalent to using the ideal lubricant, not having any viscosity. This process is far from reality, so this model can be used for the lower estimate of the energy consumption in milling.

The consumption of mechanical energy in the form of the differential expression can be written

$$dE_m = -F\,dr, \tag{1.37}$$

in which the force F, acting on the ends of the cylinder, is defined as

$$F = \sigma_b S_m. \tag{1.38}$$

Here σ_b is the ultimate strength of the material. The area of the deformed cylinder S_m can be determined from the condition of conservation of matter:

$$S_m = \frac{\pi r_H^3}{r}.$$ (1.39)

Substituting (1.39) into (1.38), then into (1.37) and integrating over r within the range $r_H - r$, give

$$E_m = \pi \sigma_b r_H^3 \ln \frac{r_H}{r}.$$ (1.40)

From (1.40)

$$\bar{E}_m = \frac{\pi \sigma_b r_H^3}{\lambda} \ln \frac{r_H}{r_H},$$ (1.41)

in which r_H for 1 kg of a substance is defined as

$$r_H = \sqrt[3]{1/\pi\rho}.$$

Using (1.41) and (1.29) in (1.28) gives the formula

$$\bar{E}_M = \frac{\eta_M}{\lambda}\left(\frac{3\sigma_s}{\rho r} + \pi\sigma_b r_H^3 \ln \frac{r_H}{r}\right),$$ (1.42)

suitable for estimating the lower boundary of energy consumption in milling of a ductile material with an ideal lubricant.

1.7. Comparison of energy consumption in different methods of powder production

The results of calculations of energy consumption in different processes in the dispersion of tin, obtained by means of (1.4), (1.5), (1.15), (1.26), (1.27), (1.36) and (1.42), are shown in Fig. 1.4. The values of coefficients of efficiency η_g, η_c, η_f, η_F, η_M are taken equal to unity. The coefficient D in (1.26) was determined using experimental data [6]. The coefficient β in (1.27) is taken as 10. Tensile strength of tin $\sigma_b = 3 \cdot 10^7$ N/m² at 20°C is taken from [10]. The value of σ_τ in (1.36) is equal to 0.6 σ_b. The physical properties of substances are taken from [11].

Comparison of the curves shown in Fig. 1.4 indicates that the least economical process for the dispersion of ductile materials to particles with a radius greater than 10^{-3} m is the process of evaporation and condensation (curve 2), and in the range of radii $(10^{-6}–10^{-3})$ m – the milling process (curve 6). However, as shown by curve 7, the milling process has a reserve in the application of lubricant. In the range of particle radii $(10^{-6}–3 \cdot 10^{-5})$ m the processes of sputtering by gas and melting of the rotating rod are roughly equivalent, but less advantageous than dispersion of the melt layer.

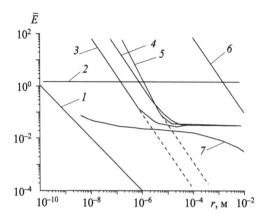

Fig. 1.4. Energy consumption in the dispersion of tin in different processes. 1 – surface energy; 2 – evaporation–condensation; 3 – spraying of a film with a 'focus' of the jets, 4 – melting the rotating rod, 6 – shear milling, 7 – milling with and ideal lubricant.

When $r < 10^{-7}$ m the most energetically favourable method is the method of evaporation and condensation.

Thus, in terms of energy savings in the process of dispersion of substances agents, the recommended process for obtaining powders with a particle size less than $2 \cdot 10^{-7}$ m is the process of evaporation and condensation, $2 \cdot 10^{-7}$–$4 \cdot 10^{-5}$ m – the process of sputtering of the melt film, $4 \cdot 10^{-5}$–10^{-2} m – the processes of spraying and of melting the rotating rod. Conducting the process of obtaining the powders by spraying immediately after the process of melting the material, which eliminates the cost of secondary melting of the material, with the size greater than $2 \cdot 10^{-7}$ m, gives the most energetically favourable technology. This is illustrated by dashed lines (see Fig. 1.4), obtained by calculations using (1.26) and (1.27), excluding the contribution of energy to melting.

Conclusions

In conclusion, the promising directions of development of methods of dispersion will be described. The equations (1.4), (1.5), (1.15) (1.26), (1.27), (1.36) and (1.42) can be used to calculate the physical efficiency of the considered technologies. The calculation results in the form

$$\eta_g = (\bar{E}_\sigma / \bar{E}_g)100 \ \%, \ \eta_f = (\bar{E}_\sigma / \bar{E}_f)100 \ \%, \ \eta_c = (\bar{E}_\sigma / \bar{E}_c)100 \ \%,$$

$$\eta_M = (\bar{E}_\sigma / \bar{E}_M)100 \ \%$$

for the dispersion of tin are shown in Fig. 1.5. In particular, the numerical values of the efficiency(%) in dispersion to particles with a radius of 10^{-5} m are as follows: for the evaporation–condensation method $8.4 \cdot 10^{-4}$, dispersion of the melt layer with gas $3.8 \cdot 10^{-2}$, spraying with a 'focus' of jets $2.1 \cdot 10^{-2}$, for melting the rotating rod $2.7 \cdot 10^{-2}$, for shear milling $4.8 \cdot 10^{-7}$, for

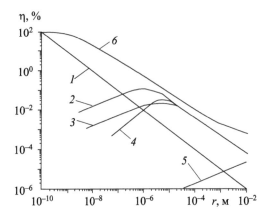

Fig. 1.5. Efficiency of various processes of dispersion of tin. 1 – gas-phase method; 2 – spraying with a 'focus' of jets; 3 – spraying a thin film of liquid; 4 – centrifugal atomization; 5 – milling with shear; 6 – milling with an ideal lubricant.

milling with ideal lubrication $7.1 \cdot 10^{-2}$. These values of efficiency demonstrate the effectiveness of only the dispersion process, not of the powder production process, as the calculations do not take into account the energy loss in real processes.

It should be noted that only the dispersion methods based on the use of physical processes were considered. In addition, all calculations were performed for the dispersion of a particular metal – tin. However, these results may be sufficient for planning future studies. Selection of a specific dispersing agent can not fundamentally affect the qualitative conclusions from the calculations:

1. Different methods of dispersion can be used for various ranges of the characteristic dimensions of the elements of the dispersed system. In the size range where the specific surface energy of the system is small compared to any characteristic specific energy of the system, for example, the latent heat of fusion, energetically favourable are the methods of comminution of the melt, and especially the methods, including the process of accumulation of the surface energy of matter prior to the comminution process, for example, the formation of a thin film.

2. In the range of sizes where the specific surface energy of the dispersed system is commensurate with the heats of phase transitions, the most suitable method for producing for the dispersed system is the evaporation and condensation method.

3. In the size range of elements of the disperse system, where the specific surface energy of the system is commensurate with the latent heat of phase transformation, the dispersion system becomes metastable. Therefore, prospective studies of processes of dispersion must be accompanied by finding ways to stabilize the ultrafine systems.

4. The presented results of comparative assessment of the specific surface energy can specify the size range of element of the disperse system of 10^{-8} m or less, where we should expect its high physical and chemical activity.

Probably, in this size range the costs to development of the dispersion methods are meaningless, since the high activity of the obtained systems requires tremendous effort in stabilization. In this case, the promising methods for studying the processes of dispersion should be accompanied by the search for ways to use the ultrafine system during its formation.

The above findings have important methodological significance, they determine the further content of this monograph. According to the second conclusion, the focus of the work is devoted to the study of condensation processes. Condensation on the surface and condensation in the volume have some fundamental differences. Heterogeneous condensation on the surface (see chapter 3) occurs at a temperature approximately equal to the known and controlled temperature of the substrate, mainly in the migration mechanism of transport of atoms or molecules of matter on the condensation front to their places in the structure of the condensate. Homogeneous condensation (see chapter 4) is characterized by autoregulation of temperature at the condensation front and by the mainly convective transport of vapour to the condensation front. Therefore, the studies of these two processes are reported separately. A common focus in these studies is that in both cases attention is given to the investigated processes, leading to the formation of disperse systems in a wide size range of disperse particles. In accordance with the first conclusion, the chapters 9 and 10 discuss some possibilities for increasing the surface energy of a substance in the liquid state, which radically improves the methods of dispersing the material in the molten state. The conclusions 3 and 4 have initiated studies of the processes leading to the formation of composite materials consisting of nanoparticles in a conservation organic matrix. These studies are described in Chapters 6 and 7.

List of symbols

A – the ratio of dynamic pressure of the gas and the Laplace pressure in a droplet;

B – coefficient;

c_s, c_l – the specific heats of solid and molten state, J / (kg $\cdot$ K);

D – empirical coefficient;

E_σ – the surface energy of droplets in the sum equalling 1 kg, J/kg;

E_m – The cost of mechanical energy in milling material, J;

$\overline{E}_\sigma$ – Surface energy, attributed to the latent heat of vaporization;

$\overline{E}_g$ – The relative energy required to evaporate 1 kg of the substance;

$\overline{E}_c$ – The relative energy required to produce 1 kg of powder by centrifugal atomization;

$\overline{E}_k$ – The relative kinetic energy of the emitted droplets;

$\overline{E}_l$ – The relative energy required to melt 1 kg of the substance;

$\overline{E}_f$ - The relative energy expended on spraying 1 kg of melt with the 'focus' of jets;

$\overline{E}_F$ - The relative energy expended on spraying 1 kg of the film of the melt;

$\bar{E}_m$ - A relative increase in the surface energy in milling;

$\bar{E}_M$ - The relative energy consumption during milling of ductile materials;

e - the surface energy of a spherical droplet, J;

F_c - the centrifugal force acting on the droplet, N;

F_σ - the surface tension force, N;

F_τ - shear force during the deformation of ductile materials, N;

l - the shear surface, displacement;

M - mass of the material, kg;

m - mass of the droplet, kg;

r - radius, m;

r_H - radius of a piece of material before milling, m;

r_a - atomic radius, m;

r_l - latent heat of fusion, J / kg;

R_0 - outer radius of the rod, m;

S_g and S - cross-sectional area of nozzles for spraying the gas and melt, m²;

S_H and S_τ - surface of a piece of material before and after mlling, m²;

S_m - The area of the deformed piece of material, m²;

T_0, T_p, T_g - initial temperature, melting point and vaporization temperature,   K;

u - velocity of melt droplets, m / s;

u_g - velocity of the dispersing gas, m / s;

v - the peripheral speed at the edge of the end of the rod, m / s;

$We_c = \rho v^2 R_0 / \sigma$ - the Weber number;

β - coefficient of gas saving;

λ - latent heat of vaporization, J / kg;

η_g - energy efficiency of the gas-phase process;

η_c - energy efficiency of the centrifugal process;

η_f - Energy efficiency process of 'focusing' of jets;

η_F - Energy efficiency the sputtering process of the film;

η_M - Energy efficiency of grinding of ductile materials;

ρ - density, kg/m³;

σ - the surface tension, N/m;

σ_s - specific surface energy of solid surface, Nm;

σ_b - ultimate strength of the material, N/m²;

σ_τ - shear strength of the material;

$\chi = G_g/G$ - ratio of gas flow to the flow rate of the melt.

Indices:

H, 0 - initial;

a - atom;

b - the ultimate strength of the material;

c - centrifugal;

f - 'focus' of the jets;

F - the processes in a liquid film;

g - gas phase;
k - kinetic;
l - melting;
m - milling;
s - the solid state;
σ - the surface.

Literature

1.   Fedorchenko I.M., Andrievsky R.A., Fundamentals of powder metallurgy, Kiev: Izdvo AN USSR, 1963, 420.

2.   Borodin V.A., et al., Atomization of liquids, Moscow, Mashinostroenie, 1967, 204.

3.   Powder metallurgy and sprayed coatings, Antsiferov V.I., et al., Moscow, Metallurgiya, 1987.

4.   Gratsianov Yu.A., et al., Metal powders from melts, Moscow, Metallurgiya, 1970.

5.   Nichiporenko O.S., et al., Dispersed metal powders, Kiev, Naukova Dumka, 1980.

6.   Powder metallurgy of special purpose materials, Eds. G. Bark, B. Weiss, Moscow, Metallurgiya, 1977.

7.   Frishberg I.V., et al., Gas-phase method of powder production, Moscow, Nauka, 1978.

8.   Altunin Yu.F., et al., Poroshk. Metall., 1970, No. 2, 120–126.

9.   Zholob V.N., Koval' V.P., Poroshk. Metall., 1979, No. 6, 13–16.

10.   The mechanical properties of rare metals, Ed. L.D. Sokolov, Moscow, Metallurgiya, 1972.

11.   Tables of physical quantities, Ed. I.K. Kikoin, Moscow, Atomizdat, 1976.

Formulation of experimental studies of homogeneous and heterogeneous vacuum condensation of metal vapours

This section describes the methods developed in the laboratory of Thermophysics of Microfine Systems of the Institute of Thermophysics of the Siberian Branch of the Russian Academy of Science to solve specific problems – condensation processes leading to the formation of disperse systems.

2.1. The object of research and conditions

The simplest case of formulation of experimental studies of heterogeneous condensation is the situation when a solid surface at a temperature T borders with a vapour medium with pressure p_f and temperature T_f. From the vapour region a flow of particles travels to the surface of unit area

$$j_f^+ = p_f (2\pi mkT_f)^{-1/2}. \tag{2.1}$$

A flat surface with the vapour condensate emits to the vapour phase a flow of particles

$$j_f^- = p_\infty (2\pi mkT)^{-1/2}. \tag{2.2}$$

In (2.1) and (2.2) m is the mass of vapour particles; k is the Boltzmann constant; p_∞ is the saturation pressure at the surface temperature.
The ratio j_f^+ and j_f^- characterizes the supersaturation at the vapour–condensate interface:

$$S = p_f T^{1/2} / (p_\infty T_f^{1/2})$$

or in the isothermal case $S = p_f / p_\infty$.

Such characteristic of the vapour–solid surface system is sufficient for research of stages preceding condensation – adsorption and nucleation. To study the processes of condensation at the condensation coefficient, commensurate with unity, this definition of supersaturation is not sufficient to characterize the solid–vapour system. In this case, we should take into account the mean mass motion of the vapour to the surface with a velocity u. At vapour velocities u much smaller than the velocity of thermal motion of the vapour particles v, j_f^+ can be determined using the approximate formula

$$j_f^+ = nu + \frac{1}{4}nv, \qquad (2.3)$$

where n is the density of the vapour particles.

Equation (2.3) shows that to characterize the condensation conditions and define we need to know the parameters of the vapour flow. The apparent simplicity of the formulation of experiments to study heterogeneous condensation hides the need to diagnose the vapour near the condensation surface. The vast majority of experiments on the condensation of metal films on a solid surface is realized by the scheme: the source of metal vapour in the form of an open crucible with the melt a substrate set at a certain distance. Such a scheme under the assumption of free-molecule evaporation from the surface of the metal melt at a certain geometry can be used to calculate the parameters of the vapour flow in the space around the source. Such calculations have been widely discussed. In the case of other schemes of evaporation – electric discharge, magnetic discharge, laser, electron beam – the calculation of the parameters of the vapour flow is significantly impeded by the complex geometry and non-isothermal conditions. When using such schemes it is compulsory to measure the parameters of the vapour flow.

2.1.1. Metagus-2 vacuum equipment

As shown by long-term experience, a vacuum chamber designed for experimental studies of vacuum condensation processes leading to the formation of disperse systems must meet certain specific conditions.

1. High vacuum is not required, as impurities of foreign gases and vapours take part in the formation of metallic dispersed systems.

2. Multipurpose design for the possibility of carrying out various experiments with different materials and different process parameters.

3. The ability to quickly re-equip using various objects of research and diagnostic methods.

4. Provide easy access to the object under study for both the diagnosis of the object and for the rapid evacuation of the product.

5. Ease of maintenance: cleaning, repair and replacement of the pumping the system.

6. The pumping system must provide fast access to the operating mode.

a   *b*

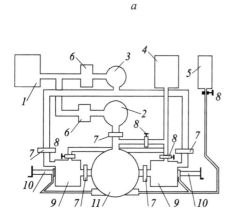

Fig. 2.1. Diagram (a) and appearance (b) of the vacuum system. 1 – the main mechanical pump, 2 – vapour–oil pump in the middle chamber, 3 – vapour–oil pump of the outer chambers, 4 – backing mechanical pump, 5 – mechanical pump of manipulators, 6 – solenoid valve, 7 – slide, 8 – gate 9 – outer chambers; 10 – manipulator, 11 – chamber.

Figure 2.1 shows the diagram of the vacuum chamber, along with the pumping system, developed in accordance with the stated requirements and using the experience gained while working with simpler installations.

The vacuum system consists of three vacuum chambers 9 and 11, connected in series by vacuum slide gates 7. The outer chamber 9 are equipped with manipulators that can be used to transfer samples from chamber to chamber. The pumping system consists of two high-vacuum pumps 2 and 3 with liquid nitrogen traps with a maximum high vacuum of about $1.33 \cdot 10^{-3}$ Pa, three backing mechanical pumps 1, 4 and 5. The pump 1 is designed to create fore-vacuum of vapour–oil pumps, pump 4 – for pre-pumping, the pump 5 is used to ensure the vacuum density of the manipulators. The pumps are connected vacuum leads the remote-controlled gates 6 and 7 and valves 8 with manual control. It provides several options for joint or independent pumping of the three chambers. One of the outer chambers is equipped with a bag gateway for introducing samples into the vacuum chamber without breaking the vacuum (see Fig. 2.1 b).

To install experimental equipment, the middle chamber has three hatches and a stationary metal evaporator with a maximum temperature of 2000 K; in the outer chamber there are five hatches, which greatly facilitates the organization of various methods of diagnosis. The three cameras can be used together in a single experiment or independently for different experiments.

The name of the Metagus installation comes from the abbreviation of the words 'metal gas dynamic installation', which means that it is an analog of the vacuum wind tunnel but working with metal vapours. Its uniqueness lies in the fact that it does not require a powerful pump down system. Metal vapours easily condense on the chamber walls. The performance of such a tube as regards the consumption of the working gas is limited only by the attached vapour source.

2.1.2. Source of metal vapours

All of the above methods of evaporation have one thing in common – a substance evaporates in open space of the vacuum chamber. There is no guarantee that the substance in the open space vaporizes to the molecular level. A possible transfer of substance vapours with the inclusions of a cluster phase into the area of the experiment makes the parameters of the vapour phase uncertain. The vapour source usually radiates to the surrounding space electromagnetic and electric fields or a flux of high-energy particles. This may cause an additional distortion of the condensation conditions.

In addition, most of the known sources are based on the non-stationary heating mode. This fact greatly complicates the study of homogeneous and heterogeneous condensation, since the difficult to measure parameters must be studied on average for some period of time. Therefore, in the formulation of experimental studies of heterogeneous condensation we should formulate specific requirements on the vapour source:

1. The source must generate a flow of vapour with a controlled and regulated content of the clusters.

2. The configuration of the vapour flow from the source should be easy for calculating the parameters of the known classical schemes.

3. Associated emission of the vapour source into the surrounding area and the area of the substrate must be controlled and regulated.

4. The possibility of steady-state operation of the source.

Such requirements are satisfied by the source of metal vapours, designed for the research on homogeneous and heterogeneous condensation [1]. The principal and at the same time the simplest feature of this source is the possibility of superheating the vapour. The scheme of the source is shown in Fig. 2.2. Evaporation of the metal takes place in a molybdenum evaporator 3. The vapour of the evaporated substance is fed into the molybdenum superheater

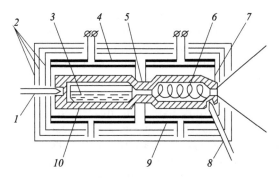

Fig. 2.2. Scheme of the vapour source. 1, 8 – thermocouples, 2 – screens, 3 – evaporated substance, 4, 9, – coaxial heaters, 5 – vapour line, 6 – superheater, 7 – sonic nozzle, 10 – evaporator.

6, filled with molybdenum chips. Superheated vapour is fed to the nozzle 7. A free supersonic jet of superheated metal vapour forms at the exit of the nozzle. Design features of the source: self-contained coaxial heaters of the evaporator and superheater 4 and 9, their thermal uncoupling by the vapour line 5. Sufficiently large surface area ratios of the melt in the evaporator, the area of the channel of the vapour line and the critical section area of the nozzle provides the possibility of determining the stagnation pressure of the vapour flow p_0 on the saturation curve at the evaporator, the stagnation temperature T_0 with respect to the temperature of the superheater and nozzle prechamber. A large enough surface of metal evaporation in the evaporator, compared with the area of the nozzle throat, eliminates the possibility of metal boiling in the evaporator and minimizes the formation of liquid droplets. Sufficiently large orifices of the vapour line and the superheater in comparison with the critical nozzle section eliminate the appearance of a hydraulic pressure drop between the evaporator and the prechamber of the nozzle. Features of trapping of microdroplets in the superheater are discussed separately based on the research presented in Chapter 11. The use of the screens 2 minimizes heat leakage from the source into the surrounding space and, most importantly, into the study area of heterogeneous condensation using a vapour jet. In the application of tantalum heaters and molybdenum screens the source has the following main characteristics:

Power consumption, kW	to 5
The maximum temperature, K	1800
The radiation into the region of the jet at $T_0 = 1350$ K, W/m²	11
The maximum metal charge, ml	20
Time to steady state at $T_0 = 1350$ K, h	1
The maximum temperature difference between superheater and crucible, K	400
The location of the source	vertical, horizontal

The diagram of the electric power circuit of the heaters is shown in Fig. 2.3. Stand-alone heaters of the evaporator and the superheater are fed from the secondary windings of the power step-down transformers. Thyristor power control is realized by a programmable thermostat Proterm-100. The input signal for control is supplied from the thermocouples embedded in the evaporator and superheater bodies.

Features of the pilot study using the described vapour source can be explained by the p–T diagram shown in Fig. 2.4 in logarithmic coordinates. The stagnation parameters p_0, T_0 (point 0) are reached in the nozzle prechamber. In comparison with the saturation curve 1, where K is the triple point, the point 0 is in the superheated vapour region. At pressure in the vacuum chamber p_b and temperature T_b the adiabatic expansion of the metal vapour through the nozzle and in the jet behind the nozzle takes place. The process is shown in Fig.

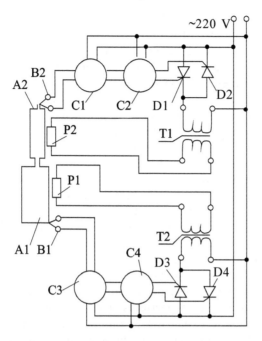

Fig. 2.3. Power supply. A1 – evaporator; A2 – superheater, B1, B2 – thermocouples, C1, C3 – thermoregulators R-133, C2, C4 – Amplifiers U-13, D1, D2, D3, D4 – thyristors, T1, T2 – transformers, P1, P2 – heaters.

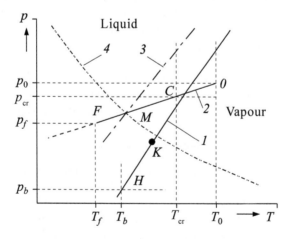

Fig. 2.4. p–T diagram of the process under investigation of heterogeneous condensation of the directional vapour flow. 1 – saturation curve, 2 – adiabate, 3 – the curve of the critical supersaturation, 4 – boundary of translational 'freezing'.

2.4 by the adiabate. Point C represents the vapour parameters in the critical section of the nozzle. For a given nozzle geometry, each point of the adiabate corresponds to a particular section of the vapour flow. In the jet behind the nozzle in a cross-section with the parameters p_f, T_f there is the solid surface used as a substrate in the investigation of heterogeneous condensation. For this section at a certain geometry of the nozzle under the assumption of adiabatic expansion of vapour we can all vapour parameters, including particle density n_p, the Mach number of the flow M_p, average mass velocity u_p, temperature T_f. At a constant geometry of the nozzle and the position of the substrate by changing the mode of the vapour source we can move the point 0 in the coordinates p–T. Point F moves by the same distance as point 0, which means free variation of the parameters of the vapour, incident on the substrate. For example, moving point 0 along the isotherm shown in the Figure changes p_0 at T_0 = const, which corresponds to changes of the flow incident on the vapour substrate at a constant temperature. Knowing the parameters of the vapour at the point F, we can easily calculate its supersaturation with respect to the given conditions on the substrate (point H).

We should discuss some features of this formulation of the study of heterogeneous condensation associated with the possible emergence of homogeneous condensation. In Fig. 2.4 curve 3 denotes the critical supersaturation for the start of homogeneous condensation. If the point F is located outside the segment MC, the jet may show homogeneously formed clusters and condensed phase particles. Curve 4 shows the conditional boundary of the adiabatic expansion of the vapour in the continuous gas-dynamic mode while moving along the point 0 along the isotherm T_0. The section of the adiabate MF corresponds to collisionless expansion of vapour. The appearance of a homogeneous condensate in the jet is determined by the curve (3 or 4) first intersected by the adiabate. If the adiabate intersects first curve 4, the condensate cannot form in the jet and part of the curve MF corresponds to the adiabatic collisionless acceleration of the vapour particles in a vacuum. If the adiabate first intersects the curve 3, the substrate at the point F will interact with the vapour flow.

Thus, the vapour source described here allows the study of heterogeneous condensation as both purely molecular and clustered vapour in a wide range of supersaturation, temperature, and the velocity of the vapour.

2.1.3. The structure of the magnesium vapour jet

Experiments with the production and study of magnesium vapour jets [2] were conducted using the circuit shown in Fig. 2.5, at the nozzle temperature of 1273 K, which determines the stagnation temperature T_0 of the vapour. The temperature of the evaporator with the melt was maintained at 1073 K, which corresponds to the stagnation pressure p_0 = 4256 Pa, defined by the saturation pressure for magnesium. The diameter of the outlet of the sonic nozzle d_{cr}, used in most experiments, was equal to $4 \cdot 10^{-4}$ m. The values of the Reynolds and Knudsen numbers, defined by the parameters at the nozzle exit, with p_0 = 4256 Pa and T_0= 1723 K, were Re = 84, Kn = 0.04. The viscosity of magnesium vapour and the mean free path for the calculation of Re and Kn were calculated from the kinetic equations. The pressure in the vacuum chamber during experiments was maintained equal to $1.33 \cdot 10^{-2}$ Pa.

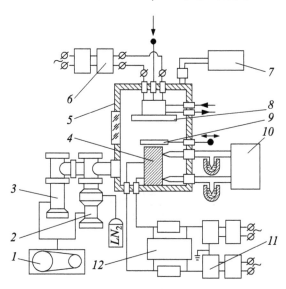

Fig. 2.5. The experimental setup. 1 – backing pump, 2 – high vacuum pump, 3 – booster pump, 4 – vapour source, 5 – chamber, 6 – power supply, 7 – vacuum gauge, 8 – plate, 9 – screen, 10 – potentiometer, 11 – transformer, 12 – control of the power source.

The structure of the jet of magnesium vapours was studied by measuring specific flow ρu. Here ρ, u is the density and velocity of the vapour. The method consists in condensing magnesium vapours on a copper plate 8 (see Fig. 2.5) installed across the jet. The plate was mounted on a heat exchanger equipped with electric heating from the power source 6 and cooled by liquid nitrogen, which allowed her to set the temperature in the range 77–500 K. The movable flap 9 was used to eliminate the transient regimes of operation of the source and control the time of interaction of the jet with the plate. In this procedure, the important parameter is the temperature of the plate. In the interaction of magnesium vapour with the plate having a temperature of less than 430 K the resultant condensate takes the form of a cracked and exfoliated film. At the temperature of the plate in the experiments of 430–470 K the efficiency of condensation is close to unity, and the adhesion of the condensate is sufficient for the formation on the plate of a dense solid film whose thickness can be measured. During the condensation of magnesium on the plate (1–5 min), the maximum film thickness was varied in the range of 0.3–1.5 mm.

Transverse profiles of the specific consumption, measured at various distances x from the nozzle, normalized by the values on the axis of the jets, are shown in Fig. 2.6. Curves 1–4 correspond to x/d_{cr} = 30, 50, 70, 100. There is a rough similarity of the profiles in the coordinates ρu, y/x, which indicates that in the experimental conditions the magnesium vapour jet can be modelled by a spherical source like the gas jet [3].

The resulting profiles of specific consumption are significantly narrower than the profiles corresponding to the expansion of a monatomic ideal gas. The

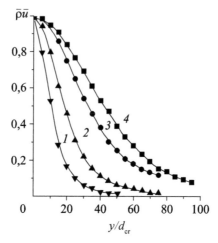

Fig. 2.6. Transverse profiles of the relative specific consumption in the magnesium vapour jet. $p_0 = 4256$ Pa, $T_0 = 1273$ K, $d_{cr} = 4 \cdot 10^{-4}$ m, $x/d_{cr} = 30$ (1), 50 (2), 75 (3), 100 (4).

difference in the profiles is due, apparently, to the presence of condensation processes in the jet of magnesium vapour, narrower profiles form by the impact of condensation on the expansion of the vapour in the jet. The measured specific consumption is the sum of specific consumptions of vapour and condensed phases. The share of the condensate near the axis of the jet can be increased due to better conditions for condensation and slight dispersal separation of the particles in the condensate formed at small distances from the nozzle outlet.

We should expect that the processes of condensation in the jet are sensitive to the effects of energy from outside. Then the changes in the jet under the influence of external energy could confirm the presence of condensation processes. The impact of energy on the vapours in the jet without the mechanical disturbances of its structure may have the form of a glow discharge. Profiles were measured in the conditions of excitation of high-frequency discharge (HF) in the jet. To do this, electrodes made of steel wire with a diameter of 2 mm were placed in the periphery of the vapour jet. The frequency and power of the discharge were respectively 40.6 MHz and 20 W.

It is established that the transverse profile of specific flow in the jet with a glow discharge is in better agreement with the calculated data for the isentropic expansion of a monatomic gas. It can be assumed that the heating of magnesium vapour wit HF discharge leads to evaporation of part of the condensate in the jet. The fact of matching theory and experiment is an indirect confirmation of the effect of condensation on the structure of the jet.

Calculation of the flow of Mg vapour in the jet behind the sonic nozzle was carried out in the approximation of the inviscid gas with no thermal conductivity with the condensation process taken into account [4]. It was believed that the process of condensation in the jet is described by the

Becker–Doring–Frenkel'–Zel'dovich–Frenkel' theory, and the condensate fraction is calculated using the algorithm proposed in [5].

The following physical properties of magnesium were used in the calculations: the surface tension σ = 0.569 N/m, heat of vaporization L = $5.44 \cdot 10^6$ J/kg, the density of liquid magnesium ρ_L = $1.59 \cdot 10^3$ kg/m³, the curve of the phase transition p = $1.33 \cdot 10^2 \cdot 10^{8,82-7741/T}$ Pa; the heat capacity of liquid Mg c_L = $3.31 \cdot 10^4$ J/(K·kmol) and vapour Mg c_p = $2.08 \cdot 10^4$ J/(K·kmol).

The calculation assumed a constant value of surface tension σ = const = 0.569 N/m at T = 924 K, independent of temperature. The value of σ and its temperature dependence strongly affect the nucleation rate, so $I = AI_D$, where I_D is the rate of nucleation according to the classical capillary nucleation theory. To obtain meaningful results in the calculations, the value of A was varied from 1 to 10^{10}.

Figure 2.7 shows the profiles $\overline{\rho u}$ at a distance x/d_{cr} = 75 from the nozzle outlet, normalized for the axis values obtained in the experiment (curve 1) and also the calculated profiles for the flow regime of magnesium vapour without condensation (γ = 1.67, curve 2) and in the presence of condensation in the jet (curve 3). It is seen that the calculated profiles 2 and 3 are much flatter. By varying the calculation parameters it became clear that the experimentally obtained profiles $\overline{\rho u}$ can not match the calculated profiles obtained in this model. Based on the above, the following assumption is made: the difference is due to the fact that in this discharge mode an important role is played by slippage of the condensate particles; the area of formation of these particles is at a distance of one or two diameters from the nozzle. If we assume that after reaching the maximum rate of nucleation the condensate particles move in straight lines, we obtain the profile 4 in Fig. 2.7; the half-width of this profile is close to the half-width of the experimental profile. In addition, it appears that part of the substrate (region A in Fig. 2.8) approximately corresponds to the corner into which all the condensate 'falls' assuming complete slipping.

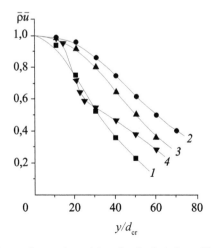

Fig. 2.7. Comparison of experimental and calculated profiles of specific consumption.

The sharp decline in the profile 4 corresponds to the transition from the field on which the condensate 'falls' to the region in which only the vapour phase contributes to $\overline{\rho u}$. The experimental profile shows a smooth transition from one region to another. This difference is probably due to the fact that the actual trajectory of the particles is not straight, although the slippage is rather large.

Thus, it seems likely that the experimentally obtained profiles $\overline{\rho u}$ are the result of condensation in the jet. However, in this flow regime it is difficult to rely on the agreement between calculation and experiment, as calculations use the continuum model, the applicability of which is limited to a very small region near the nozzle exit at Kn = 0.04.

Note that the described calculations are qualitative in nature, since the available experimental data are valid for the discharge mode for which the theoretical model clearly makes a substantial error. Nevertheless, an amendment to the model (the assumption of slippage), which seems to be correct in terms of physics of the process, makes it possible to obtain a qualitative agreement with the experimental data.

2.1.4. The interaction of a metal vapour jet with a solid surface

Condensation processes in the jet, of course, affect to some extent the process of its subsequent condensation on the surface, so the review of the processes of condensation on the surface can give information about the structure of the jet. A more detailed study of heterogeneous condensation is described in Chapter 3. However, changes in the method of measuring the specific flow rate produced some interesting results that characterize the metal vapour jet.

The scheme of these experiments is shown in Fig. 2.8. The massive copper plate 2 is installed across the Mg vapour jet which flows from a sonic nozzle 1. Additional plate 3 was mounted perpendicularly on a copper plate to create areas of shading the copper plate from the vapour flow.

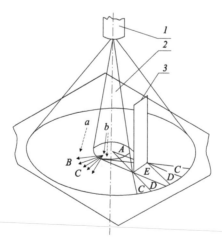

Fig. 2.8. Diagram of the jet. 1 – nozzle, 2 – copper plate, 3 – plate mounted perpendicular to the copper plate.

Experiments with the condensation of magnesium vapours on the copper surface with artificial shading in the temperature range 300–370 K showed a number of structural features of the jet. First, there was a characteristic radial orientation of the crystals of larger size formed by condensation of the jet in region *B*. Second, the central, axial part *A* of a film condensed on the plate 2, examined at a magnification of 90 showed a relatively larger surface roughness than area *B*. Third, all the rough places in the area *B*, for example parts of the film separated from the plate, were covered by individual particles with a typical size of about 0.01 mm. Fourth, vertical barriers 3, mounted on the plate 2 perpendicular to its surface, give a shadow image. Zone *E* is free of condensation of magnesium, it contains only a few magnesium particles. Zones *D* are covered by a deposit of condensate particles with a characteristic size of about 0.01 mm. In areas *C* there is a condensate consisting of a dense flat film with embedded individual particles. The nature of the shadow image behind the obstacle shows that the peripheral portion of the plate is covered with magnesium, coming from two sources. One source is nozzle 1, the other – the area *A*.

These features suggest the following picture of the interaction of the Mg vapour jet with the plate. The paraxial part of the jet, bounded by the conical surface with half-angle at the vertex of 17°, is the flow of the gas phase, conventionally represented by the dashed arrow *a*, and carried the particles of the condensate shown by the arrow *b*. The coefficient of adhesion of the condensate particles on the plate seems to be less than unity. Part of the condensate particles, impinging on the surface, are partially reflected by it, and perhaps destroyed on impact. The resulting particle debris, shown in Fig. 2.8 by the arrows *c,* fly along the surface of the plate.

Thus, close to the plate in the region *B* there are two flows: vapour phase, as shown in Fig. 2.8 in the form of dashed arrow *a* with the velocity vector directed from the nozzle, and fragments of condensate particles having on average the radial direction. Due to the collisional interaction of the particles formed by these streams flow of material coming in the radial direction of the condensation surface at an angle to it. This may be due to the radial orientation of the crystals on the plate and the shadow picture of the obstacles. The accumulation of the particle debris at low obstacles is explained by settling of the particles on the surface of stagnant areas.

2.1.5. The mechanical properties of condensate films

Of interest are the mechanical properties of films obtained by condensation of Mg vapour jets on the solid surface. In this regard, measurements were taken of their microhardness in the places corresponding to the condensation of the axial part of the jet. Microhardness was determined by the Vickers method in Epitip-2 equipment. The load applied to the samples was 50 g, loading time 10 s. The measurements were performed on films obtained in different experimental conditions. The results are shown in Table 2.1.

Table 2.1. Microhardness of films produced in different conditions

$T_N = 1273$ K, $T_0 = 1073$ K, $x/d_{cr} = 100$	H_v
Sonic nozzle	40.9
Supersonic nozzle	36.2
Sonic nozzle with HF discharge	73.6
Microsection of cast magnesium	36.1

The table shows that the films obtained by condensing jets have increased microhardness compared with the original cast magnesium microhardness. The properties of the film are especially strongly affected by the presence of a glow discharge.

Attention should be given to noteworthy features of the films obtained on the surface of the plate under the condensing vapour jet. In most experiments the area B is bordered by a strip of darker colour condensate consisting of a deposit of ultrafine particles. This result is apparently explained by the fact that in the peripheral part of the vapour jet condensation occurs in the residual gas in the chamber. This process is similar to that used in the gas-phase method for obtaining ultrafine powders [6]. Precipitation of magnesium powder on all surfaces inside the vacuum chamber is also connected with the interaction of the Mg vapour jet with the residual gas in the chamber. Moreover, it has been noted that the deposition of magnesium occurs mainly on the heated surfaces.

2.1.6. The residual atmosphere in the chamber

When investigating the heterogeneous condensation in vacuum the qualitative and quantitative composition of the atmosphere of residual gases in the chamber is of considerable importance. Continuous monitoring of the residual gases content in the chamber with vapour–oil pumping is extremely difficult due to the fact that mass spectrometers are usually unstable in such a system. Prolonged use of the mass spectrometer in a vacuum chamber with the presence of heavy organic compounds is accompanied by contamination of its scanning and detecting units, and the spectrometer loses its sensitivity. Therefore, only a qualitative study of the atmosphere in the chamber and the effectiveness of application of the liquid nitrogen freezing trap was carried out.

A monopole mass spectrometer MX-7304 was used. The ionizer of the mass spectrometer was mounted on the axis of the Mg vapour jet in the body, cooled with liquid nitrogen. In the cooled body there was an opening for feeding in a molecular beam of the Mg vapour jet in the area of the ionizer of the mass spectrometer. Measurements of the atmosphere in the vacuum chamber were taken with the activated magnesium vapour source. Figures 2.9 and 2.10 show the mass spectra of the residual atmosphere in the chamber: the first figure – in the absence of liquid nitrogen in the frozen traps of the chamber and the

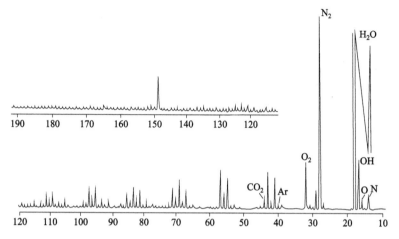

Fig. 2.9. The mass spectrum of the residual atmosphere of the vacuum chamber without freezing with liquid nitrogen.

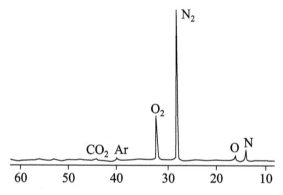

Fig. 2.10. The mass spectrum of the residual atmosphere of the vacuum chamber with freezing for the trap of the ionizer.

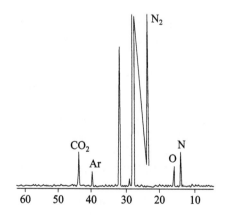

Fig. 2.11. The mass spectrum of the residual atmosphere of the vacuum chamber with the heater of the vapour source switched on.

Table 2.2. Molar composition of the residual atmosphere of the vacuum chamber in three different conditions

Mode	p, 10^3 Pa	N	O	OH	H_2O	N_2	O_2	Ar	CO_2	$(CH_2)_nH_m$	I, rel. units
Without cooling and source	13.3	0.98 (4.65)	0.63 (2.97)	3.9	22.1	15.1 (70.8)	3.6 (16.8)	3.5 (1.62)	0.68 (3.14)	52.8	862 (184.9)
Cooling with liquid nitrogen without source	8	4.9	2.5	–	–	70.7	21	1.4	Traces	–	100.5
Cooling with liquid nitrogen with activated heater of the source	8	4.21	2.53	–	–	69.4	17.9	1.7	4.25	–	106.8

Comment. The last column of the table gives the total intensities of the spectrum lines in relative units. The values in the brackets are the results of inspection of air components.

mass spectrometer and a total pressure of $1.33 \cdot 10^{-2}$ Pa, the second – during the freezing of liquid nitrogen traps of the chambers and the mass spectrometer at a total pressure in the chamber of $8 \cdot 10^{-3}$ Pa. In Fig. 2.10 the mass spectrum was obtained with the heater of the metal vapours switched off, in Fig. 2.11 when the heater of metal vapours switched on. The results of processing of mass spectra, assuming the equal size of the ionization cross section and the equal multiplicity of the components are, shown in Table 2.2.

According to the results given in the table, we can draw the following conclusions:

1. Comparison of the first and second rows of the table shows that the change of pressure in the chamber during cooling down liquid nitrogen traps is approximately proportional to the integrated intensity of the spectral lines belonging to the non-condensable component of air. This means that the liquid nitrogen of the trap chamber with a factor of ≈ 0.6 reduces the total pressure in the chamber without changing the composition of the residual atmosphere. More than half of the molar composition of the atmosphere are the fragments of hydrocarbons. The composition of the atmosphere as regards the non-condensable component of air remains constant.

2. The liquid nitrogen trap of the mass spectrometer can almost completely stop penetration of organic components into the ionizer, and so this setting can be used for the diagnosis of the vapour jet.

3. Comparing the second and third rows of the table indicates the appearance of an additional gas flow from the source of vapour with the

heaters switched on. In this stream there is an increased content of carbon dioxide formed, apparently, as a result of the pyrolysis and oxidation of hydrocarbons in the vapour source heaters.

The approximate constancy of the content of the main components of the residual atmosphere (organic matter, water, molecular nitrogen and hydrogen) allows experiments to investigate the heterogeneous condensation without a detailed continuous monitoring of atmospheric composition. In the results presented here it was considered sufficient to measure the pressure in the chamber.

Elevated levels of carbon dioxide emitted by the heaters of the source suggest that the high-temperature tantalum heaters of the source are protected from oxidation by the presence of vapours organic substances in the residual atmosphere. Indeed, during the ten-year operation of vapour sources of the construction described there was no failure of these devices due to oxidation of the heaters.

2.1.7. The choice of simulation metal for experiments

The specificity of the research of the formation of disperse systems imposes the following requirements on the metal used in the experiments:

1. High enough saturated vapour pressure at moderate temperatures for varying the parameters of the vapour flow in a wide range.

2. Resistance of the vapour source materials in the desired temperature range to molten metal.

3. Study of the basic physical properties of the metal in three phases: liquid, solid and vapour.

4. Moderate chemical activity of the metal to the media of the experimental vacuum units, which makes it possible, if desired, to investigate the role of chemical reactions in the tested process.

5. The practical applicability of disperse systems of the investigated metal.

In accordance with the requirements formulated in this paper, magnesium was selected as the basic model metal.

Table 2.3 shows the values of the vapour pressure of magnesium. The saturation curve for magnesium is described in most detail in [7]. The physical properties of magnesium are presented in [8] and in Table 2.4. The chemical properties of magnesium are given in [9].

2.2. Calorimetry of heterogeneous condensation process

In heterogeneous condensation the latent heat can cause heating of the condensation front and changes in the conditions of structure formation.

Table 2.3. Vapour pressure of magnesium

p, 133 Pa	10^{-11}	10^{-10}	10^{-9}	10^{-8}	10^{-7}	10^{-6}	10^{-5}	10^{-4}	10^{-3}	10^{-2}	10^{-1}	1	10	10^2	10^3
T, K	388	410	432	458	487	519	555	600	650	712	781	878	1000	1170	1400

Table 2.4. Physical properties of magnesium

Parameter	Value
Isotopic composition in nature	Mg^{24}(78.6 %), Mg^{25}(10.11 %), Mg^{26}(11.29 %)
Configuration of outer electrons	$3s^6$
Ionization energy $Mg^0 \rightarrow Mg^+ \rightarrow Mg^{2+} \rightarrow Mg^{3+} \rightarrow Mg^{4+} \rightarrow Mg^{5+}$, eV	7.64; 15.03; 80.12; 109.29; 141.23
Hexagonal, close-packed crystal lattice, Å	$a = 3.2028$, $c = 5.199$
Atomic mass, conv. units	24.312
Atomic radius, Å	1.6
Density 99.9% Mg, kg/m³	$1.739 \cdot 10^3$
Melting point, K	924
Boiling point, K	1380
Heat of melting, J/kg	$3.614 \cdot 10^5$
Heat of evaporation at 1380 K, J/kg	$5.249 \cdot 10^6$
Specific heat capacity at 293 K, J/(kg K)	1038
Specific heat conductivity, W/(m K)	15

Nevertheless, the thermal effects in condensation have been studied insufficiently. In [10] direct measurements were taken of the thermal effects in the condensation of bismuth and indium in order to assess their potential impact on the conditions of structure formation of the condensate. Measurements were taken with a film thermocouple. Marked thermal effects associated with thermal radiation of the vapour source, the heat of condensation and the enthalpy of vapour were observed. It was found that the main contribution to the thermal effects comes from the heat of condensation. The magnitude of the measured thermal effects did not exceed 8–12 K. Study [11] discussed the role of thermal effects in the observed shifts in of the condensation mechanisms in the experiments.

Theoretical analysis of the thermal problem in the process of heterogeneous condensation was carried out in [12]. The temperature fields on the substrate and the layer of the condensate were calculated, taking into account the incident and reflected components of the heat flux.

Separate determination in the experiments of the main heat flux components makes it possible to extract quantitative information about the process of condensation from the analysis of the thermal state of condensation front, in addition to determining the condensation conditions. This is especially important when we study the condensation of the clustered vapour flow. By combining independent measurements of the rate of condensation and the thermal effect of condensation we can judge the composition of the deposited flux. Therefore, the aim of this work was to study the problem of developing methods for measuring the heat flux and surface temperature, suitable for use in demanding conditions. The basis was the transient method of measuring the heat flows into the body in the regular mode.

A differential calorimeter in the form of a metal plate suspended on a wire heater and a thermocouple is placed in a metal vapour jet. The metal vapour jet, supplied to the calorimeter, can be interrupted by a movable flap.

2.2.1. Measurement of the heat content of vapour in the jet

Heating of the body placed in a stream of metal vapour is described by the equation

$$c_d m_d \left(\frac{dT}{d\tau} \right)_1 = P^+ - P^-(T), \qquad (2.4)$$

where c_d, m_d is the specific heat and mass of the sensor; P^+ is the power delivered to the sensor by the vapour jet; $P^-(T)$ is the power of the heat losses of the sensor. Changes in the temperature of the sensor in time are shown conventionally in Fig. 2.12 as section 1. At the time τ_b the vapour flow is blocked by a flap. Cooling of the body under the assumption that the losses depend only on its temperature can be described by the equation

$$c_d m_d \left(\frac{dT}{d\tau} \right)_2 = -P^-(T). \qquad (2.5)$$

The solution of this equation is graphically depicted in Fig. 2.12 as part of the curve 2. Subtraction of (2.5) from (2.4) at the same temperatures T gives

$$c_d m_d \left[\left(\frac{dT}{d\tau} \right)_1 - \left(\frac{dT}{d\tau} \right)_2 \right] = P^+. \qquad (2.6)$$

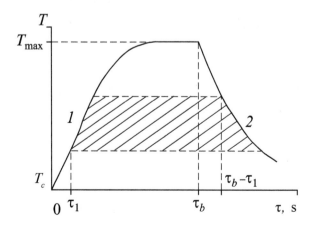

Fig. 2.12. Time dependence of temperature. 1 – heating; 2 – cooling. The shaded area – steady indications of the calorimeter.

Equation (2.6) can be used to determine the power of heat exchange of the body with a jet of metal vapour from the experimentally measured dynamics of the temperature of the body, similar to that shown in Fig. 2.12. Processing of the experimental curve at different T can help to determine the power P^+ as a function of the body temperature during steady-state conditions in the jet or if $(T_{max} - T_c) / T_c \ll 1$ – power of unsteady heat transfer.

The calorimeter, used in the experiments, was made on the basis of a vacuum thermocouple pressure transducer PMT-2. At the junction of the thermocouple and the heater of the probe there was placed a copper plate with a thickness of 50 μm with the dimensions of about 2×2 mm, making a total mass of the calorimeter $m_d = 1.71 \cdot 10^{-6}$ kg, the area in plan of $F_d = 2.635 \cdot 10^{-6}$ m². The heat flux to the calorimeter can be determined by the formula

$$q^+ = \frac{c_d m_d}{F_d} \left[\left(\frac{dT}{d\tau} \right)_1 - \left(\frac{dT}{d\tau} \right)_2 \right].$$

An example of measurements of q^+ on the magnesium vapour jet axis arising from the sonic nozzle with a critical diameter $d_{cr} = 0.4$ mm at a distance from the nozzle $x = 227$ mm is shown in Fig. 2.13. The figure shows that within 8 s after applying the jet to the calorimeter measurements become unstable, and at $\tau_1 > 9$ s the readings stabilized: $q^+ = 40; 36.5, 37.5$ W/m² for points 1, 2, 3, respectively.

For these same conditions Fig. 2.14 shows the specific heat flux of cooling of the calorimeter, calculated from the measurements by the formula

$$q^- = \frac{c_d m_d}{F_d} \left(\frac{dT}{d\tau} \right)_2.$$

A characteristic feature of all measurements is the presence of the linear part of the dependence of q^- on $(T - T_c)$. Since the measurements were carried

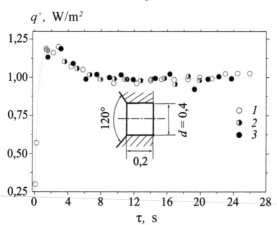

Fig. 2.13. An example of measurements with the calorimeter of the specific heat flux in relation to time.

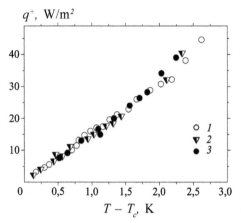

Fig. 2.14. The specific heat flux of cooling the calorimeter. The experimental points correspond to Fig. 2.13.

out under high vacuum in the absence of convective heat transfer, the linear dependence of q^- on $(T - T_c)$ indicates the nature of the conductive cooling of the calorimeter in its wire harness in the regular mode:

$$c_d m_d \left(\frac{dT}{d\tau} \right)_2 = \frac{(T - T_c)\lambda_h}{l_h} F_h.$$

Here λ_h is the thermal conductivity of the material; l_h is length; F_h is the total cross section of the wire suspension. The copper plate – the calorimeter – is suspended on two platinum wires with a diameter of 0.1 mm, length 20 mm and a chromel–copel thermocouple with a diameter of 0.07 mm. Because of the relatively high thermal conductivity of the plate in comparison with chromel and copel it is possible to estimate cooling of the calorimeter by the thermal conductivity of the platinum wire. Under the assumption of the Fourier number

$$\text{Fo} = \lambda_h \tau / \left(c_d \rho_d l_h^2 \right) = 1,$$

the estimate of the characteristic time of establishment of the steady-state heat flux of leakage is calculated as

$$\tau_h = c_d \rho_d l_h^2 / \lambda_h = 20 \text{ s}.$$

A similar value the time of establishment of heat transfer in a copper plate gives

$$\tau_d = c_d \rho_d l_d^2 / \lambda_d = 0.04 \text{ s}.$$

The results of these evaluations show that the instability of the readings at the first seconds after the arrival of the flow to the calorimeter (see Fig.

2.13) can be explained by the unsteady heat transfer in the device. Such non-stationary measurements are also observed at the junction of the parts of the curves 1 and 2, shown in Fig. 2.12. This is confirmed by the increase in the scatter of the points in Fig. 2.14 in the region of maximum $T - T_c$. In this connection it is expedient to use the results which lie in the shaded area in Fig. 2.12.

Figure 2.13 shows that the random measurement error in the steady-state heat transfer region is no more than 5%. The measurement accuracy can be improved by statistical processing of this section. Experience with 10 points of measurement in the shaded area shows that the measurement error can be reduced to ±1%.

The dependence $q^-(T - T_c)$, obtained at some specific values, can be used as a calibration curve for measurements of q^+ at lower T_{max}. The maximum error of these measurements is well within ±1%.

If we use this calorimeter in an environment where the temperature region $T_c - T_{max}$ is located in the range of possible condensation of metal vapours on the plate of the calorimeter, we should be aware that m_d and c_d are variable. The above processing of the measurement results is valid only if the mass of condensate on the plate is $m_c \ll m_d$. The condensate can be from the plate by heating to a temperature at which sublimation of the condensate becomes significant, but the sublimation of the plate is negligible. Experience in measurements in magnesium jets shows that the evaporation of magnesium and complete cleaning of the plate surface at a pressure in the vacuum chamber of $p_b = 5 \cdot 10^{-3}$ Pa can be carried out efficiently by heating to a temperature of 700 K.

2.2.2. Measurement of specific flow in the metal vapour jet

Figure 2.15 shows the curves 3 and 4 of heating the calorimeter at a constant heating power respectively after measurements of q^+ in the mode of condensation of substances and the clean calorimeter. The characteristic defect

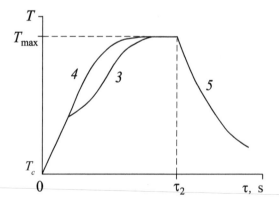

Fig. 2.15. The curves of thermal cycling in the calorimeter: in the evaporation mode of the condensate (curve 3) and the clean calorimeter (curve 4); 5 – cooling after switching off the heater at time τ_2. Curves 1 and 2 – see Fig. 2.12.

in the heating curve 3 indicates the evaporation of the condensate produced on the plate of the calorimeter during the measurement of q^+.

The curves 3 and 4 can be used to calculate the mass of the condensate available on the calorimeter before heating, assuming that the equilibrium evaporation of the condensate. The equation of the curve 4 can be written as in (2.4)

$$c_d m_d \left(\frac{dT}{d\tau}\right)_4 = P_h - 2\varepsilon_4 \sigma_0 f \left[\left(\frac{T_4}{100}\right)^4 - \left(\frac{T_c}{100}\right)^4\right], \qquad (2.7)$$

where p_h is the heater power; ε_4 is the emissivity of the plate of the calorimeter without the condense; σ_0 is the Stefan–Boltzmann constant; f is the form factor of the radiative heat transfer. In (2.7), the heat loss from the calorimeter plate is in the form of the radiation component.

The equation of curve 3 can be written as

$$c_d m_d \left(\frac{dT}{d\tau}\right)_3 = P_h - 2\varepsilon_3 \sigma_0 f \left[\left(\frac{T_4}{100}\right)^4 - \left(\frac{T_c}{100}\right)^4\right] F_d - L\frac{dm_c}{d\tau}, \qquad (2.8)$$

where ε_3 is the effective emissivity of the calorimeter plates with the condensate in the general case – a variable during evaporation; L is the latent heat of sublimation at the vaporization temperature. Subtracting (2.8) from (2.7) at the same time and integrating the resulting expression gives the equation

$$m_c = \frac{c_d m_d}{L}(T_4 - T_3) + \frac{2F_d f \sigma_0}{L}\left\{\left(\frac{T_c}{100}\right)^4 \int_0^\tau \left[\varepsilon_3(T) - \varepsilon_4\right] d\tau - \right.$$
$$\left. - \int_0^\tau \varepsilon_3(T)\left(\frac{T_3}{100}\right)^4 d\tau + \varepsilon_4 \int_0^\tau \left(\frac{T_4}{100}\right)^4 d\tau\right\}. \qquad (2.9)$$

However, due to the indeterminacy of $\varepsilon_3(T)$ the expression (2.9) is not suitable for the processing of the experimentally obtained curves 3 and 4 (see Fig. 2.15). At the same time in the experiments noted that curves 3 and 4 are the same before the start of evaporation. This indicates equality $\varepsilon_3(T) = \varepsilon_4$, which greatly simplifies the expression (2.9):

$$m_c = \frac{c_d m_d}{L}(T_4 - T_3) + \frac{2\varepsilon_4 F_d f \sigma_0}{L}\int_0^\tau \left[\left(\frac{T_4}{100}\right)^4 - \left(\frac{T_3}{100}\right)^4\right] d\tau.$$

The value of $\varepsilon_4 f$ can also be determined from experiment. From (2.7) if $T_4 = T_{max}$, $(dT/d\tau) = 0$, it follows

$$P_h = 2\varepsilon_4 F_d f \sigma_0 \left[\left(\frac{T_{max}}{100}\right)^4 - \left(\frac{T_c}{100}\right)^4\right]. \qquad (2.10)$$

Substitution of (2.10) in (2.7) yields

$$\varepsilon_4 f = \frac{c_d m_d \left(\dfrac{dT}{d\tau}\right)_4}{2 F_d \sigma_0 \left[\left(\dfrac{T_{max}}{100}\right)^4 - \left(\dfrac{T_c}{100}\right)^4\right]}. \qquad (2.11)$$

Thus, the experimentally obtained curves 3 and 4 (see Fig. 2.15) with the aid of (2.10) and (2.11) can be used to calculate the mass of the condensate evaporated from the plate of the calorimeter, and the specific local vapour flow $\rho u = m_c / \tau_1 \alpha_c F_d$, where τ_1 is the time of condensation on the calorimeter, α_c is efficiency of condensation; ρ, u is the density and velocity in the flow. Indeterminacy of α_c caused the measured value to be interpreted as $\alpha_c \rho u$.

2.2.3. The results of testing methods

In the process of adjustment of thermal measurements, using a source of thermal radiation with a power of about 1 W, it was possible to determine the minimum sensitivity of the calorimeter. It was 0.1 W/m². Later this value was used to determine the boundary of minimum pressure in the vapour jet available in the measurements. It was also found that at this sensitivity the isothermal nature of the structural elements of the installation around the calorimeter is very important. In particular, the greatest obstacle to measuring is the radiation of the source nozzle, open in the camber. The effects of changing the atmosphere inside the chamber, associated with varying modes of operation of the vapour source, are also important. In connection with this the calorimeter was placed in a thermally stabilized housing, which reduces the influence of Mg vapours scattered in the chamber. The radiation from the vapour source was reduced by stopping the flow with a hole 20 mm in diameter, made in a water-cooled plate installed between the source and the shutter. For reasons of stabilizing the stray radiation from the vapour source, falling to the calorimeter through the diaphragm, the measurements were performed at a constant temperature of the nozzle. Radiation was initially measured at the beginning and end of the experiment in the absence of magnesium vapour jets and was taken into account in determining the heat flux $J = q^+ - q_R^+$. It was also noted that to obtain stable reproducible results, the modified calorimeter must be trained by multiple measurements.

Under the assumption of complete thermal accommodation of the vapour atoms incident on the calorimeter, the measured enthalpy of the flow can be approximately represented as the enthalpy flow $J = \rho u [C_p (T_0 - T_a) + L (1 - \alpha_c)]$, where c_p is the specific heat flux at the measured point; α_c is the percentage of the possible condensate. According to [8] L is approximately an order of magnitude higher than $c_p(T_0 - T_a)$ at the given T_0. Therefore, for qualitative analysis of the measurement results it is sufficient to use $J = \rho u L / (1 - \alpha_c)$.

Measurements of the specific consumption of magnesium in the jet using a calorimeter by the method described above are associated with the following constraints.

1. Reduction of the heating power in evaporation of magnesium built up on the calorimeter increases the role of the second term in (2.10). This, in turn, reduces the accuracy of measurements due to increased errors in integration.

2. Increasing the heater power and hence the heating rate leads to non-equilibrium evaporation. This can result in an error associated with incorrect consideration of the heat expended on evaporation.

It was established by experiments that at the heating rate, exceeding 240 K/s, and the thickness of the magnesium deposited in the calorimeter of few microns it is possible to separate Mg from the calorimeter in the form of parts of the film up to hundreds of microns thick. Therefore, the experiments employed heating rates of less than 150 K/s. In addition, to reduce random errors, the exposure time of magnesium accumulated on the calorimeter was chosen to ensure that the evaporation was performed on the same amount of magnesium regardless of the experimental conditions. For example, the exposure time of the calorimeter at $p_0 = 6.6 \cdot 10^3$ Pa was 5 min, at $p_0 = 1.3 \cdot 10^4$ Pa it was 2.5 min, which corresponds to the thickness of the Mg layer deposited on the calorimeter of 0.8 μm. Almost complete evaporation of the magnesium layer at a heating rate of the calorimeter of 150 K/s occurred within ≈ 1 s, which corresponds to the fraction of the second term in equation (2.8) of about 30%. Given that neglect of the second term greatly simplifies the processing of the measurement results, this is acceptable for obtaining relatively accurate results.

2.3. A technique for measuring the effectiveness of the condensation

Measurements of the effectiveness of condensation are important both for understanding the mechanism of heterogeneous condensation and for evaluating the effectiveness of the process during the formation of the condensate useful for practice. Finally, the value of the efficiency of condensation may be closely related to the type of structure of the resultant condensate.

The conventional method for determining the efficiency of condensation is based on the ratio of the measured rate of condensation to the specific consumption of the vapour supplied to the surface. However, in actual experiments, the determination of the specific vapour consumption represents a significant challenge. The absence of a simple but non-perturbing method of measuring the specific flow makes requires the use of relative values of the effectiveness of condensation.

This section sets out the methodology of determining the absolute values of the efficiency of condensation, which does not require knowledge of the parameters of the oncoming flow. The diagram of the device with which this technique can be implemented is shown in Fig. 2.16.

In front of the extended surface of the substrate 2, on which it is required to measure the efficiency of condensation there is the wire 1 mounted in the

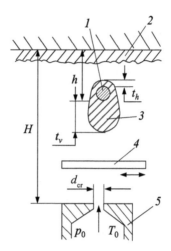

Fig. 2.16. The scheme of the device for measuring the coefficient of condensation. 1 – wire; 2 – condensate on the substrate; 3 – condensate on the wire; 4 – flap; 5 – nozzle of the vapour source.

parallel direction with the diameter much smaller than the spacing. In a real experiment it is a tantalum wire with a diameter of 0.2 mm at a distance of $h = 6$ mm. Near the vapour source there is a movable flap pair 4 for interrupting the vapour flow. The characteristic size of the condensate 3 on the wire can be used to calculate from the efficiency of condensation. The thickness of the condensate t_v in front at the wire can be determined by the formula

$$t_v = \alpha_v \, j \Delta \tau / \rho, \tag{2.12}$$

where α_v is the efficiency of condensation at the front point; j is the specific vapour consumption; $\Delta \tau$ is the condensation time; ρ is the density of magnesium in solid form. With a sufficient length of the substrate the thickness of the condensate in the rear stagnation point of the wire can be determined by the formula

$$t_h = (1 - \alpha_v) \, \alpha_h \, j \Delta \tau / \rho, \tag{2.13}$$

where α_h is the efficiency of condensation in the rear frontal point. Due to the small diameter of the wire and the size of condensation on it in comparison with h the shading of the vapour flow by the wire can be ignored. The combination of (2.12) and (2.13), assuming that $\alpha_v = \alpha_h$, gives the equation

$$\alpha_v = 1 - t_h / t_v. \tag{2.14}$$

In general, when $\alpha_v \neq \alpha_h$, which may be when the efficiency of condensation depends on the flow density, (2.14) is valid only in the region where $\alpha_v \rightarrow 0$ and $\alpha_v \rightarrow 1$, and between α_v and α_h there is a relationship

$$\alpha_v = \alpha_h \, / \, (t_h \, / \, t_v + \alpha_h),$$

which can be used to process the experimental results, assuming that the dependence of α_h on j $(1 - \alpha_v)$ is identical to the dependence of α_v on j. Then, if at least one value of α_{v1} at j_1 is known, for example in the region of small α_v, the value for another value of j can be determined by solving the equation

$$\alpha_v^2 - \alpha_v \left[\left(1 - j_1 \, / \, j\right)\left(t_h \, / \, t_v\right) + 2 + \alpha_{v1} - j_1 \, / \, j \right] + \left(1 + \alpha_{v1} - j_1 \, / \, j\right) = 0. \quad (2.15)$$

(2.15) was derived using a linear interpolation of intermediate points between α_v and α_{v1}.

2.3.1. A technique for measuring the efficiency of condensation and orientation of the structures of vacuum condensates on a gradient substrate

A technique for investigation of the heterogeneous condensation process using a substrate with a temperature gradient was first proposed and used in [13, 14] by L.S. Palatnik and co-workers. Subsequently, the researchers of this school obtained with the help of this method some unique experimental results [15–20], revealing the main regularities of the initial stage of heterogeneous condensation and the properties of vacuum condensates.

In this paper, the possibility of the method of the gradient substrate are augmented by reasons set out above. The wire used to measure the efficiency of condensation is attached in front of the substrate in the form of an incomplete ring. A temperature gradient is created in the wire by heating and cooling the ends of the wire with liquid nitrogen. In the section of the wire where the temperature of the wire coincides with the temperature of the extended substrate we can measure the absolute value of the efficiency of condensation. Measurements of the thickness of the condensate in the front point of the wire in other sections and its comparison with the thickness of the substrate in the isothermal section allow us to calculate the efficiency of condensation along the entire length of the wire and, consequently, in a wide temperature range.

When the gradient wire ring is installed in the vapour flow without the rear mounted substrate, the experiment can be carried out to determine the local efficiency of condensation in the cross-sectional contour of the wire, and, therefore, in dependence on the angle of the flow.

Tests of this technique have shown that in addition to the local efficiency of condensation it is possible to measure the characteristics of the structure of the condensate – the grain size and orientation of the structure. The measurement results are discussed in section 3.4.

2.4. The possibility of using electron-beam diagnostics to measure the density of magnesium vapours

This section discusses the possibility of using electron-beam diagnostics to measure the density of magnesium vapour. We present the results of several experiments. The difficulties and prospects of development of this diagnosis are discussed.

In connection with the study of the formation of disperse systems, naturally there is interest in supersonic jets of metal vapours – a new and rather unusual object of study. For the study of the supersonic jet of metal vapour in vacuum it is necessary to know the density field in the jet. Measurements of the density field in supersonic jets of gas in vacuum by electron-beam (EB) diagnostics are well known [21]. The purpose of the study is to show the possibility of using EB diagnostics to measure the density in a supersonic jet of magnesium vapours.

The EB method for measuring the density can be briefly described as follows [21]: a narrow collimated beam of fast electrons intersects the flow of gas (vapour of the metal) and excites emission in it. The radiation intensity of the excited gas is proportional to its density. By measuring the intensity of any spectral line we can find the gas density. This technique is used to estimate the density of atomic and molecular gases. It is known [21] that in the EB diagnostics fairly stringent requirements are imposed on the spectral line, namely:

1) sufficient intensity;

2) the short lifetime of the upper level of the transition corresponding to a given spectral line;

3) a small amount of absorption of this line in the studied environment;

4) high excitation energy of the line.

Based on the available literature data [22, 23], we can try to find the right line. Study [22] contains information on the cross sections of excitation of lines of Mg I and Mg II by the electron impact in the energy range from excitation threshold to 300 eV, study [23] presents spectral tables. Diagnosis was carried out on lines of magnesium ion $\lambda = 4481.3, 4481.1$ Å (doublet) and 2795 Å, with the maximum cross section of excitation by electron impact. This choice is due to the fact that for these lines the conditions 3) and 4) are better satisfied than for the lines of the magnesium atom.

2.4.1. Determination of the range of applicability of electron-beam diagnostics of density

Let's see how the conditions 1) and 2) are satisfied for these lines. From these estimates, we can define a range of densities where EB diagnostics is applicable.

The radiation intensity of a spectral line excited by an electron beam is determined by the formula

$$J_{ki} = A_{ki} N_k \hbar v_{ki},$$

where J_{ki} is the intensity of lines corresponding to the transition $k \rightarrow i$; A_{ki} is the Einstein coefficient for the transition $k \rightarrow i$; N_k is the density of particles in the excited state k; $\hbar$ is Planck's constant; v_{ki} is the frequency of the transition $k \rightarrow i$.

In the absence of influence of gas-kinetic collisions

$$N_k = nn_e v_e \sigma_{0k}(E_e) \bigg/ \sum_{i<k} A_{ki},$$

where n is the density of gas atoms; n_e is the density of electrons in the beam; v_e is the velocity of the electrons in the beam; σ_{0k} is the cross section for excitation of level k from the ground state of electrons with energy E_e.

No influence of the gas-kinetic collisions means that the lifetime of the level (excited state of a particle) is much less than the time between collisions, ie $\tau_{\text{lifetime}}/t_c \ll 1$. As a rule, it is assumed that $\tau_{\text{lifetime}}/t_c \leq 0.1$. From this criterion we can found the upper limit of applicability of the EB diagnostics of density, given that

$$\tau_c = \lambda_a / v_a = \left(\sqrt{2} n \sigma_a \sqrt{\frac{8kT}{\pi m}} \right)^{-1},$$

where λ_a is the mean free path of the atoms; v_a is the average velocity of the gas atoms; σ_a is the section of the gas-kinetic collisions, T is the temperature of the gas.

As a result, we obtain $n \leq \dfrac{1}{10\tau_{\text{lifetime}}} \dfrac{1}{4\sigma_a} \sqrt{\dfrac{\pi m}{kT}}$. For $\lambda = 4481$ Å the lifetime of the upper level is equal to $4.55 \cdot 10^{-9}$ s [24], for $\lambda = 2795$ Å $\tau_{\text{lifetime}} = 3.7 \cdot 10^{-9}$ s [25, 26]. Consequently, at $T = 1000°C$ for $\lambda = 4481$ Å $n \leq 4.6 \cdot 10^{16}$ cm^{-3}, or $p \leq 800$ Pa, for $\lambda = 2795$ Å $n \leq 5.7 \cdot 10^{16}$ cm^{-3} or $p \leq 984$ Pa. These are the maximum densities at which the following relationship still holds:

$$J_{ki} = \left(A_{ki} \bigg/ \sum_{i<k} A_{ki} \right) nn_e v_e \sigma_{0k}(E_e) \hbar v_{ki}.$$

The lower limit of applicability of EB diagnostics is determined by the sensitivity of the apparatus. To find it, we compare the intensity of magnesium vapour and of the well studied nitrogen in the same conditions: at equal densities, with the same parameters of the apparatus of the complex – the spectrometer and the electron beam gun. Then from the known lower limit for nitrogen we find the lower limit for magnesium:

$$\frac{J_{ki}^{\text{Mg}}}{J_{ki}^{\text{N}_2}} = \frac{\sigma_{0k}^{\text{Mg}}(E_e)}{\sigma_{0k}^{\text{N}_2}(E_e)} \cdot \frac{v_{ki}^{\text{Mg}}}{v_{ki}^{\text{N}_2}} \cdot \frac{A_{ki}^{\text{Mg}}}{A_{ki}^{\text{N}_2}} \cdot \frac{\displaystyle\sum_{i<k} A_{ki}^{\text{N}_2}}{\displaystyle\sum_{i<k} A_{ki}^{\text{Mg}}}.$$

Table 2.5. Determination of possible application of spectral lines for EB diagnostics at $T = 1000°C$

Wavelength, Å	τ_L, s	I_{Mg}/I_{N_2}	Lower limit		Upper limit	
			n, cm^{-3}	p, Pa	n, cm^{-3}	p, Pa
4481	$4.55\cdot10^{-9}$	$7.6\cdot10^{-3}$	$4.8\cdot10^{14}$	1.33	$4.6\cdot10^{16}$	800
2795	$3.70\cdot10^{-9}$	$7.1\cdot10^{-2}$	$5.6\cdot10^{13}$	0.133	$5.7\cdot10^{16}$	984

The electron beam energy, for example, is 10 keV. The excitation cross section of the band (0,0) 1 of the negative system of nitrogen $\sigma_{0k}^{N_2}$ (10 keV) $= 9\cdot10^{-19}$ cm^2 [27]. The remaining parameters for nitrogen are: $\lambda_{ki} = 3914$ Å, $A_{0k}^{N_2} = 1.24\cdot10^7$ Hz, $\tau_{\text{lifetime}} = \dfrac{1}{\sum\limits_{i<k} A_{ki}^{N_2}} = 6.6\cdot10^{-8}$ s.

Since not all experimental data are available for some parameters of magnesium, there are some difficulties in determining them. Basically, this refers to the excitation cross sections of lines $\lambda = 4481$ and 2795 Å. To determine them, we use the interpolation formulas from [28] and the values of cross sections near the excitation threshold from [22]. Then, for $\lambda = 4481$ Å we obtain σ_{0k} (10 keV) $= 0.64\cdot10^{-20}$ cm^2, for $\lambda = 2795$ Å σ_{0k} (10 keV) $= 3.7\cdot10^{-20}$ cm^2. Hence, for $\lambda = 4481$ Å the radiation intensity ratio of magnesium and nitrogen is $J_{ki}^{Mg}/J_{ki}^{N_2} = 7.6\cdot10^{-3}$, and for the line $\lambda = 2795$ Å it is $7.1\cdot10^{-2}$.

Electron-beam diagnostics of nitrogen can be carried out in the density range $4\cdot10^{12} \le n \le 4\cdot10^{15}$ 1/cm^3. At $T = 300°C$, this corresponds to the pressure range $1.33\cdot10^{-2} \le p \le 13.3$ Pa. Consequently, the lower limit of applicability of the EB diagnostics for magnesium vapour is as follows: for the line $\lambda = 4481$ Å $n \le 4.8\cdot10^{14}$ cm^{-3}, the line $\lambda = 2795$ Å $n \le 5.3\cdot10^{13}$ cm^{-3}. The results are presented in Table 2.5.

Thus, the above estimates show that the EB diagnostics is applicable for measuring the density of magnesium vapour. However, the range of densities, which can be measured using EB diagnostics in magnesium vapours, is significantly smaller than in nitrogen.

2.4.2. Measurements of spectra

To test these assumptions and estimates, spectral measurements were taken in magnesium vapour. The emission spectrum, excited by the high-energy electron beam in a supersonic jet of Mg vapour, was recorded. The measurements were performed in a low-density gas dynamics tube. The experimental setup is shown in Fig. 2.17. The metal vapour source 2 is placed inside the vacuum chamber 1. The design of the evaporator is described in [29]. The evaporator received power from step-down transformers OSU-20. The supersonic jet 6 formed during discharge of the magnesium vapour from the evaporator into vacuum. The vacuum chamber is equipped with the electron beam gun 3, which generates an electron beam 4 with an energy of 10–20 keV and a maximum

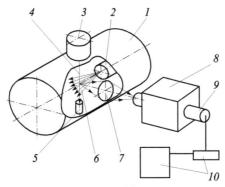

Fig. 2.17. The experimental setup for electron-beam studies. 1 – vacuum chamber; 2 – the vapour source; 3 – the electron beam gun; 4 – electron beam; 5 – electron collector; 6 – vapour jet; 7 – lens; 8 – monochromator; 9 – photomultiplier; 10 – amplifier and recorder of the photomultiplier signal.

current of 1 mA. The electron beam current is measured with the collector 5. The supersonic jet intersects the electron beam 4, which excites the glow of magnesium vapour. This radiation is collected by the optical system 7 at the entrance slit of the spectrometer 8. The spectral instrument is a MDR-3 monochromator. The radiation, passing through the spectral instrument, is recorded with a FEU-79 photomultiplier 9. The signal from the photomultiplier is recorded by the system 10, consisting of an PMT-05 and a KSP-4 recording potentiometer.

Experimental conditions: the electron beam energy 20 keV, beam current 0.8 mA, beam diameter 1 mm; spectrometer dispersion 13 A/mm, slit width 1 mm, scan speed 200 A/min. The dark current of the photomultiplier was $5 \cdot 10^{-11}$ A. The optical system consists of two quartz lenses, the gain of the system is 0.33, the relative solid angle $5.5 \cdot 10^{-3}$. The transmission of the system of registration of radiation is from 3000 to 6000 A. Evaporator mode: $T_0 = 860°C$, $p_0 = 2260$ Pa. The experiment used a sonic nozzle, the diameter of the throat $d_{cr} = 0.54$ mm. The nozzle–electron beam distance $x/d_{cr} = 25$. The pressure in the vacuum chamber during the experiment $6.65 \cdot 10^{-2}$ Pa.

The spectral measurements recorded 13 lines (Fig. 2.18). The comparison spectrum for interpretation was the mercury spectrum. All fixed lines are identified within the experimental error of ± 13 Å, as the Mg I and Mg II lines. The results of spectral measurements are presented in Table 2.6 which also includes the spectroscopic data [23–25, 29] and data on the excitation cross sections of these lines [22]. The table shows that the intensity of noise in the longer wavelength range is higher. This is due to the background created by the glow of the hot evaporator.

2.4.3. Defocusing of the electron beam by the electromagnetic field of the vapour source

The experiments revealed the influence of current supply of the evaporator on

Table 2.6. Radiation spectrum of Mg vapours excited by the electron beam

No	λ_comparison, Å (Δλ = ±13 Å)	λ_true, Å	E_upper	E_lower	J_table	J_real, 10⁻¹¹ A	J_noise, 10⁻¹¹ A	Signal/noise	Degree of ionization	Transition	Lifetime τ_L, s	Q_max 10¹⁹, cm²	E_max, eV
1	5703	5711.1	6.52	4.35	30	17	22	0.77	I	$3p^1P^o$–$5s^1S$	$5.5\cdot10^{-7}$	52	8.7
2	5531	5528.4	6.59	4.35	40	33	18	1.8	I	$3p^1P^o$–$4d^1D$	$6\cdot10^{-9}$	150	10.3
3	5172	5183.6	5.11	2.72	45				I		$1.6\cdot10^{-9}$	1400	6.7
		5172.7	5.11	2.72	44	42	9	4.7	I	$3p^3P^o$–$4s^3S$	$2.85\cdot10^{-9}$	840	6.7
		5167.3	5.11	2.72	42				I		$8.3\cdot10^{-9}$	260	6.7
4	4730	4730	6.97	4.35	10				I	$3p^1P^o$–$6s^1S$	$1.2\cdot10^{-6}$	6.4	9.0
		4739.7	14.18	11.57	5	1.5	6.2	0.24	II	$4d^2D$–$8f^2F^o$	–	–	–
		4739.6	14.18	11.57	6				II	$4d^2D$–$8f^2F^o$	–	–	–
5	4700	4703	6.98	4.35	30	8.6	6.2	1.4	I	$3p^1P^o$–$5d^1D$	$4.3\cdot10^{-9}$	69	11.1
6	4567	4571.1	2.71	0.00	28	2.1	5.9	0.36	I	$3s^2S$ –$3p^3P^o$	$1.2\cdot10^{-3}$	13	4.55
7	4474	4481.3	11.63	8.86	13	0.5	5.7	0.09	II	$3d^2D$–$4f^2F^o$	$4.55\cdot10^{-9}$	5.7	40
		4481.1	11.63	8.86	14				II		$5\cdot10^{-9}$		
8	4347	4354.5	7.19	4.35	6	2.2	5.7	0.39	I	$3p^1P^o$–$7s^1S$	$2.3\cdot10^{-6}$	2.4	12
		4351.9	7.19	4.35	20				I	$3p^1P^o$–$6d^1S$	$37\cdot10^{-7}$	25	12
9	4168	4177.1	7.31	4.35	2	0.7	5.7	0.12	I	$3p^1P^o$–$7d^2D$	–	–	–
		4167.3	7.31	4.35	15				I	$3p^1P^o$–$8s^2S$	–	16	12.3
10	4065	4057.5	7.4	4.35	10	0.24	6.3	0.04	I	$3p^1P^o$–$8d^2D$	–	6.8	12.6
		4054.7	7.4	4.35	2				I	$3p^1P^o$–$9s^1S$	–	–	–
11	3843	3838.3	5.95	2.72	20–20				I	$3p^1P^o$–$3d^3D$	$2.6\cdot10^{-10}$	130	7.8
		3832.3	5.95	2.71	20–				I		$4.3\cdot10^{-10}$	71	7.8
		3829.4	5.95	2.71	20	4.1	6.3	0.65	I		$9.1\cdot10^{-10}$	32	7.8
		3850.2	12.08	8.86	36				II	$3d^2D$–$5p^2P^o$	$7.8\cdot10^{-7}$	–	–
		3848.2	12.08	8.86	7 8				II		$6.4\cdot10^{-7}$	–	–
12	3348	3336.7	6.43	2.72	20	0.64	6.3	0.1	I	$3p^3P^o$–$5s^3S$	$8.3\cdot10^{-9}$		
13	3106	3096.9	6.72	2.72	24				I		$4.2\cdot10^{-9}$		
		3093	6.72	2.71	22	0.52	6.3	0.08	I	$3p^3P^o$–$4d^3D$	–	36	8.5
		3104.8	12.86	8.86	8				II	$3d^2D$–$8f^2F^o$	–	–	–
		3104.7	12.86	8.86	9				II		–	–	–

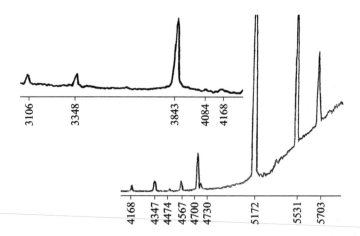

Fig. 2.18. The radiation spectrum of magnesium vapour excited by an electron beam. MDR-3, FEU-79, the gap of 1 mm, the electron energy of 20 keV.

the electron beam. The evaporator was powered with a current of 300 A and voltage up to 4 V. Under the action of electric and magnetic fields of current the electron beam moved in the space along an elliptical trajectory. Ellipse parameters: major axis length 80 mm, minor axis length 5 mm.

The impact of supply current on the electron beam will be estimated. It is assumed that the current is ring-shaped with radius R = 50 mm. The electron beam and the ring current are located in one plane. Movement of the electron beam perpendicular to this plane is determined by the of the magnetic field of current, and displacements in the plane by the electric field. The magnetic field is defined by the formula

$$H = 2I / cR_b,$$

where I = 300 A is the current of the heater of the evaporator; c = $3 \cdot 10^{10}$ cm/s is the speed of light; R_b = 100 mm is the distance from the centre of the ring current to the electron beam. Then the magnetic field intensity in the beam region is $H \approx 6$ Oe. The electrons flying at velocities v_e perpendicular to magnetic field lines of force will move in circular orbits with radius R_e = $mv_e c/(eH)$. The velocity of the electrons at E_e = 20 keV is v_g = $8.4 \cdot 10^7$ m/s. Then

$$R_e = \frac{mv_e c}{eH} = \frac{9.1 \cdot 10^{-28} \cdot 8.4 \cdot 10^9 \cdot 3 \cdot 10^{10}}{4.8 \cdot 10^{-10} \cdot 6} \approx 0.88 \text{ m}.$$

Taking into account the experimental geometry, the motion in a circular orbit with such a radius in the collector region gives the deviation equal to 25 mm. The maximum energy that an electron can acquire in the electric field of the supply current of the evaporator is equal to 4 eV, which corresponds to the electron velocity v_e = $1.2 \cdot 10^6$ m/s. The time during which the electron travels the gun–collector distance L = 60 cm is τ_b = $L/v_e \approx 7.1 \cdot 10^{-9}$ with. Hence, the deviation is l = $v\tau / 2 \approx 4.3$ mm. These estimates correlate well with the observations in the experiments. Consequently, the displacement of the electron beam is caused by the influence of the current of the evaporator.

To test the effect of defocusing the electron beam under the influence of electromagnetic field of the evaporator on the line intensity, measurements were made on three of the strongest lines λ = 5703, 5531 and 5172 Å. It was found that when the supply currents of the evaporator are switched for a short period of time the line intensities behave as follows: the intensity of the line λ = 5703 Å increased 10 times, the line λ = 5531 Å 1.5 times, whereas the intensity of the line λ = 5172 Å decreases by 10 times. The complicated behaviour of the lines suggests that their intensity is influenced at the same time by several factors. Firstly, the geometrical defocusing, and secondly, the effect of low-energy electrons. When the power supply of the evaporator is switched the electron beam under the influence of electric and magnetic fields of these currents begins to move in space and as a result the electron beam partially leaves the field of observation. The value of the detected signal decreases. When the power supply is switched on the number of low-energy electrons increases due to the fact that the residence time of the secondary electrons in the beam region increases. In fact, the magnetic field of 6 Oe the electron with an energy of 4 eV has the Larmor radius of 13 mm. This fact certainly hampers the diffusion of low-energy electrons from the beam region. The factor determined by the influence of low-energy electrons plays evidently a dominant role for the line $\lambda = 5172$ Å. This line has the lowest excitation threshold, equal to 5.11 eV. Its cross-section in the energy range of the excitation threshold is one of the biggest among the lines of Mg I. In addition, it decreases rapidly with increasing energy [22]. All this indicates that the sharp drop in the intensity of the line $\lambda = 5172$ Å when the current supply is switched off comes from the influence of low-energy electrons. Thus, the dominant factor for the lines $\lambda = 5703$ and 5531 Å is the defocusing of the electron beam. So when the current supply of the evaporator is switched on the intensity of these lines increases.

2.4.4. The choice of spectral lines

Based on these results and taking into account the signal, noise, and the effect of defocusing, diagnosis was carried out on the lines Mg I with $\lambda = 5531$, 4700 and 3843 Å. The lifetimes of the upper levels of the transitions, corresponding to these lines, are quite small, $\tau_{lifetime} \leq 6 \cdot 10^9$ s [25]. These lines are not absorbed in the magnesium vapour as they are not resonant. However, the excitation energy of these lines is low and, therefore, the influence of secondary electrons can be significant. All these lines are quite intense. The magnesium ion line $\lambda = 4478$ Å (4481 Å), originally selected for the diagnosis, has a relatively low intensity. However, the experimentally obtained intensity of the lines can not be used for estimates, since the effective opening of the spectrometer slit was 1×6 mm, and in the measurement of density for the locality of 1 mm^3 the required size of the spectrometer slit should be the size 0.3×0.3 mm.

Thus, when measuring the density the strength of the signal compared to that obtained in this experiment falls sixty times. This means that even the strongest lines are not suitable for EB diagnostics. Nevertheless, there are at least three ways to increase the intensity of the lines. First, it was found that the optical system that collects radiation from the volume is not matched with the spectral instrument. The result is that the signal strength decreases by 60 times. To address this shortcoming, we can fully compensate the loss of intensity arising from the need to localize measurements by reducing the spectral gaps. The second way to increase the intensity of the lines is the suppression of defocusing. Installing a magneto-soft screen between the vapour source and the electron beam can significantly reduce the influence of electromagnetic fields on the beam. The third possible way is to increase the beam current. If we use all options to increase the intensity of the lines, the selected lines, as well as the magnesium ion line $\lambda = 4474$ Å can be used for EB density measurements. Only one alignment of the optical system allows the line $\lambda = 5531$ Å to be used for measurements up to $x/D_{cr} \leq 86$, provided that the signal/noise ratio is ≥ 0.1.

We compare the results of calculation and experiment. Comparison can be made only for the line $\lambda = 4481$ Å. It is estimated in section 9.1 that the minimum density that can be measured is $4.8 \cdot 10^{14}$ $1/cm^3$. In the experiment, the Mg vapour density in the area of the electron beam, computed under the assumption of isentropic flow in the supersonic jet, was $3.8 \cdot 10^{13}$ $1/cm^3$. This discrepancy may be caused by both the fact that in reality the equipment used has a higher sensitivity and the fact that the isentropic calculation gives understated density values compared with the actual experiment.

Thus, the spectrum of magnesium vapour, excited by electron impact, contains fairly intense lines $\lambda = 5531$, 4700, 4478 and 3843 Å for the EB measurements of the density of the vapour. These lines also satisfy certain other requirements of the lines for EB diagnostics. However, the final conclusion on the applicability of the EB diagnosis of magnesium vapour can be done only after further research, in particular, the influence of secondary electrons and additional measures to eliminate defocusing the electron beam by the electromagnetic field of the metal evaporator.

Conclusions

1. Experimental studies of heterogeneous condensation, to examine the processes of formation of disperse systems on solid surfaces, were formulated In addition to using the known methods, new methods of creating the flow of vapour and its diagnosis were proposed.

2. A metal vapour source with a given and regulated superheating was constructed for studies of heterogeneous condensation of superheated vapours on solid surfaces and for studies of homogeneous condensation in rapidly expanding vapours in a vacuum.

3. A calorimetric technique is described, which can be used both for the study of the dynamics of metal vapours in a jet expanding into a vacuum in terms of real effects of homogeneous condensation and for the study of heterogeneous condensation.

4. A new method for absolute measurements of the efficiency of the condensation of metal vapours for the study of heterogeneous condensation in a wide range of angles of incidence to the flow on the substrate and temperatures, as well as under the strong influence of the residual atmosphere in the chamber.

5. The possibility of using electron-beam diagnostics to measure the density of magnesium vapour has been investigated. It is shown that the spectrum of magnesium vapours, excited by electron impact, contains fairly intense line $\lambda = 5531$, 4700, 4478 and 3843 Å with which to measure the density of these vapours. These lines also satisfy certain other requirements of the lines for EB diagnostics. However, the final conclusion of the applicability of the EP diagnostics of magnesium vapour can be done only after further research, particularly on the effect of secondary electrons, and after installing a device

for shielding the electron beam from the electromagnetic field of the vapour source.

List of Symbols

A - coefficient;

A_{ki} - Einstein coefficient for the transition $k \rightarrow i$, Hz;

c - specific heat, J / kg;

c - the speed of light, m/s;

D_{cr} - the diameter of the critical section, m;

E_e - electron energy, eV;

F_o - Fourier number;

F_d - area of the sensor surface, m²;

F_h - total cross section;

f - the form factor;

H - magnetic field strength, e;

$\hbar$ - Planck's constant, J·s;

I - current heating the evaporator, A;

I - nucleation rate, J / kg;

J - flux of enthalpy, W·m⁻²;

J - the intensity of radiation W/m²;

j - flux of particles, m⁻²·s;

Kn - Knudsen number;

k - Boltzmann constant, J/K;

L - distance an electron gun, electron collector, m;

L - latent heat of vaporization, J / kg;

l_h - length, m; deviation of the electron under the influence of the electromagnetic field, m;

M - Mach number;

m - mass of the atom, kg;

m_d - the mass of the sensor, kg;

N_k - the density of particles in the excited state k, m⁻³;

n - the partial vapour density, m⁻³;

n_e - the density of electrons in the beam, m⁻³;

P - power, W;

p - pressure, N/(m²·Pa);

q - heat flux, W·m⁻²;

Re - Reynolds number;

R_b - the distance from the center of the ring current to the electron beam, m;

R_e - the Larmor radius, m;

S - supersaturation;

T - temperature, K;

t - thickness, m;

u - average mass velocity, m/s;

v - the arithmetic mean thermal velocity of atoms, m/s;

v_e - velocity of the electrons in the beam, m / s;
v_a - the average velocity of the gas atoms, m / s;
x, y - linear coordinates, m;
α_k - efficiency condensing;
α_c - fraction of the condensate;
σ_a - section of the gas-kinetic collisions, m^2;
ε - the emissivity;
λ - wavelength, Å;
λ_a - the mean free path of atoms, m;
λ_h - thermal conductivity of the material;
v_{ki} - frequency of the transition $k \rightarrow i$, c^{-1};
ρ - density of vapour, kg/m^3;
σ - surface tension, N / m;
σ_0 - Stefan-Boltzmann constant;
σ_{0k} - cross section for excitation of level k of the ground state atom, m^2;
τ - time, s;
$\tau_{vr.zh}$ - lifetime, s;
T_c - the time between collisions in the gas, c;
τ_b - time of flight of an electron from the gun to the collector with.

Indices:

0 - inhibition, the initial state;
1-4 - number of plots;
+ - Coming;
- - Outgoing;
∞ - saturation;
vr.zh - lifetime;
to - room;
cr - critical;
N - heater;
a - air;
d - the sensor;
e - the electron;
f - pairs;
h - the rear critical point;
k, i - number of energy levels;
max - the maximum;
Mg - magnesium;
N_2 - nitrogen;
p - vapoury;
R - radiation;
v - the frontal point.

Literature

1.   Bochkarev A., Gaiskii N.V., Zolkin A.S., Condensation of Mg vapour during expansion into vacuum and on the surface, Phase transitions in pure metals and binary alloys, Novosibirsk, Publishing House, ITP SB RAS, 1980, 133–145.
2.   Bochkarev A., et al., as Ref. 1.
3.   Ashkenas H., Sherman P.S., The structure and utilization of supersonic jets in low density wind tunnel, In: Rarefied Gas Dynamics. N.Y., Acad. Press, 1966, V. 11, 84–105.
4.   Skovorodko P.A., Influence of homogeneous condensation in a free jet on the molecular beam intensity, Some problems of hydrodynamics and heat transfer. Novosibirsk, Publishing House ITF SO AN SSSR, 1976, 106.
5.   Davydov A.M., Mekh. Zhid. Gaza, 1971, No. 3, 66–73.
6.   Frishberg I.V., et al., Gas-phase method to obtain powders, Moscow, Nauka, 1978.
7.   Rosebery F., Handbook of vacuum engineering and technology, Moscow, Energiya, 1972.
8.   Physical Encyclopedic Dictionary, Sov. entsiklopediya, 1963, V. 3, 48.
9.   Encyclopedic Dictionary of Chemistry, Sov. entsiklopediya, 1983, 308.
10.   Komnik Yu.F., Fiz. Tverd. Tela, 1963, V. 5, 90.
11.   Komnik Yu.F., Physics of metal films, Moscow, Atomizdat, 1979.
12.   Miki R., Chang D.J., Li P., Thin Solid Films, 1987, V. 150, 259Sov. entsiklopediya,267.
13.   Palatnik L.S., Komnik Yu.F., Dokl. AN SSSR, 1959, V. 124, No. 4. 808–811.
14.   Palatnik L.S., Komnik Yu.F., Fiz. Met. Metalloved., 1960, V. 10, No. 4, 632–633.
15.   Palatnik L.S., Gladkikh N.T., Dokl. AN SSSR, 1961, V. 140, No. 3, 567–570.
16.   Palatnik L.S., et al., Fiz. Tverd. Tela, 1962, V. 4, No. 1, 202–206.
17.   Palatnik L.S., Gladkikh N.T., Fiz. Tverd. Tela, 1962, V. 4, No. 2, 424–428.
18.   Palatnik L.S., Gladkikh N.T., Fiz. Tverd. Tela, 1962, V. 4, No. 8, 2227–2332.
19.   Palatnik L.S., et al., Fiz. Met. Metalloved., 1963, V. 15, No. 3, 371–378.
20.   Kosevich V.N., et al., Fiz. Tverd. Tela, 1964, V. 6, No. 1, 3240–3246.
21.   Bochkarev A.A., et al., in: Experimental methods in rarefied gas dynamics. Novosibirsk, Publishing House ITF SO AN SSSR, 1974.
22.   Aleksakhin I.S., et al., Optika Spektroskopiya, 1973, V. 34, No. 6, 1053–1061.
23.   Striganov A.R., Sventitskii N.S., Tables of spectral lines of neutral and ionized atoms, Moscow, Atomizdat, 1966.
24.   Grimm G., Spectroscopy of plasma, Moscow, Atomizdat, 1969.
25.   Corliss Ch., Bozman W. Transition probabilities and oscillator strengths of 70 elements, Springer-Verlag, 1968.
26.   Smith G., Phys. Rev., 1966, V. 145, 26.
27.   Hiroh M.N., et al., Phys. Rev. A, 1970, V. 1, P. 1615.
28.   Vainshtein L.A., et al., Excitation of atoms and broadening of spectral lines, Moscow, Nauka, 1979.
29.   Zolkin A.S., et al., Producing magnesium vapour jets, Proceedings of the Conference of Young Scientists, ITF SO AN SSSR, Novosibirsk, 1980.

3

The formation of disperse systems on a solid substrate in heterogeneous condensation

The formation of disperse systems on a solid substrate by condensation from a foreign vapour phase takes place in several stages. Sorption processes begin upon contact of the vapour medium with the surface or when applying a molecular beam to the surface. With time a stationary concentration of adsorbate particles forms on the surface and the size of the particles is determined by the parameters of the vapour environment, the properties of the substrate and the external physical conditions. If the vapour environment is supersaturated, the adsorbed phase can also become supersaturated. Due to thermal migration of particles and fluctuations agglomerates of particles of the adsorbed phase form continuously on the surface. The state of the agglomerates varies depending on the size of the particles. The agglomerates which have not reached the critical size of stability in the adsorbed and vapour phases start to decay. The agglomerates that are larger than the critical size are stable and able to grow. The result of this second stage, called nucleation or nucleation, is the formation of a disperse system as the particles in the condensate of submicron dimensions distributed on the substrate surface. In the third stage the growth of nuclei and the birth of additional ones leads to the formation of a continuous coating on the substrate surface by the condensate of the vapour phase. Depending on the combination of the surface properties and the properties of the condensing vapour we can obtain a variety of disperse systems. In the case of vapour condensation by the vapour–liquid mechanism the formation of liquid nuclei over time transforms to the liquid-drop condensation on the surface, and in the case of wetting of the last –to film condensation. In condensation by the vapour–liquid–crystal mechanism the formation of liquid nuclei as they grow leads to crystallization of the solid phase and production of a solid film on the substrate. Condensation by the vapour–crystal mechanism is characterized by the formation of solid nuclei, followed by filling the entire surface by the solid condensate of the vapour phase. Depending on the properties of the substrate surface with respect to the condensate, greatly varying films of the solid condensate can be produced. On single-crystal substrate with a lattice parameter close to the lattice parameter of the condensate, a low-intensity effect of disturbing factors may produce single-crystal films (molecular epitaxy). Polycrystalline substrates usually show the formation of a polycrystalline solid condensate film with the grain size, determined by the surface properties and condensation conditions. Quasi-amorphous solid condensates can be produced as a result of condensation at high supersaturation.

The most studied are the processes of production of epitaxial single-crystal films because they are used most extensively in many areas of new technology, such as nuclear research, microelectronics, microwave engineering, optics, and superconductivity research. Advances in this field have been achieved by the displacement of film production technology to the range of high vacuum, the optimal temperature, and high-purity conditions. Nature, in view of its complexity and multiparametric nature, is 'stingy' as regard the creation of single-crystal thin films. Therefore, the success of epitaxial technology has provided a number of unique discoveries and a 'jump' in scientific and technological progress. Of course, most researchers of the condensation processes have studied the epitaxial growth of island and single-crystal films. The field of condensation in the vapour medium in the 'dirty' conditions, when a variety of components and parameters leads to the formation of a complex, usually polycrystalline disperse system, remained poorly understood. But it is this region in which complex systems form and is rich in the physical and chemical phenomena. Technical applications of the systems here are determined by the need to develop new materials with unique electrical, magnetic, thermal, mechanical and chemical properties. The uniqueness of these properties results primarily because from the commensurability of the effects of surface and bulk phenomena, which fundamentally distinguishes such dispersions from massive bodies. The need for further research in this area is also dictated by the fact that nature provides *in vivo* as a rule such complex disperse systems. Therefore, knowledge of the processes involved here is one of the priorities of the natural sciences.

There are several unresolved fundamental problems preventing a complete description of the process of heterogeneous condensation in the general case.

1. The study of sorption processes in the interaction of vapour or gas phase with the boundary of the condensed solid or liquid phase. Its difficulties are, first of all, the uncertainty of the specific physical constants when describing the thermal accommodation of gas phase particles on the surface, the probability of 'sticking' the particles on the surface, the type and magnitude of the binding energy of the particles to the surface, the surface migration relationships.

2. The study of sorption processes in the presence of the real properties of the surface (polycrystallinity, roughness, physical, and chemical composition) and under the action of external physical factors (electromagnetic, electrical, magnetic, thermal fields).

3. Nucleation processes on solid surfaces, taking into account its real properties. There is currently no accurate quantitative description of nucleation processes, even in simplified conditions of one-component vapours and an ideal single-crystal surface.

4. The mechanisms of growth of nuclei on a solid surface. Difficulties in solving this problem are, first of all, the uncertainty in the quantitative description of ways to transport of material of the vapour phase to the condensation front.

5. The formation of structures of solid condensates. At the present time based on the huge amount of experimental data available there are at best qualitative charts of modes for the formation of various structures for the condensation of specific substances on specific surfaces.

The materials in this chapter are presented so as to cover at least part of each of these problems. At the same time, the completeness of the exposition of these issues is regulated by the authors' desire not to deviate from the focus of the work – the creation of qualitatively feasible ways of obtaining a dispersed system. The study is based both on the literature experimental data and the results obtained by the authors.

3.1. Critical phenomena of condensation in the presence of impurities

Beginning with the work of Knudsen [1], Langmuir [2] and Wood [3], the effect of the critical condensation temperature above which condensation does not occur on the substrate was explained and extensively studied. L.S. Palatnik and Yu.R. Komnik [4, 5] found the second critical temperature at which the structure of vacuum condensates substantially changes. Finally, the study [6] presented data on the temperature boundaries at which there are significant changes if the efficiency coefficient of condensation and the structure of the condensate. The results were obtained for the condensation of a number of metals: Au, Cu, Ag, Be, Co, Ni, Fe, Ti, Pt, Cr, In, Sn, Pb, Bi. For the condensation of Zn and Cd on copper substrates at temperatures ranging from 193 to 273 K in vacuum of $1.33 \cdot 10^{-2} - 1.33 \cdot 10^{-4}$ Pa it was found that the condensate is formed only in low-temperature (I) and high-temperature (III) areas. Between the regions I and III there is a 'breakdown' of condensation, i.e. condensate is virtually absent in the range of the density of the vapour flow less than $5 \cdot 10^{-4}$ kg/(m²·s). In addition, in the regions I and III there are various structures of the condensate. In the low-temperature region I the structure of the condensate is globular, and in the region III it is fine-grained. When the density of the vapour flow is $5 \cdot 10^{-3} - 10^{-2}$ kg/(m²·s) the metal vapours condense also in the intermediate region II with the formation of single crystal and spheroidal particles. It was shown that the temperature region II, where the condensation is not observed, depends on the density of the molecular beam and the properties of the substrate. Due to the change in the structure of the condensate in the characteristic temperature region the authors of [7–9] concluded that it is necessary to relate the observed effects with the change of the vapour–liquid–solid condensation mechanism to the vapour–solid mechanism. Correlations of the characteristic temperature of 'collapse' with the melting point of the condensing metal were determined. However, the fluctuations of this temperature in the range 0.2–0.9 of the melting point for the condensation of various metals have left the question open.

According to Komnik [10], a possible reason for the existence of characteristic temperature regions for condensation is the presence on the substrate of adsorbed layers of oil vapours and gases. However, in this work the effect of the pressure of the residual atmosphere on the collapse effects is not shown.

The authors of [6–9] observed the characteristic friability of the condensate on the borders of the zone of 'collapse' of condensation. This suggests that the condensation modes near the zone of 'collapse' are of special interest for the processes of formation of dispersed structures. Therefore, we carried out experiments aimed at establishing a connection of this zone with the appearance of dispersed structures of the condensate.

3.1.1. The influence of the residual atmosphere on the 'breakdown' of magnesium vapour condensation

To study the effect of the pressure of the residual atmosphere in the vacuum chamber on the condensation process, measurements were taken of the thickness of the condensate h_c and the condensation rate $(\rho u)_c \sim dh_c/dt$ using a quartz balance. Measurements were carried out in continuous deposition on a quartz probe with a smooth variation of the magnesium vapour flow, and in the mode of a stationary flow of magnesium vapour by opening shutters in front of the quartz sensor for a time sufficient to obtain steady readings. The steady readings were consistent with results of measurements in the continuous mode. Measurements of the steady condensation rate at a substrate temperature of 294 and 349 K depending on the specific density of the magnesium vapour flow are shown in Fig. 3.1. The specific flow density was varied by varying the pressure p_0 at the vapour source. The results of measurements of the rate of condensation on the substrate at a temperature $T_S = 294$ K at various pressures of residual gas in the chamber p_b are indicated by the curves 1–4. As can be seen from the figure, the condensation rate increases rapidly with an increase of the specific density of magnesium vapour. This corresponds apparently to the exit from the zone of 'collapse' of condensation. In this case the characteristic values of the specific flow density or pressure in the source p_0 are approximately proportional to the pressure of the residual atmosphere in the chamber. In the region of 'collapse' of condensation an increase of the specific flow of the magnesium vapour results at first in a slow increase in the condensation rate and $(\rho u)_c$ then assumes a constant value. For curve 3, this value is equal to $2 \cdot 10^{-9}$ kg/(m²·s), for curve 4 $3 \cdot 10^{-9}$ kg/(m·s).

Such a characteristic behaviour of the condensation rate is apparently due to the change in the residual atmosphere. In Fig. 3.1 b curves 3', 4' show the change in pressure in the chamber p_b depending on the flow of magnesium vapour in the respective regimes. The initial part, where the condensation rate

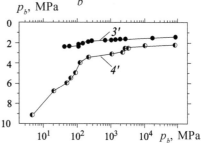

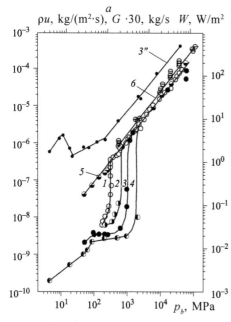

Fig. 3.1. The condensation rate of magnesium vapour $(\rho u)_c$ on the substrate depending on the vapour pressure at the source. Pressure in the vacuum chamber p_b, Pa: $1 - 7.2 \cdot 10^{-4}$; $2 - 9.3 \cdot 10^{-4}$; $3 - 1.3 \cdot 10^{-3}$; $4 - 2.1 \cdot 10^{-3}$; $T_S = 294$ K; $5 - 2.6 \cdot 10^{-4}$; $T_S = 349$ K; $6 -$ magnesium vapour flow G through the evaporator nozzle; $3'' -$ heat flux W at the condensation surface; $3'$, $4' -$ chamber pressure p_b in measurement of the condensation rate for curves 3, 4.

increases and pressure decreases the impurities are probably absorbed from the chamber by the porous condensate and a certain amount of the impurity reacts chemically with magnesium to produce heat. Curve 3" shows the values of the heat flow into the substrate for the curve 3, obtained by the method described in chapter 2.2.

Experiments to measure the condensation rate at various magnesium vapour flows, carried out at a substrate temperature of 349 K, showed no 'collapse' (curve 5). In this case, the substrate is pre-warmed at 473 K for 30 min, which resulted in efficient cleaning to remove impurities and to achieve complete condensation of magnesium.

Control measurements were carried out of the Mg vapour flow from the source using a directional Faraday trap with restricted access of the molecules of the residual atmosphere. The measurement results, described by the curve 6, are, in fact, a characteristic of the vapour flow used in the experiments. The equidistant curves 5 and 6 indicate the absence of any features of condensation on the heated substrate.

Characteristic forms of the readings of the quartz balance when measuring the mass of the condensate in dependence on time after opening the shutter are shown in Fig. 3.2. The time of closing the shutter is marked on the curves by point A. The measurements refer to different values of p_0 for curve 4, Fig. 3.1. In Fig. 3.2 a curve 1 shows the characteristic shape of the change in the thickness of the condensate during the time corresponding to the condensation rate in the steady state $(\rho u)_c = 5.63 \cdot 10^{-10}$ kg/(m·s²). Curve 2 corresponds to $(\rho u)_c = 2.95 \cdot 10^{-9}$ kg/(m·s²). The measurements represented by curve 1 in Fig. 3.2 b were obtained in the transition region with a sharp increase in the

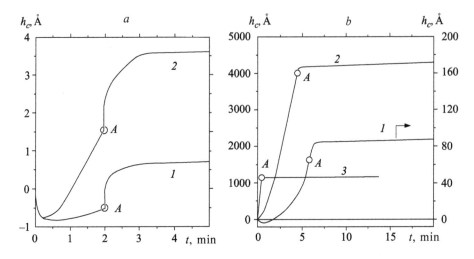

Fig. 3.2. Changes in the thickness of the condensate h_c over time (as indicated by the quartz balance over time). a – p_0 = 17.3 (1), 199.5 Pa (2), b – p_0 = 2194 (1), 2460 (2) and 1.633·10⁴ Pa (3).

condensation rate $(\rho u)_c$ = 1.18·10⁻⁸ kg/(m·s²). Typical readings after reaching full condensation are shown by curves 2 and 3 (see Fig. 3.2 b) $(\rho u)_c$ = 2.08· 10⁻⁶; 1.11·10⁻⁵ kg/(m·s²). The nature of time readings of the quartz balance at different pressures in the vapour source is fundamentally different. At relatively high pressures (curve 3 in Fig. 3.2 b) the readings are stabilized in a time of about 2 s, which apparently corresponds to the adsorption of magnesium atoms and the formation of the condensation front on the substrate. Long-term relaxation of the readings of the quartz balance after the closure of the shutters leads to the adsorption and absorption of residual gases of the working chamber with a fresh Mg condensate.

When reducing the vapour pressure at the source by about an order of magnitude the time to obtaining stable indications of the quartz balance also increases approximately by one order of magnitude (curve 2, Fig. 3.2 b), the curves 1 and 2 show characteristic peaks of signals at the time of opening and closing the shutter. These bursts can be explained by desorption of the molecules of the residual atmosphere from the substrate when exposed to light radiation from the nozzle of the source when opening the throttle and adsorption at closing. The longer relaxation time of the signal after closing the shutter may indicate a more permeable, possibly porous Mg condensate in this mode, which increases its absorption capacity for the residual atmosphere [11].

With a further slight decrease of the magnesium vapour flow from the source (Fig. 3.2 b, curve 1) the nature of the interaction of the vapour flow with the surface of the substrate changes. During the time of a few minutes there is no formation of a solid condensation front and no establishing of the complete condensation mode. Condensation occurs only on certain islands. In the region of 'collapse' of condensation (see Fig. 3.2 a, curves 1 and 2) the commensurability of the sorption–desorption 'surges' with the main signal suggests that these processes hinder the development of condensation. Lack of long-term relaxation of the signal after closing the shutter demonstrates that the substrate does not contain a significant amount of the condensate. After repeated measurements in this mode, the substrate surface is covered with a loose dark-coloured deposit.

Figure 3.3 shows the experimentally determined temperature region where the Mg condensate is practically absent. The pressure in the chamber corresponds to a sharp increase in the condensation rate in Fig. 3.1. With increasing pressure in the chamber (points 1–4, see Fig. 3.3) condensation starts at a higher specific flow of the magnesium vapour. In addition, the temperature range of the zone of 'collapse' of condensation increases with increasing pressure in the chamber.

As a result, it was found that the temperature limits of the region of 'collapse' of condensation in the coordinates of $T_S - p_0$ depend on the pressure of residual gases in the chamber, and the cause of this relationship is shown – the processes of sorption–desorption of non-condensable impurities on the substrate.

3.1.2. The efficiency of condensation near the boundaries of 'collapse'

The efficiency of condensation of magnesium was measured in the range of the substrate temperature T_S = 113–673 K by the method described in section 2.3.1. The pressure of the residual atmosphere in the chamber was p_b = 4.25·10⁻³ Pa and the stagnation pressure of the vapour jet in the source p_0 = 239.4 Pa, which corresponds to the specific consumption of vapour 3·10⁻⁶ kg/(m²·s). The results are shown in Fig. 3.4. The measurement results show that at T_S > 643 K condensation of magnesium does not occur. There is also a range of temperatures 232–304 K with an immeasurably low rate

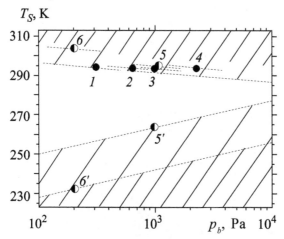

Fig. 3.3. The temperature ranges of the condensation of magnesium on the substrate (shaded) at various pressures of the residual gas in the chamber. The points 1–6 correspond to p_b, Pa: 7.2·10⁻⁴; 1·10⁻³; 1.3·10⁻³; 2.1·10⁻³; 1.4·10⁻⁴; 4.2·10⁻³, points 5′, 6′ shows the lower boundary of the zone of 'collapse' of condensation.

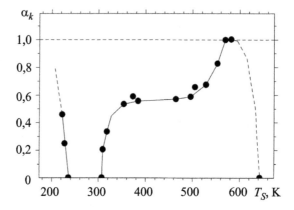

Fig. 3.4. The efficiency of magnesium vapour condensation on a copper substrate, depending on the substrate temperature T_S.

of condensation of magnesium. When approaching the borders of this region there is a sharp decrease in the condensation coefficient. The condensate on the substrate close to a temperature of 304 K has a polycrystalline fine-grained structure, and at temperatures near 232 K globular formations appear. In the temperature range 373–573 K the condensate structure is columnar changing to equiaxed at a higher temperature. The structures were studied on transverse sections of a gradient wire substrate in a Neophot-21 microscope. The results agree qualitatively with the observational data [6] for Zn and Cd. The temperature interval 232–304 K most likely corresponds to the zone of 'collapse' of condensation.

Also measured was the efficiency of condensation when the specific consumption of vapour at a constant temperature of the wire substrate $T_S = 300$ K was varied, at a vapour stagnation temperature of 1350 K. Figure 3.5 shows photos of transverse sections of the magnesium condensate produced by successive deposition in three different modes.

These condensation zones I, II, III correspond to $p_0 = 1.3 \cdot 10^3$; $2.7 \cdot 10^3$; $6.6 \cdot 10^3$ Pa, $(\rho u)_c = 4.9 \cdot 10^{-6}$, $1.35 \cdot 10^{-5}$, $4.5 \cdot 10^{-4}$ kg/(m²·s). The pressure in the

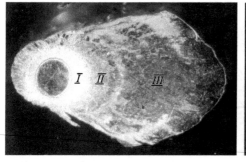

Fig. 3.5. Cross-sections of the condensate produced on the wire, ×96 and ×1000 for the left and right images, respectively.

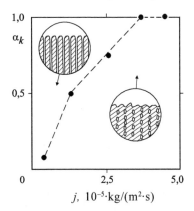

Fig. 3.6. The dependence of the efficiency of condensation and the type of structure on the specific vapour consumption.

vacuum chamber was maintained at $2 \cdot 10^{-4}$ Pa. The dependence of the efficiency of condensation, calculated for the regions from (2.14) and (2.15), is shown in Fig. 3.6. The characteristic cross-sectional dimensions of the condensate on a wire were changed in an optical microscope with a maximum error of less than 1 μm.

The anisotropy and porosity of vacuum condensates of metals having a decisive influence on their transport thermophysical properties, are usually associated with the presence of foreign microparticles on the substrate surface, penetration of the macroparticles of evaporated substance into the condensate, diffusion and coalescence during condensation, and features of the microrelief of the substrate [12]. The main regularities of manifestation of a particular structure of the condensate – conical polycrystals, columnar structure, equiaxed crystals – depending on changes in the substrate temperature and the condensation rate were determined [13–17]. In this study, in addition to the known facts, it was attempted to establish the relationship between the structure of the condensate and the efficiency of condensation.

Analysis of the structure in different zones showed the following facts.

1. Condensation at $(\rho u)_c = 4.9 \cdot 10^{-6}$ kg/(m²·s) occurs at a relatively low condensation coefficient and produces a pronounced columnar structure of the condensate. In the frontal part of the condensate, facing the vapour source, columnar crystals in some places grow into each other with the formation of closed porosity. Due to the fact that at the low efficiency of condensation the front and rear parts of the condensate were obtained at a constant temperature of the wire and at approximately the same flows of condensed matter, the difference observed in the structure can not be explained within the framework of [13–16]. Apparently, in addition to temperature and the condensation rate the structure of the condensate is influenced by additional factors. These factors in this experiment are the radiation of the vapour source and the fact that the frontal part of the condensate is bombarded with magnesium atoms from the source with higher energy than the energy of the atoms reflected from the substrate.

2. When p_0 is increased in parallel with an increase in the efficiency of condensation the structure in the frontal part of the condensate becomes equiaxed. Moreover, this transition occurs as a result of increasing the number of bridges between the columnar crystals, especially those directed at the vapour source. The surface of the condensate is formed by the outputs of intergrown columnar crystals.

3. The direction of growth of the columnar crystals is oriented with respect to the directions of the incident flow and the position of the surface on which they are formed. Their rate of growth does not depend on the misorientation relative to the incident flow within ±12°. The crystal can bend during the growth process.

More than the ten-fold change in the efficiency of condensation at the relatively small change in the vapour flow and the corresponding transition of the structure of the condensate from columnar to equiaxed suggest some radical changes in the condensation conditions. We can assume that at a constant vapour temperature T_0 = 1350 K increasing p_0 creates suitable conditions for the homogeneous condensation of magnesium vapours in the flow that leads to the formation of magnesium clusters and, consequently, changes in the conditions of condensation on the substrate. However, the study of homogeneous condensation is the subject of a separate study in the next chapter. Here we provide another possible explanation of the formation of disperse structures of vacuum condensates, based on the effect of capillary condensation of impurities [17].

3.1.3. Role of capillary condensation in the formation of disperse systems

We assume that the atmosphere of the vacuum chamber in which the process of obtaining a vacuum condensate is realized, contains vapours of a substance with a partial pressure p_b. In the case of vapour–oil vacuum pumps, it may be the vacuum oil. Pressure p_b in this case is determined by the saturation curve of oil at a chamber temperature of $p_{h\infty}(T_b)$. The degree of supersaturation of the vapours $\sigma = p_b/p_{h\infty} - 1$ determines the possibility of their condensation with the formation of nuclei of critical size [18]:

$$r_* = 2\gamma\omega/(kT_b \ln (1 + \sigma)). \qquad (3.1)$$

Here γ is surface energy; $\omega = m/\rho$ is the atomic volume in the condensed phase; m is the mass of the molecule; k is the Boltzmann constant; T_b is the temperature of the vacuum chamber. Graphical representation of (3.1) for γ = 0.03 N·m^{-1}, m = 7.524·10^{-25} kg, ρ = 0.9·10^3 kg/m^3, T_b = 300 K is shown in Fig. 3.7. Curve 1 is used to determine the minimum radius of curvature of the stable nucleus of the condensed phase in terms of supersaturation ($\sigma > 0$, $p_b > p_{h\infty}$). Curve 2, which formally follows from (3.1) with $\sigma < 0$, $p_b < p_{h\infty}$, corresponds to an atmosphere of dry vapour with respect to the flat substrate and is usually not considered. Its physical meaning is the principal possibility of the existence of a stable nucleus of the condensed phase with a concave surface, as conventionally shown on the left in Fig. 3.7. This means

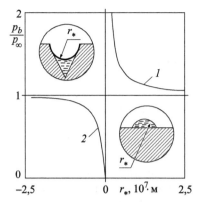

Fig. 3.7. Dependence of the relative equilibrium pressure for the critical nuclei of vacuum oil at a temperature of 300 K. 1 – convex nuclei, 2 – concave nuclei.

that all the irregularities of the substrate in the form of depressions with a radius of curvature smaller than that defined by the formula (3.1) are filled with the condensed phase of the impurities. The same can be attributed to the junction of the growing crystals of the vacuum condensates. That is, apparently, the joint of the neighbouring crystals of the vacuum condensate formed in the vacuum chambers with vapour-oil pumping, is filled with the oil vapour condensate as the crystals grow. The probable mechanism of filling the joint with of the crystals with oil is the surface diffusion of adsorbed molecules in the direction opposite to the gradient of surface curvature [18].

It is easy to trace the qualitative agreement between our model of growth of columnar crystals and the available experimental data [12, 13]. Increasing the substrate temperature is directly dependent on $p_{h\infty}$, $\sigma \to -1$, $r_* \to 0$, which means reducing the gap between the crystals and increasing the extent of their intergrowth. Increase of the specific flow of the main condensing component causes a decrease in the concentration of adsorbed oil vapours on the crystal surface and hence leads to difficulties in filling the joint in the crystals with the impurity which also leads to the growth of crystals into each other. Increasing the energy of atoms of the main flow or the additional energy impact on the substrate, such as radiation of the evaporator, have apparently the same effect through desorption of the impurity. The desorption of impurity atoms and molecules from the crystal surfaces should lead to an increase in the efficiency of condensation of the basic substance, which also correlates with the observed experimental facts. The pressure of the condensate of the oil vapours filling the gaps between the columnar crystals which has a concave meniscus shape with a radius of curvature determined by the conditions in the chamber, is $p = p_b - 2\gamma/r_*$. Negative pressure $p - p_b$, due to the Laplace difference, may be the cause of the appearance of compressive stresses in the vacuum condensate. In the case of condensation on a flat surface these stresses can cause cracking and deformation of the condensate film, which is often observed in the experiment.

It should be noted that the role of the impurities that cause the discrete nature of the vacuum condensates produced by the described mechanism can be fulfilled by any substance, provided that the substrate temperature is less than the critical temperature. At cryogenic temperatures, a similar role can be performed by the gas, which is confirmed by experiments [16]. At higher temperatures this role can be played by the impurity of a more volatile metal with a low-melting point metal, such as mercury in relation to magnesium [14]. A small admixture of a more fusible metal may create precipitation of a eutectic in the joints of the crystals where the negative curvature is greatest. It should, however, incur a correlation between the structure of the condensate and the temperature of the substrate, referred to the melting point of the base metal. The latter fact has been established and studied in [13, 19].

The foregoing suggests that the discrete structure of the vacuum condensates can be qualitatively explained on the basis of mass transfer between the vapour and the substrate, the diffusion transport of matter on the surface, processes of segregation of impurities and inhomogeneities of the surface curvature.

3.2. Model of the kinetics of sorption processes on a solid surface, taking into account chemical reactions and condensation

In this section, the abnormal phenomena such as 'collapse' in condensation, experimentally investigated in [6, 20] and in section 3.1, are explained by modelling of sorption processes.

The first attempt at a theoretical study of the sorption exchange between the medium surrounding the solid and the layer of molecules adsorbed on the surface was made by Langmuir [21]. The development of this idea in the form of the BET model [22] is widely known and used to date [23] for the analysis of steady states of the surface–medium systems. However, the BET model and the assumption of an infinite number of layers of adatoms used in the model are exactly satisfied only at the critical point of the adsorbed substance. For states below the critical point analysis of the experimental data using the BET model requires the choice of fitting parameters, determined, for example, by the quality of the real surface and the possible influence of capillary condensation. Attempts to study the dynamics of adsorption–desorption, especially of the multicomponent medium, are rare [24]. The problem of the process of adsorption–desorption on the surface of a solid of a one-component vapour, taking into account surface migration of adsorbed particles in the trap and in the volume of the body was considered by Ya.E. Geguzin and S. Kaganovsky [18].

3.2.1. Formulation of the kinetic adsorption–desorption problem

The following assumptions were made for the kinetic adsorption–desorption problem:

1. Adhesion of the adsorbed particles occurs only on the surface area free of adatoms with a constant adhesion coefficient. Thermal accommodation of the particles on the surface takes place instantaneously.

2. The density of vacancies of adsorption on the solid surface remains constant regardless of the possible coverage of the surface of the body by products of chemical reactions.

3. The density and capacity of the traps and sinks for the adatoms are constant. Traps are located on the surface evenly with a characteristic length smaller than the mean free diffusion path of the adsorbed particle on the surface.

4. The probability of ejection of the adsorbed particles by foreign particles from the gaseous medium or by any other effect, such as the influence of electromagnetic radiation, does not depend on the degree of filling the surface with adsorbed particles. The knocked-out particle does not remain on the surface.

Adsorption takes place on a solid surface, placed in an atmosphere of a mixture of gases or vapours. The rate of change of surface concentrations of atoms or molecules of the i-th component of the gas mixture containing N components, neglecting surface diffusion and considering only the reactions of synthesis and decay, can be written in the form of the equation

$$\dot{\xi}_i(\tau) = P_i\left(1 - \frac{1}{n_a}\sum_{j=1}^{N}\xi_j\right) - \frac{\xi_i}{\tau_i} - \sum_{l=1}^{L}\alpha_{li}I_l\xi_i - \sum_{q=1}^{Q}k_q\xi_i^{\,n}\xi_j^{\,m} + \sum_{b-1}^{Q}k_b\xi_{nm} - \sum_{h=1}^{H}\eta_{hi}(\xi_i - \xi_{hi}),$$

$$(3.2)$$

where τ is time; P_i are the specific densities of fluxes of the particles of components on the solid surface; n_a is the density of adsorption vacancies, adopted the same for all types of particles; τ_i is the average lifetime of the particles of i-th component on the surface until their spontaneous re-evaporation; α_{li} is the probability of desorption of particles of the i-th component L by various energy effects: electromagnetic radiation, electron or ion bombardment, impact of particles of another component (in this case $I_l = P_j$, $L = N-1$). The fourth and fifth terms on the right side of (3.2) reflect the absorption and the appearance of the adsorbed particles of the i-th component in Q decay and synthesis reactions with rate constants k_q and k_b; ξ_{nm} is the concentration of adsorbed molecules of the reaction products of synthesis. The last term of the equation in the linear approximation reflects the interaction of adsorbed particles with the probability η_{hi} with traps and sinks H of different types: condensation on the body surface, capillary condensation in the cavities, pores and other surface defects, the absorption of particles in the volume of the body. Here ξ_{hj} is the concentration of adsorbed molecules when the adsorbate is in equilibrium with filled traps.

Equation (3.2) must be supplemented by Q-equations for the kinetics of the adsorbed reaction products and H $(N + Q)$-equations for the dynamics of the filling of traps or sinks. The system of $(H+1)\cdot(N+Q)$-equations can be studied by the techniques used in chemical kinetics.

The model of the sorption processes described in general terms was used by the authors in a few simpler concrete tasks. First, the model was used for theoretical analysis of experimental data on the condensation of vapour jets of magnesium on a copper substrate in a vacuum chamber. The analysis was performed using the equation (3.2) for the sorption of three components: magnesium, oxygen and vacuum oil of the chamber, taking into account the oxidation of magnesium, the equation for the sorption of magnesium oxide and the equations describing the flow of adsorbate oil into the capillary traps of the substrate [25]. The results of analysis were used to explain complex experimental data, indicate the mode of oxidation of magnesium with oxygen with a deficiency of magnesium vapours and/or oxygen, specify the mode of nucleation of metallic magnesium on a copper substrate, the mode of condensation of magnesium to produce a columnar structure and the trapping of oil in the gaps between the columns.

Second, experimental data [26] for the nucleation of water vapour in the presence of various He pressures have been analyzed. The analysis was performed using two equations of the type (3.2) for the adsorption of water molecules and helium atoms on the pre-critical aggregates of water. The experimental data for a wide range of temperatures and pressures of helium have been summarized in the coordinates 'the critical supersaturation – occupation of the adsorption vacancies of water aggregates by the He adatoms. It is shown that the occupation of the adsorption vacancies by the impurities requires an increase in the effective supersaturation in order to achieve the required nucleation rate [27].

The sorption model (3.2) was also used to substantiate the mechanism of chain nucleation [28]. The essence of this mechanism is that the capillary condensation of the impurities frees additional adsorption vacancies near the already existing nuclei, which increases the concentration of the adsorbate and stimulates the production of nuclei near the existing ones.

Thus, the simplest theoretical approach in the single-layer approximation may be used to take into account the peculiarities of the nucleation of the condensed phase of the adsorbate.

The simplest examples available for analysis will now be discussed.

3.2.2. Adsorption of two-component gas mixture with a binary chemical reaction on the surface

In the case of recording only the irreversible binary reaction of the particles of the two components in the mutual desorption action by bombardment with foreign particles and desorption by radiation the sorption kinetics can be described by six equations:

$$\dot{\xi}_1(\tau) = P_1(1 - (\xi_1 + \xi_2)/n_a) - \xi_1/\tau_1 - \alpha_{12}P_2\xi_1 - \alpha_1 I\xi_1 + \eta_1(\xi_{h1} - \xi_1) - k_q\xi_1\xi_2;$$

$$\dot{\xi}_2(\tau) = P_2(1 - (\xi_1 + \xi_2)/n_a) - \xi_2/\tau_2 - \alpha_{21}P_1\xi_2 - \alpha_2 I\xi_2 + \eta_2(\xi_{h2} - \xi_2) - k_q\xi_1\xi_2;$$

$$\dot{\xi}_{12}(\tau) = K_q\xi_1\xi_2 - \xi_{12}/\tau_{12} + \eta_{12}\,(\xi_{h12} - \xi_{12}); \qquad\qquad (3.3)$$

$$\dot{\xi}_{h1}(\tau) = \eta_1\,(\xi_1 - \xi_{h1})/V_1; \quad \dot{\xi}_{h2}(\tau) = \eta_2\,(\xi_2 - \xi_{h2})/V_2; \quad \dot{\xi}_{h12}(\tau) = \eta_{12}\,(\xi_{12} - \xi_{h12})\,/\,V_{12},$$

where ξ_{12} is the concentration of adsorbed particles of the products of the binary chemical reaction on the solid surfaces; τ_{12} is the lifetime of the surface to re-evaporation; η_{12} is the probability flow of particles into the trap with a specific capacity V_{12}. For the analysis of equations (3.3) it is important to agree on the degree of saturation of the traps for autonomous for both components ξ_{h1}, ξ_{h2} and reaction products ξ_{h12}. This issue is related to the prior history of the solid. If the body up to the time $\tau = 0$ were held infinitely long in the saturated atmosphere of the 1-*th* component without active desorption by the external influence at a constant temperature, then the first equation for $\tau \to \infty$ gives

$$\xi_{h1} = P_{h1} / (P_{h1} + n_a/\tau_1),$$

where $P_{h1} = p_{h1} (2\pi m_1 kT)^{-1/2}$, p_{h1} is the saturation pressure; m_1 is the molecular mass of the particles of the 1st component; k is the Boltzmann constant; T is temperature. In this case $\xi_{h2} = \xi_{h12} = 0$. Similarly, if the body was in the saturated atmosphere of the second component, then $\xi_{h2} = P_{h2} / (P_{h2} + n_a/\tau_2)$, $\xi_{h1} = \xi_{h12} = 0$. For generality, it should be assumed that ξ_{h1}, ξ_{h2}, ξ_{h12} are non-zero, and to simplify the task they are functions of temperature alone, which corresponds to an infinitely large capacity of traps or sinks, at the influence of diffusion inside the body and at the condensation of the respective components. Solution of a particular problem requires special refinement of the values of ξ_{h1}, ξ_{h2}, ξ_{h12}.

The first two equations of (3.3) do not contain the variable ξ_{h12} and can be considered independently. Making them dimensionless them with $\Theta_i = \xi_i/n_a$, the average lifetime $\tau_0 = (\tau_1 + \tau_2) / 2$, $t = \tau/\tau_0$ and introduction of the notations $a_1 = \tau_0 (1/\tau_1 + \alpha_{12}P_2 + \alpha_1 I_1)$; $a_2 = \tau_0(1/\tau_2 + \alpha_{21}P_1 + \alpha_2 I_2)$; $b_1 = P_1\tau_0/n_a$; $b_2 = P_2\tau_0/n_a$; $c_1 = \eta_1\tau_0$; $c_2 = \eta_2\tau_0$; $K = k_q n_a \tau_0$ leads to a system of non-linear differential equations:

$$\dot{\Theta}_1(t) = b_1 - (a_1 + b_1 + c_1) \Theta_1 - b_1\Theta_2 - K\Theta_1\Theta_2 + c_1\Theta_{h1},$$
$$\dot{\Theta}_2(t) = b_2 - (a_2 + b_2 + c_2) \Theta_2 - b_2\Theta_1 - K\Theta_1\Theta_2 + c_2\Theta_{h2}. \qquad (3.4)$$

The system (3.4) has two stationary points, defined by the expressions

$$\dot{\Theta}_{1C}^{\pm} = A \pm \{A^2 + [b_1(a_2 + c_2(1 - \Theta_{h2})) + c_1\Theta_{h1}(a_2 + b_2 + c_2)]/[K(b_2 - b_1 - a_1 - c_1)]\}^{1/2},$$

$$\dot{\Theta}_{2C}^{\pm} = [b_1 + c_1\Theta_{h1} - (a_1 + b_1 + c_1)\Theta_{1C}^{\pm}]/(b_1 + K\Theta_{1C}^{\pm}),$$

where $A = [K(b_1 + c_1\Theta_{h1} - b_2 - c_2\Theta_{h2}) - b_1b_2 + (a_1 + b_1 + c_1)]/[2K(b_2 - a_1 - b_1 - c_1)]$ is the intersection of the isoclines:

$$\Theta_1 = (b_1 + c_1\Theta_{h1} - b_1\Theta_2) / (a_1 + b_1 + c_1 + K\Theta_2),$$
$$\Theta_2 = (b_2 + c_2\Theta_{h2} - b_2\Theta_1) / (a_2 + b_2 + c_2 + K\Theta_1).$$

The equation for the trajectories of system (3.4) is the expression $d\Theta_1/d\Theta_2 = F(\Theta_1, \Theta_2)$, obtained by dividing the first equation by the second

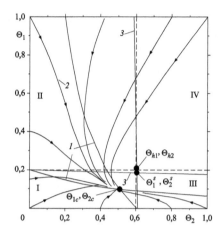

Fig. 3.8. The phase portrait of system of equations (3.4). $a_1 = 99$; $a_2 = 1$; $b_1 = 20$; $b_2 = 9$; $K = 2 \cdot 10^3$; $\Theta_{h1} = 0.2$; $\Theta_{h2} = 0.6$. 1 isocline for $K \neq 0$; 2 – the trajectory of solutions, 3 – isoclines at $K = 0$. Dashed line – the boundaries of the regions I–IV.

one. For all positive a_1, a_2, b_1, b_2, c_1, c_2, K in the first quadrant of the phase plane Θ_1, Θ_2 there is a stable node with coordinates Θ_{1c}, Θ_{2c}. An example of the phase portrait in the first quadrant is shown in Fig. 3.8. The other stationary point is in the second quadrant at $a_1 + b_1 + c_1 < b_2$, $a_2 + b_2 + c_2 > b_1$; in the third quadrant at $a_1 + b_1 + c_1 > b_2$, $a_2 + b_2 + c_2 > b_1$; in the fourth quadrant at $a_1 + b_1 + c_1 > b_2$, $a_2 + b_2 + c_2 < b_1$ and does not constitute substantial interest for the physical interpretation of the system (3.4), as the bifurcations of the solutions occurring during displacement of the second stationary point do not change qualitatively the phase portrait in the first quadrant.

In Fig. 3.8 the dashed lines, dividing the phase space into four characteristic regions, show the saturation state of the individual components relative to the body surface, which corresponds to $\Theta_{hi} = b_{hi}/(a_i + b_{hi})$, $a_i = \tau_0/\tau_i$, $b_{hi} = P_{hi}\tau_0/n_a$. If the stable node is in region I, then at $t \to \infty$ the values of the concentration of the adsorbed particles on the surface of the body are below the equilibrium values for saturation for the individual components and so condensation does not occur. The position of the stable node in the areas II and III corresponds to the establishment of the concentration of adsorbed particles of the first or second component, the position of the node in the area IV – to the simultaneous condensation of two components. A more detailed analysis of the phase portrait shows further three points, having a physical meaning.

1. Some paths may twice intersect the saturation line, which is equivalent to the appearance of the condensate for some time, followed by evaporation.

2. To begin the condensation of one of the components in the presence of the atmosphere of another vapour or gas we require the specific flow of this component that is stronger than in high vacuum.

3. With the decrease of the impurity flow of the vapour or gas condensation occurs in the conditions of a lower concentration of adsorbed impurity particles on the surface of the condensate, which can lead to a change in its morphology.

In the case of solutions of (3.3) in a dynamic space, such as numerical methods, we can calculate practically useful characteristics of the process:

– instantaneous values of the coefficients of condensation of the components:

$$\alpha_{r1} = c_1 \, (\Theta_1 - \Theta_{h1}) \, / \, b_1, \; \alpha_{r2} = c_2 \, (\Theta_2 - \Theta_{h2}) \, / \, b_2 \qquad (3.5)$$

and the reaction products:

$$\alpha_{r12} = c_{12} \, (\Theta_{12} - \Theta_{h12}) \, (m_1 + m_2) \, / \, (b_1 m_1 + b_2 m_2);$$

– instantaneous value of the total condensation coefficient:

$$\alpha_c = [(\alpha_{r1} b_1 m_1 + \alpha_{r2} \, b_2 m_2) \, / \, (b_1 m_1 + b_2 m_2)] + \alpha_{r12}; \qquad (3.6)$$

– instantaneous mass composition of the condensate produced at time t, if the products of chemical reactions remain on the surface:

$$c_{ri} = \alpha_{ri} b_i m_i \, / \, [\alpha_{r1} b_1 m_1 + \alpha_{r2} b_2 m_2 + \alpha_{r12} \, (b_1 m_1 + b_1 m_2)]. \qquad (3.7)$$

Substituting into (3.5)–(3.7) the steady-state values Θ_{1c} and Θ_{2c} we can calculate steady-state values α_{ri}^C, α_{r12}^C, α_c^C, c_{ri}^C, c_{r12}^C. To calculate the integral values of these characteristics for a finite time interval, the equations (3.5)–(3.7) must be integrated and averaged in this interval of time.

3.2.3. Analysis of the problem in the absence of chemical reactions

In the special case $K = 0$ the equations (3.4) are transformed into a system of inhomogeneous differential equations:

$$\dot{\Theta}_1(t) = b_1 - (a_1 + b_1 + c_1)\Theta_1 - b_1\Theta_2 + c_1\Theta_{h1},$$
$$\dot{\Theta}_2(t) = b_2 - (a_2 + b_2 + c_2)\Theta_2 - b_2\Theta_1 + c_2\Theta_{h2}. \qquad (3.8)$$

The roots of the characteristic equation of system (3.8) are given by

$$\beta^\pm = (a_1 + b_1 + c_1 + a_2 + b_2 + c_2)/2 \pm [(a_1 + b_1 + c_1 - a_2 - b_2 - c_2)^2/4 + b_1 b_2]^{1/2}.$$

The solution of (3.8) has the form

$$\Theta_1(t) = G_1 \exp(-\beta^+ t) + G_2 \exp(-\beta^- t) + \Theta_1^S,$$
$$\Theta_2(t) = G_1 \left[(\beta^+ - a_1 - b_1 - c_1)/b_1\right] \exp(-\beta^+ t) + G_2 \left[(\beta^- - a_1 - b_1 - c_1)/b_1\right] \exp(-\beta^- t) + \Theta_2^S,$$

$$(3.9)$$

where

$$\Theta_1^S = \frac{b_1[a_2 + c_2(1 - \Theta_{h2})] + (a_2 + b_2 + c_2)c_1\Theta_{h1}}{b_1(a_2 + c_2) + (a_1 + c_1)(a_2 + b_2 + c_2)},$$

$$\Theta_2^S = \frac{b_2[a_1 + c_1(1 - \Theta_{h1})] + (a_1 + b_1 + c_1)c_2\Theta_{h2}}{b_1(a_2 + c_2) + (a_1 + c_1)(a_2 + b_2 + c_2)} \qquad (3.10)$$

– steady-state values of the extent of surface coverage by adsorbed particles. The integration constants G_1, G_2 can be determined using the initial conditions $t = 0$; $\Theta_1 = \Theta_{01}$, $\Theta_2 = \Theta_{02}$:

$$G_1 = \left[(\beta^- - a_1 - b_1 - c_1)(\Theta_1^S - \Theta_{01}) - b_1(\Theta_2^S - \Theta_{02})\right]/(\beta^+ - \beta^-),$$

$$G_2 = -\left[(\beta^+ - a_1 - b_1 - c_1)(\Theta_1^S - \Theta_{01}) - b_1(\Theta_2^S - \Theta_{02})\right]/(\beta^+ - \beta^-). \qquad (3.11)$$

The isoclines of system (3.8) are shown in Fig. 3.8 by the curves 3. Their intersection with the coordinates of steady-state values Θ_1^S, Θ_2^S is situated in region III. It is evident that the absence of chemical reaction between the components under the same conditions leads to the possibility of condensation of 2nd component.

In the special case when $c_1 = c_2 = 0$, the system of equations (3.8) was considered in [24] in which, unfortunately, an error was made in dealing with the solution of type (3.9) for the second component and in determining the integration constants of the type (3.11). Condition $c_1 = c_2 = 0$ corresponds to the absence of particle traps, and $\Theta_{h1} = \Theta_{h2} = 0$ corresponds to the action of always empty traps or condensation at high supersaturation. For this case, the rate of accumulation of mass of the adsorbed gases per unit surface area of the body in dimensionless form can be written as

$$dM/dt = \left(\dot\Theta_1 m_1 + \dot\Theta_2 m_2 + c_1\Theta_1 m_1 + c_2\Theta_2 m_2\right)/m_0, \qquad (3.12)$$

where m_0 is the mass scale, for example, $m_0 = (m_1 + m_2)/2$. The first two terms of (3.12) reflect the contribution to the increase in mass of the adsorbed particles on the surface of the body, the third and fourth – the contribution of possible condensation. The results of calculations by formulas (3.9)–(3.12) for the case $c_1 = c_2 = 0$, i.e. in the absence of condensation of both components, are shown in Fig. 3.9. The calculations were performed with initial conditions $t = 0$, $\Theta_1 = \Theta_{01} = 0$. The different curves correspond to different Θ_{02}. Curves 2 and 3 were obtained at $t = 0$, $dM/dt = 0$, $d^2M/dt^2 = 0$, which corresponds to the initial conditions:

$$\Theta_{02} = \Theta_{02}^* = 1 - a_2/(b_1 m_1/m_2 + a_2 + b_2),$$

$$\Theta_{02} = \Theta_{02}^{**} = 1 - a_2/\left[(a_1 + b_1 + a_2 + b_2) - (a_1 a_2 + a_1 b_2 + a_2 b_1)/(b_1 m_1/m_2 + a_2 + b_2)\right].$$

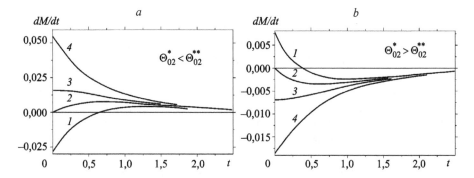

Fig. 3.9. Change of the rate of adsorption with time. $a_1 = a_2 = 1$; $b_1 = 0.2$; $b_2 = 1$; $c_1 = c_2 = 0$; $m_1 = 24$. a – $m_2 = 18$, curves 1–4 correspond to $\Theta_{02} = 0.57, 0.5588, 0.5538, 0.534$; b – $m_2 = 28$, curves 1–4 correspond to $\Theta_{02} = 0.536, 0.5394, 0.542; 0.548$.

3.2.4. The physical interpretation of models

The dependence of dM/dt versus time (see Fig. 3.9) illustrates the variety of readings of the quartz balance in the measurement of the deposition rate in the absence of chemical reactions and condensation on the sensor. The possible condensation on the sensor of the quartz balance makes it even more difficult to measure them over time. Figure 3.10 b shows the calculated dM/dt for various conditions. Curves 1–3 show changes of the data obtained by the quartz balance at different specific flow rates of the vapour entering the vapour sensor. For small flow rates of the vapour to the sensor, situated in the saturated atmosphere of the 2nd component, there are negative dM/dt values, corresponding to the initial desorption of the 2nd component.

For curve 1 we see that when closing the shutter, interrupting the vapour flow at time A, the transfer of the quartz balance again to the saturated in

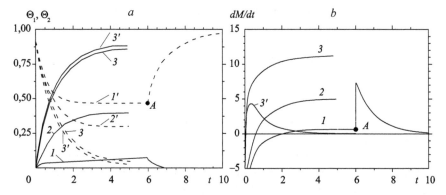

Fig. 3.10. The dependence of the concentration of adsorbed particles (a) and deposition rate (b) on time. Calculation by formulas (3.9) and (3.12) with $a_1 = a_2 = 1$; $b_1 = 1$; $K = 0$; $c_2 = 0$; $\Theta_{01} = \Theta_{h1} = 2 \cdot 10^{-3}$; $\Theta_{02} = 0.997$. Curves 1–3 correspond to $b_1 = 0.2$; 2, 20; $c_1 = 0.555$; curve 3' – $b_1 = 20$; $c_1 = 2 \cdot 10^{-2}$; Figure a solid lines Θ_1, dashed Θ_2.

atmosphere of the 2nd component causes a spike in the readings. At high specific flow rates the negative deposition rates at $t = 0$ and spikes at closing the shutter decrease and disappear. Curve 3' in comparison with curve 3 illustrates the complicated nature of the readings of the balance at a much lower condensation coefficient.

All types of curves presented in Fig. 3.10 b were observed when measuring the rate of condensation in our experiments with the quartz balance. Figure 3.10 a shows the results of calculations of the relative concentrations of the components in the sensor for the corresponding curves in Fig. 3.10 b. It is seen that the adsorbate particle concentration is strongly dependent on the specific density of the flow of vapour to the surface, and weakly on the condensation process.

To analyze the above experimental results, it should be noted that the magnesium used in the experiments is active with respect to oxygen and the water vapour of the vacuum chamber. When evacuating the vacuum chamber by means of the vapour–oil diffusion pump the atmosphere of residual gases also contains a significant amount of oil vapours inert to magnesium. In this case, the experimental data should be analyzed using the system of equations:

$$\dot{\Theta}_1(t) = b_1 - (a_1 + b_1 + c_1)\Theta_1 - b_1\Theta_2 - b_1\Theta_2 + K\Theta_1\Theta_2 + c_1 C,$$

$$\dot{\Theta}_2(t) = b_2 - (a_2 + b_2 + c_2)\Theta_2 - b_2\Theta_1 - b_2\Theta_3 + K\Theta_1\Theta_2 + c_2\Theta_{h2},$$

$$\dot{\Theta}_3(t) = b_3 - (a_3 + b_3 + c_3)\Theta_3 - b_3\Theta_1 - b_3\Theta_2 + K\Theta_1\Theta_2 + c_3\Theta_{h3}, \qquad (3.13)$$

$$\dot{\Theta}_{12}(t) = K\Theta_1\Theta_2 - a_{12}\Theta_{12} - c_{12}(\Theta_{12} - \Theta_{h12}),$$

$$c_3 = Bc_1(\Theta_1 - \Theta_{h1}).$$

The first two equations of the system (3.13) reflect the dynamics of the relative concentrations of chemically interacting magnesium and the oxidant (oxygen or water), the fourth – the dynamics of the products of their reactions, and the third – the dynamics of surface filling by the oil adsorbate. The fifth equation shows the relationship of the probability of the flow of the oil adsorbate into the trap at a rate of condensation of 1st component, which means physically describes the capture of oil molecules stages by growth stages in condensation of magnesium or capillary condensation in the intergranular space of the magnesium condensate.

Calculations using the system (3.13) were carried out under the assumption that at $\Theta_i < \Theta_{hi}$ the value $c_i = 0$, i.e., condensation of i-th component takes place only in case of saturation of the adsorbed phase. The calculation results are shown in Fig. 3.11.

Curves 1–3 refer to the range of low b_1, where $\Theta_1 \ll \Theta_3$, $\Theta_2 \ll \Theta_3$, $\Theta_1 < \Theta_{hi}$ and deposition of the reaction products takes place on the surface with the rate $c_{12}(\Theta_{12}^S - \Theta_{h12})$. Curves 1', 1", 2', 2" relate to the area $\Theta_1 > \Theta_{h1}$, $\Theta_2 \ll \Theta_1$, where the condensation of Mg occurs at a rate of $c_1(\Theta_1^S - \Theta_{h1})$. The curves marked with the numerals with one stroke were obtained by neglecting the capture of molecules by the magnesium condensate, two strokes – $\Theta_{h3} = 0$,

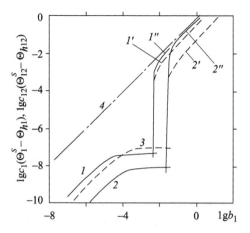

Fig. 3.11. The deposition rate depending on the specific flow density of the 1st component. The calculation of the system of equations (3.13), $a_1 = 1$; $a_2 = 10^{-4}$; $a_3 = 1$; $K = 10$; $\Theta_{h1} = 2 \cdot 10^{-3}$; $\Theta_{h2} = 0.2$; $\Theta_{h3} = 0.997$; $\Theta_{h12} = 0$. For curves 1–3 $c_1 = c_2 = c_3 = 0$; $b_1 = 1$; $b_2 = 10^{-7}$; 10^{-7}; 10^{-6}; $b_3 = 1, 10, 10$. For curves 1', 2' $c_1 = 10$; $c_2 = 0$; $b_2 = 10^{-7}$; 10^{-6}; $b_3 = 1$; 10.

which means empty traps for oil molecules formed during the condensation of magnesium. Curves 1, 1', 1" were obtained at the oil vapour flow ten times smaller b_3 to the surface in comparison with the curves 2, 2', 2". Curve 3 corresponds to a tenfold increase in the flow of oxidizer b_2 in comparison with curve 2. The supersaturation in the vapour phase with respect to magnesium b_1 / b_{h1}, at which intense condensation of Mg on the surface begins, for curves 1', 1" and 2', 2" is equal to 2 and 10.

Comparison of simulation results presented in Fig. 3.11 with the experimental data shown in Fig. 3.1 shows their qualitative agreement.

3.2.5. Critical temperatures of heterogeneous condensation

The conditions of 'collapse' or the beginning of heterogeneous condensation, corresponding to the transition of the stationary point $(\Theta_{1c}, \Theta_{2c})$ in Fig. 3.8 from region to region and illustrated in Fig. 3.11 by the change in the specific vapour consumption b_1, can be analyzed by putting $\Theta_{1c} = \Theta_{hi}$ or $\Theta_1^S - \Theta_{hi}$. For a two-component gas medium and without considering chemical reactions, from (3.10) we obtain the 'collapse' condition of the 1st component:

$$b_1 = \frac{\Theta_{h1} a_1 \left(a_2 + b_2 + c_2\right)}{a_2 \left(1 - \Theta_{h1}\right) + c_1 \left(1 - \Theta_{h1} - \Theta_{h2}\right)}. \tag{3.14}$$

Note that the linear relationship between b_1 and b_2 in (3.14) at a 'collapse' of condensation was observed in the experiments described above. With the help of (3.14) we can analyze the conditions of 'collapse' of condensation when the temperature changes. When only thermal desorption is considered with the assumption of diffusive flow of adatoms in the active centres of condensation and approximation of the saturation pressures of the components according to [29], the parameters a_i, b_{hi}, c_i can be expressed as

$$a_i = \tau_0 v_\perp \exp\left(-\varepsilon_i / kT\right);$$

$$b_{hi} = \tau_0 \left(v_\perp n_a^2 \left(2\pi m_i kT\right)^{1/2} / \kappa_i\right) \exp\left(-3\varepsilon_i / kT\right);$$

$$c_i = 2\left(\tau_0 n_{4i} \, v_\| / \pi n_a^2\right) \exp\left(-u_{Di} / kT\right). \tag{3.15}$$

Here $v_\perp$, $v_\|$ are the frequencies of thermal vibrations in the transverse and longitudinal surface directions; ε_i is the energy of one bond of the adatoms; κ_i is the effective condensation coefficient of atoms of the vapour phase on the surface of the condensed phase; n_{4i} is the adsorption vacancy concentration.

After substituting (3.15) (3.14) can be obtained by the equation

$$b_1 + A_1 b_1 \exp\left(-2\varepsilon_2/kT\right) = B_1 T^{1/2} \exp\left(-3\varepsilon/kT\right) + B_2 T \exp[-(3\varepsilon_1 + 2\varepsilon_2)/kT] +$$
$$+ B_3 T^{1/2} \exp\left[-(3\varepsilon_1 - \varepsilon_2)/kT\right] + B_4 T \exp\left[-(3\varepsilon_1 + \varepsilon_2)/kT\right] +$$
$$+ B_5 T^{1/2} \exp\left[-(3\varepsilon_1 - \varepsilon_2 + u_{D2})/kT\right] + B_6 T \exp\left[-(3\varepsilon_1 + \varepsilon_2 + u_{D2})/kT\right] + \tag{3.16}$$
$$+ B_7 T \exp\left[(2\varepsilon_2 - u_{D2})/kT\right] + B_8 T b_1 \exp\left[-(2\varepsilon_1 + \varepsilon_2 + u_{D2})/kT\right].$$

For positive b_1 equation (3.16) in accordance with the number of sign changes can have at most three roots with respect to temperature. The scheme of the graphical solution of this equation is conventionally presented in Fig. 3.12. If the right side of the equation contains exponential terms with exponents much smaller than in the second term on the left side, the dependence of the sum of all exponential terms on temperature is given by curve 2. In this case, for small b_1 there are three critical temperature T_{c1}, T_{c2}, T_{c3}. Temperatures T_{c1} and T_{c2} can be interpreted as the borders of the zone of 'collapse' of condensation found in the work of L.S. Palatnik. Temperature T_{c3}

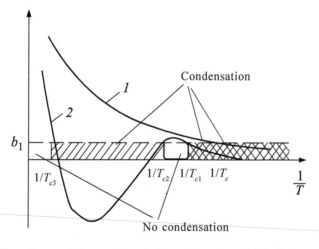

Fig. 3.12. The scheme of the graphic solution of equation (3.16). 1, 2 – various options for the sum of exponential terms,

corresponds to the high-temperature boundary of condensation when comparing the flow incoming to the surface and the evaporation flow. For large b_1 there is only this high-temperature limit of condensation. Curve 1 in Fig. 3.12 shows the sum of exponential terms in equation (3.16), when the term with the coefficient A_1 is not essential. Then the entire range of b_1 there is only one condensation temperature limit T_c. The above indicates that the existence of a zone of 'collapse' of condensation requires a combination of physical parameters and the properties of the solid system–vapour environment system.

Note that the analysis shows the dependence of T_{c3} on the residual atmosphere pressure. It was found in earlier experiments [20].

So, based on models of the kinetics of sorption processes we clarified qualitatively the known experimental results on the 'collapse' of vapour condensation on a solid surface in the presence of impurities.

3.3. Features of the initial stage of formation of columnar dispersed structures on the substrate

In section 3.1 it is shown that areas near the boundary of the region of 'collapse' of condensation are characterized by the formation of structures representing the sum of separated columnar crystals (see Fig. 3.5 b). To understand the mechanism of their formation, attempts were made to trace their evolution in time by spraying several samples at varying exposure. Two types of substrate – pieces of polished copper foil and mesh for electron-microscopic study with a deposited carbon film 500–100 Å thick, were mounted on a 72-position turret drum inside the chamber. In addition to the moving shutter, a disk shutter was placed in the chamber between the vapour source and a drum for varying exposure in spraying samples from 0.01 s to several minutes. Measures were taken to eliminate penetration of ultrafine dust from the volume of the vacuum chamber on the substrate. Sprayed samples, obtained on the foil and the carbon film, were treated respectively in the optical and electron microscopes. At a distance of 170 mm from the vapour source the mode of $p_0 = 6650$ Pa, $T_0 = 835$ K precipitation of Mg on the foil specimens, clearly visible in an optical microscope, was observed at exposure times of at least 9 s. A characteristic feature of these experiments was the initial formation of groups of islands with a diameter of about 0.3 μm, which is at the limit of resolution of light microscopy. With increasing deposition exposure the groups of islands grew, followed by filling the entire surface of the sample. The results of measuring the size of the islands in relation to the exposure time are shown in Fig. 3.13. The figure shows that the size of the islands depends on the exposure time at the stage of condensation, until the surface is not completely covered with the condensate. After filling the surface the average dimensions of the islands in remain constant.

It was noticed that the formation of the first islands on the foil occurs near the surface microdefects – microscratches, recesses, microinclusions in the structure. The surface of the condensate is filled with successive appearance of new islands near the already existing ones.

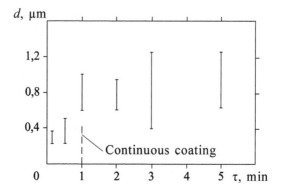

Fig. 3.13. The dependence of the particle size of the condensate on deposition time. Sonic nozzle size 0.4 mm; distance 170 mm; p_0 = 6650 Pa. Copper foil after machining, T_0 = 835 K, T_S = 298 K.

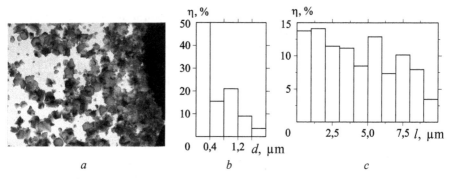

a   b   c

Fig. 3.14. Magnesium condensate on the carbon film. a – photograph at the edge of the cell; p_0 = 6650 Pa, T_0 = 1133 K, deposition time 30 s, magnification $2 \cdot 10^4$; b – particle size distribution; c – the distribution of particles over the distance from the edge of the mesh cell.

Experiments with carbon substrates showed that the appearance of the first dispersed islet cells is observed near the edges of the cells. Figure 3.14 a is an electron micrograph of one of the samples. From the picture it is clear that magnesium condensate islands have two distinct fractions which differ in size. The size distribution function of islands is shown in Fig. 3.14 b. The particles of the larger fraction are characterized by the plate shape with hexagonal faceting and preferred orientation of the islands by the plane to the surface of the carbon film. Smaller particles are distributed mainly along the periphery of large islands. Figure 3.14 c illustrates the distribution of the density of large particles in the distance from the edge of the mesh cell. Here the cell is also gradually filled with condensate particles.

The bimodality of the size distribution function of the islands can be interpreted as the result of two condensation processes. The larger islands are probably the result of their growth on the substrate. The fraction of fine particles may be due to homogeneous condensation of the vapour in the flow

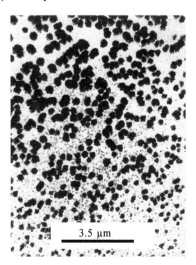

Fig. 3.15. Magnesium condensate on mica. 2 min deposition at the parameters indicated in Fig. 3.14.
Fig. 3.16. (right) Magnesium condensate on the carbon film.

before their arrival on the sample (homogeneous condensation phenomena are discussed in the next chapter). A more pronounced bimodal size distribution of condensate particles is shown in Figs. 3.15 and 3.16.

The sequence of coverage of the surface with the islands will now be clarified.

3.3.1. Model of non-stationary surface diffusion of adsorbed particles of two components

There are a number of works, e.g. [30, 31], which examined the development of nuclei of the condensed phase on the surface due to the flow of matter by migration of adsorbed particles on it. The ranges of parameters in which the diffusion mechanism of condensation is dominant were determined in [18]. Analytical solutions in the quasi-stationary approximation for the concentration profiles of adsorbed particles on the surface were obtained, with the topography of the surface taken into account. It is shown that the formation of the condensed phase on the substrate is due to diffusion of the adsorbate on the surface. In general, the appearance of the nuclei on the surface creates around them zones with a reduced concentration of the adsorbate, which leads to their uniform distribution on the surface. The regularities of growth of nuclei and formation and evolution of the condensation front were studied, and instantaneous and integral coefficients of condensation were calculated. However, in a more complicated case of two-component mixtures of condensable gases the problem was only formulated and not investigated.

This section describes special features of the formulation of the problem in the case when the condensing vapour environment contains a gas admixture in gas in the unsaturated state, and special features of nucleation can make fundamental changes in the course of the condensation process.

The kinetics of the concentration of the particles of the two-component gas mixture, adsorbed on the flat surface, in a one-dimensional formulation can be represented by the equations

$$\frac{\partial \Theta_1}{\partial t} = D_1 \frac{\partial^2 \Theta_1}{\partial x^2} + b_1 - (a_1 + b_1 + c_1)\Theta_1 - b_1\Theta_2 + c_1\Theta_{h1},$$

$$\frac{\partial \Theta_2}{\partial t} = D_2 \frac{\partial^2 \Theta_2}{\partial x^2} + b_2 - (a_2 + b_2 + c_2)\Theta_2 - b_2\Theta_1 + c_2\Theta_{h2}, \tag{3.17}$$

where D_1, D_2 are the reduced coefficients of surface diffusion of the particles of the gas mixture components; Θ_1, Θ_2 are the surface concentrations of adsorbed particles of the components made dimensionless with respect to the density of vacancies on the solid surface; t is dimensionless time; b_1, b_2 are the specific flows of the particles of the components from the gas medium on the surface; a_1, a_2 are the probabilities of desorption by re-evaporation; c_1, c_2 are the probabilities of a local exchange of the adsorbed particles on the surface. Equations (3.17) are written in the light of occupation of a part of the finite number of vacancies, when the body is held long enough in the saturated atmosphere of either component. Concentrations Θ_{h1}, Θ_{h2} correspond to the filling of surface traps in the state where the body is held long enough in a saturated atmosphere of a component.

In the stationary approximation for $\partial \Theta_1/\partial t \to 0$, $\partial \Theta_2/\partial t \to 0$, system (3.17) has the solution:

$$\Theta_1 = G_1 \exp\left(\sqrt{z^+}\, x\right) + G_2 \exp\left(-\sqrt{z^+}\, x\right) + G_3 \exp\left(\sqrt{z^-}\, x\right) + G_4 \exp\left(-\sqrt{z^-}\, x\right) + \Theta_1^S,$$

$$\Theta_2 = \left(\frac{D_1 z^+ - a_1 - b_1 - c_1}{b_1}\right)\left[G_1 \exp\left(\sqrt{z^+}\, x\right) + G_2 \exp\left(-\sqrt{z^+}\, x\right)\right] +$$

$$+ \left(\frac{D_1 z^- - a_1 - b_1 - c_1}{b_1}\right)\left[G_3 \exp\left(\sqrt{z^-}\, x\right) + G_4 \exp\left(-\sqrt{z^-}\, x\right)\right] + \Theta_2^S, \tag{3.18}$$

where

$$z^\pm = \frac{[(D_2(a_1+b_1+c_1)+D_1(a_2+b_2+c_2)] \pm \left\{[D_2(a_1+b_1+c_1)-D_1(a_2+b_2+c_2)]^2 + 4D_1 D_2 b_1 b_2\right\}^{1/2}}{2D_1 D_2}$$

and $\pm\sqrt{z^\pm}$ are the roots of the characteristic equation of system (3.17). Steady-state concentrations Θ_1^S and Θ_2^S for $x > 0$ are given by

$$\Theta_1^S = \frac{b_1\left[a_2 + c_2\left(1 - \Theta_{h2}\right)\right] + \left(a_2 + b_2 + c_2\right)c_1\Theta_{h1}}{b_1\left(a_2 + c_2\right) + \left(a_1 + b_1\right)\left(a_2 + b_2 + c_2\right)};$$

$$\Theta_2^S = \frac{b_2\left[a_1 + c_1\left(1 - \Theta_{h1}\right)\right] + \left(a_1 + b_1 + c_1\right)c_2\Theta_{h2}}{b_1\left(a_2 + c_2\right) + \left(a_1 + b_1\right)\left(a_2 + b_2 + c_2\right)}.$$

Integration constants $G_{1,2,3,4}$ should be determined from the boundary conditions of the specific task.

3.3.2. Heterogeneous nucleation and growth of nuclei in the presence of impurities

The problem formulated by equations (3.17) is considered for the initial conditions:

$$T = 0, \; \Theta_1 = 0, \; \Theta_2 = b_2 / (b_2 + a_2),$$

which is equivalent to applying a specific flow of the gas of the 1st component on the surface of the body relative to the gas of the 2nd component situated in the unsaturated component. The appearance of the particle flow of the 1st component from the gas phase causes an increase in concentration Θ_1 and decrease of Θ_2. Changes of Θ_1 and Θ_2 with time in the range $\Theta_1 < \Theta_{h1}$, $\Theta_2 < \Theta_{h2}$ assuming the equality of all points of the body surface are described by the solutions of (3.17), already cited in section 3.2.3. At $\partial^2\Theta_1/\partial x^2 = 0$, $\partial^2\Theta_2/\partial x^2 = 0$

$$\Theta_1 = H_1 \exp(-\beta^+ t) + H_2 \exp(-\beta^- t) + \Theta_1^S,$$

$$\Theta_2 = H_1\left(\frac{\beta^+ - a_1 - b_1 - c_1}{b_1}\right)\exp(-\beta^+ t) + H_2\left(\frac{\beta^- - a_1 - b_1 - c_1}{b_1}\right)\exp(-\beta^- t) + \Theta_1^S,$$

where

$$\beta^\pm = (a_1 + b_1 + c_1 + a_2 + b_2 + c_2)/2 \pm \left\{\left[(a_1 + b_1 + c_1) - (a_2 + b_2 + c_2)\right]^2 / 4 + b_1 b_2\right\}^{1/2}$$

are the roots of the corresponding characteristic equation. The integration constants H_1, H_2 are determined from the initial conditions:

$$H_1 = \frac{\left(\Theta_{02} - \Theta_2^S\right)b_1 + \Theta_1^S(\beta^- - a_1 - b_1 - c_1)}{\beta^+ - \beta^-},$$

$$H_2 = \frac{\left(\Theta_{02} - \Theta_2^S\right)b_1 + \Theta_1^S(\beta^+ - a_1 - b_1 - c_1)}{\beta^+ - \beta^-}.$$

Upon reaching Θ_1 concentration of adsorbed particles, the appropriate concentration in an environment where the body would be in a saturated environment, only the first component Θ_{h1}, there is the possibility of nucleation condensation of infinite dimensions.

With a further increase $\Theta_1 > \Theta_{h1}$ the likelihood of formation of a stable nucleus from N^* particles per unit surface area per unit time is [29]

$$J = \zeta v \Theta_1 n \, (N^*) \exp (-u_{D1}/kT), \tag{3.19}$$

where ζ is the fraction of vacancies at the border of the nucleus occupied by the particles of the first component; v is the frequency of thermal vibrations of adsorbed particles on the surface; $n(N^*) = n_a \exp [-\delta\Phi(N^*)/kT]$ is the density of aggregates containing N^* particles on the surface, forming and breaking up due to fluctuations; $\delta\Phi \, (N^*)$ is the energy barrier of aggregate formation; u_{D1} is the activation energy of diffusion. With increasing Θ_1 when the J value is about 1 per unit area of the body surface the first nucleus of the 1st component forms.

The resulting stable nucleus becomes a sink for the particles of the 1st component adsorbed on the surface. Concentration gradient Θ_1 appears along the surface. The evolution of the distribution $\Theta_i(x, t)$ can be analyzed with the help of equations (3.17). If the concentrations in calculations of the solutions of (3.9) with $J = 1$ were $\Theta_1 = \Theta^{01}$, $\Theta_2 = \Theta^{02}$, these values can be used as the initial conditions in the analysis of equations (3.17). The conditions of formation of a stable nucleus of (3.19) also correspond to the condition

$$\Theta^{01}/\Theta_{h1} = \exp(2\gamma_1\omega_1/kTr_{*1}),$$

where r_{*1} is the radius of curvature of the nucleus. If the origin of the x-coordinate is represented by the boundary of the nucleus at the surface, the initial and boundary conditions are as follows:

$$\begin{aligned} t &= 0, \, x = 0 \div \infty, \, \Theta_1 = \Theta^{01}, \, \Theta_2 = \Theta^{02}, \\ t &> 0, \, x = \infty, \, \partial\Theta_1 / \partial x = 0, \, \partial\Theta_2 / \partial x = 0. \end{aligned} \tag{3.20}$$

For examining the evolution of a single nucleus it is sufficient to consider these conditions (3.20), since the requirement of finiteness of Θ_1 and Θ_2 at $x \rightarrow \infty$ yields $G_1 = G_2 = 0$. A further increase of Θ_1 and decrease of Θ_2 due to the adsorption–desorption process on the entire surface of the body lead to a concentration gradient of Θ_1 and the emergence of the diffusion flow of particles of the 1st component to the nucleus. The development of the resultant nucleus takes place both by surface diffusion and direct contact with a particle of 1st component of the gas phase. The latter circumstance leads to a convex shape of the nucleus with the coordination of the point of junction of the boundaries of the three nucleus–body–gas medium phases in accordance with the ratio of their surface energies. In general, the border of the nucleus can be characterized by the formation of a point of the zero radius of curvature, which provides the formation of a nucleus of the condensed phase of the 2nd component, even if the concentration of the particles on the surface is much less than saturated concentration.

The appearance of the nucleus of the 2nd component at the border of the nucleus of the 1st component makes it extremely difficult to further consider the problem, since the migration of adsorbed atoms of the 1st component on the condensate the 2nd component in general may differ from migration across the surface of the body. In solving the diffusion problem we define an additional area in x, where all the parameters in equation (3.17), are changing. Further evolution of the nucleus of the 1st component should be read in conjunction with the evolution of the 2nd component. Different situations can form depending on the migration characteristics of particles of both components on the condensate of the 2nd component. If the migration on the surface of the nucleus of the 2nd component for particles of the 1st component is very difficult or does not place at all, the further development of the nucleus of the 1st component will only occur due to the vapour flow of the gaseous medium. The shape of the nucleus will be a superposition of its growth from the vapour environment and diffusive growth of the nucleus of the 2nd component.

Note that in the case of small b_1, when the flat nucleus formed on the surface of the body does not acquire a convex shape, condensation of 2nd component does not take place. This condensation mode corresponds in all likelihood to epitaxial growth.

The purpose of this study did not include a description of the form of the nucleus. Therefore, we give a qualitative description of the most interesting features of the formulation of the problem as a whole.

3.3.3. Chain mechanism of nucleation

The appearance at the boundary of the nucleus of the first component of a point with zero radius of curvature and the formation of the nucleus of 2nd component with a curved surface creates an opportunity to formulate the boundary conditions for the diffusion problem assuming that there is no migration of particles of the 1st component on the surface of the nucleus of the 2nd component:

$$t = 0, \ x = 0, \ \partial\Theta_1 / \partial x = 0, \ \Theta_2 = 0;$$
$$x = \infty, \ \partial\Theta_1 / \partial x = 0, \ \partial\Theta_2 / \partial x = 0.$$
$$t > 0, \ x = 0, \ \partial\Theta_1 / \partial x = 0, \ \Theta_2 = \Theta_{h2\infty} \exp\left(2\gamma_2\omega_2/kTr_2\right); \qquad (3.21)$$
$$x = \infty, \ \partial\Theta_1 / \partial x = 0, \ \partial\Theta_2 / \partial x = 0.$$

Conditions (3.21) imply that one should study the evolution of the concentration profiles $\Theta_1(x, t)$ and $\Theta_2(x, t)$ for the diffusion sink of adsorbed particles of the 2nd component into nuclei of the 2nd component. In the process of condensate accumulation in the nucleus of the 2nd component the curvature of its surface changes. The boundary conditions for the problem can apparently be represented by the relation

$$\Theta_2\,(x = 0)\,/\,\Theta_{h2\infty} = \exp\,(2\gamma_2\omega_2/kTr_2),$$

where r_2 is the radius of curvature defined in terms of accumulation of the condensate in the angle formed by the body surface and the surface of the nucleus of the 1st component of the junction with the conditions of contact of all phase boundaries taken into account. However, without solving this problem one can see the unusual situation that arises in the area of the body surface, adjacent to the nucleus of the condensate of the 2nd component. The reduced concentration of the adsorbed particles of the 2nd component near the nucleus changes the adsorption conditions for the particles coming to the surface of the body from the gaseous medium due to the release of some of the vacancies. The upper estimate of this concentration at the border of the second nucleus can be done by assuming that the process of changing the volume of the nucleus of the 2nd component is much slower than the establishment of diffusion profiles of concentrations of both components. This assumption is justified by two factors. First, the transition of the adsorbed particles to the condensed phase is associated with a change in the density of matter by several orders of magnitude. Second, the growth of the nucleus of the 2nd component is accompanied by the growth of the nucleus of the 1st component, thereby increasing its maximum possible volume. Solving (3.18) in the conditions

$$x = 0,\ \Theta_2 = 0,\ \partial\Theta_1\,/\,\partial x = 0,$$
$$x = \infty,\ \Theta_2 = \Theta_2^S,\ \Theta_1 = \Theta_1^S,$$

where $\partial\Theta_1/\partial x = 0$ denotes the impermeable surface of the nucleus of the 2nd component for the migration of particles of the 1st component on this surface, we can determine the integration constant:

$$G_1 = G_3 = 0,$$

$$G_2 = \frac{b_1\Theta_2^S}{\left(\sqrt{z^+/z^-} - 1\right)(a_1 + b_1 + c_1) + D_1\left(z^+ - \sqrt{z^+z^-}\right)},$$

$$G_4 = \frac{b_1\Theta_2^S\sqrt{z^+/z^-}}{\left(\sqrt{z^+/z^-} - 1\right)(a_1 + b_1 + c_1) + D_1\left(z^+ - \sqrt{z^+z^-}\right)}.$$

The concentration of the particles of 1st component at the boundary of the nucleus of the 2nd component:

$$\Theta_1\big|_{x=0} = \frac{b_1\Theta_2^S\left(\sqrt{z^+/z^-} - 1\right)}{\left(\sqrt{z^+/z^-} - 1\right)(a_1 + b_1 + c_1) + D_1\left(z^+ - \sqrt{z^+z^-}\right)} + \Theta_1^S. \tag{3.22}$$

Equation (3.22) shows that a region of the solid surface with a high concentration of adsorbed particles of the 1st component forms in the vicinity of

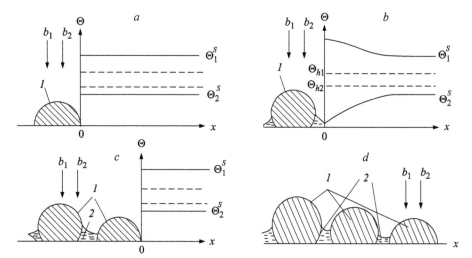

Fig. 3.17. Successive stages of nucleation. a – the formation of a nucleus of the 1st component; b – capillary condensation of 2nd component; c – formation of the 2nd nucleus of the 1st component; d – the formation of the base of the columnar structure.

the nuclei. In accordance with (3.19) this indicates an increased likelihood of the formation of the next nucleus of the 1st component in this area. Successive stages of the evolution of concentration profiles of adsorbed particles and the formation of nuclei are conventionally presented in Fig. 3.17. With further increase of the nuclei of the 1st component the gap between them is filled with the condensate of the 2nd component. The particles of the 2nd component travel here both by direct propagation from the gaseous medium and by the diffusion flow from the surface of the nuclei of the first component. So, apparently, the columnar structure of the condensate in the initial stage of condensation forms. Thus, in this section we formulate the diffusion problem of the initial stage of heterogeneous condensation in the presence of impurities in the environment. Analysis of the problem in quasi-stationary one-dimensional approximation shows the possibility of the existence of the chain mechanism of nucleation, when the formation of the nucleus through the capillary condensation of impurities stimulates the formation of subsequent nuclei following in the immediate vicinity.

Figure 3.18 shows examples of the sequential filling of the substrate with the condensate of magnesium and aluminum.

3.4. Effect of gas dynamic parameters on the formation of dispersed structures

Generalization of the experimental and theoretical studies of the formation of films of the vacuum condensate is given in the book by Yu.F Komnik [20]. The existence of critical temperatures of condensation, the change of the condensation mechanism of vapour–liquid or vapour–crystal systems in the

a   b

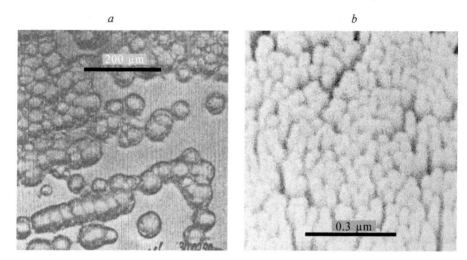

Fig. 3.18. The substrate with sequential filling with nuclei of the Mg condensate nuclei (a) on a copper foil and aluminum condensate (b) on a copper film on mica.

stage of nucleation of the film and the emergence of structure in the solid films were determined. Features of the structure of continuous films were characterized as follows. Raising the temperature of condensation promotes the enlargement of the crystallites of the film both at the stage of nucleation and growth processes due to coalescence and recrystallization. Increasing film thickness increases the average size of the crystallites and results in a linearly increase of their surface area [32]. The deterioration of the vacuum conditions leads to a reduction of the transverse dimensions of the crystallites and, at the same time, to an increase of their surface area, indicating the elongation of the crystallites. The boundaries and crystallographic planes of the crystallites which define the structure and texture of the film are oriented relative to the substrate and the direction of the vapour flow. The experimental data show that the directions of the structure and texture are located mainly between the normal to the substrate and the direction of the vapour flow [33, 34].

Thick layers of the condensate show isotropic and anisotropic microstrains and stresses. Anisotropic microstrains are oriented relative to the direction of the vapour and the normal to the substrate. To explain the origin of the stresses, models were developed based on thermal expansion, the appearance of crystal defects, changes in density at the liquid–crystal phase transition, surface phenomena on the internal and external boundaries of the crystallites, electrostatic phenomena, and the influence of the substrate. None of the models describe the experimental data for different film materials.

Annealing of the films improves the structural characteristics of the films, however, better results are obtained in the deposition of films with restructuring and bringing the structure closer to regular. Contamination of the growing film with impurities increases the degree of imperfection of the structure and weakens the relaxation processes in the resulting condensates.

Attempts were made to establish a correlation between the characteristic features of the growth of thick films and the relative condensation temperature T/T_S, where T_S is the melting point of the sprayed material. However, for example, the characteristic temperature of the appearance of structures in the films varies in different experiments in the range $(0.2–0.9)$ T_S.

The most common model of the orientation of the texture of condensates is the preferential development of crystallites favourably oriented to the vapour flow filling the entire front of condensation through segregation of differently oriented crystallites, with their mutual geometrical shadowing. This model contradicts the experimental fact that the crystallites of the film are elongated, usually along the lines of slow growth. Attempts have been made to remove this contradiction by taking into account differences in the coefficient of condensation on the different crystallographic faces due largely to the different adsorption of gas impurities on each face. A model for the formation of growth textures, based on the mechanism of spiral growth of crystals along the screw dislocation, which coincides with the direction of growth, was proposed.

It is also important to mention the model of growth of oriented structures, based on maintaining the momentum of the vapour flow by the adatoms and directed migration on the surface in the flow direction–normal to the substrate plane. Directed migration is prevented by the chaotic thermal migration of adatoms on the substrate. The direction of the incident flow is distorted by scattering on residual gas in the chamber. The superposition of these mechanisms is in qualitative agreement with experimental data [34].

In the literature we have found no attempts to quantify the mathematical description of the mechanisms of formation of structures, textures and microstrains.

Numerous experimental observations of the appearance in thick films of macrodefects as conical, faceted dome- and dome-shaped with a flat top topped inclusions are important. The origin of conical macrodefects is usually associated with the spiral growth of crystals and with a screw dislocation. The appearance of the faceted and dome-shaped inclusions in the films is explained by the penetration on the condensation front of macroparticles in the crystalline and liquid state. However, the experiments showed the development of macrodefects and formation in them of structures in the process of increasing the film thickness. This section presents the mathematical model to describe both the formation of structures in planar condensates and the evolution of macrodefects.

3.4.1. The similarity of shapes of crystallites of vacuum condensates

Study [35] presents the results of measurements of diameters of axisymmetric configurations, randomly distributed in the vacuum condensates of magnesium vapours on a polished stainless steel substrate. Spraying was conducted in a vacuum chamber at a residual gas pressure of $2.5\cdot10^{-4}$ Pa. The experiments were carried out using a vapour source with a diameter of the outlet orifice of 0.4 mm at a distance of 240 mm from the substrate. At the stagnation

a   *b*

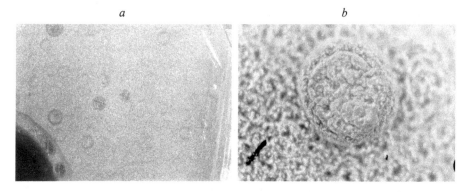

Fig. 3.19. Plan view of the condensate of the clustered magnesium vapour jet on polished stainless steel. a – magnification 60; b – magnification 400.

temperature of the vapour T_0 = 1350 K the stagnation pressure p_0 was varied in the range $10^2–10^5$ N/m², which corresponded to the specific vapour flow on the substrate j = $5\cdot10^{-7}–5\cdot10^{-4}$ kg/(m²·s). The appearance of the condensate is shown in Fig. 3.19. The results of measurements of the diameters of the dome formations, obtained for the condensate thickness h = 0–130 mm, are shown in Fig. 3.20 with the profile of one of the domes. The linear dependence of $d \sim h$ is observed in a wide range of measurement of j.

For the condensates with the thickness of about 1 mm we can make a cross section of axisymmetric inclusions. The photograph of the section with a cone-shaped crystal is shown in Fig. 3.21. Studies of sections, similar to those shown in Fig. 3.19–3.21, show that the dome-shaped axisymmetric configurations in the condensate are crystallites that develop in the condensate on the macroirregularities of the substrate or microparticles.

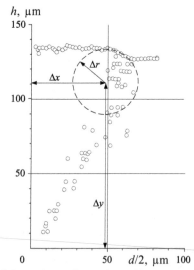

Fig. 3.20. Diameter of dome formations, depending on the thickness of the condensate and the profile of one of the domes.

Fig. 3.21. The section of the Mg condensate normal to the substrate. Fragments of the dome-shaped formations. ×100.

Study [36] described an experiment in which a wire simulates macroroughness or a particle falling on the condensation front. Interrupting the condensation process resulted in the formation of characteristic lines, illustrating the development of a crystallite in time. Figure 3.5 shows the cross section of the condensate of the vapour section of magnesium on a tantalum wire 0.2 mm in diameter, installed across the flow.

The above results allow us to conclude that in the condensation of the vapour flow on the surface the presence of asperities leads to the realization of geometrically similar shapes of the crystallites. Changes of the condensation conditions by varying the vapour flow does not affect the similarity of forms. This indicates that the property of similarity of the forms of the vacuum condensate is not determined by the condensation kinetics, and the presence of the established columnar structure in geometrically similar crystallites indicates a lack of influence of the properties of the substrate. The last circumstance suggests that the reasons for the similarity of the forms of the crystallites of the condensate are embedded in the stationary character of the gas dynamic structure of the vapour flow incident on the condensation front. At a constant geometry of the nozzle the Mach number at a certain point of the jet is maintained independently of the inhibition parameters p_0, T_0 [37]. Thus, if the gas-dynamic similarity of the supersonic jets in these experiments leads to the geometric similarity of the crystallites formed in the condensate, the model of heterogeneous condensation must take account of the Mach number of the vapour flow, i.e., the ratio of the portable and thermal motions of the vapour molecules.

3.4.2. The growth of individual crystallites by the vapour–crystal mechanism. Influence of parameters of the vapour flow

The above picture of the process of formation of vacuum condensates is sufficient to describe a model of condensation on the surface in a vacuum, which describes the observed structures of the condensates and shows the connection of the structure with the parameters of the vapour flow. In [18] the authors considered a mathematical model of unsteady crystal growth on a substrate from a spherical nucleus with condensation of the vapour flow. In [36], the kinetics of the transformation of the form of a single nucleus at condensation is discussed in a more general case, taking into account the directional flow of vapour, background vapour flow scattered in the space around the substrate, and the flow reflected from the substrate at normal incidence of the vapour flow to the substrate.

In this section we consider the more general case where the vapour flow is directed to the normal of the substrate at an arbitrary angle φ_0. It is assumed that the vapour source generates a two-dimensional flow with mean velocity u, Mach number M and Maxwell's velocity distribution of the atoms. The ratio of the mean free path of atoms in the vapour flow to the characteristic size of the growing crystal is large (Kn > 10). It is assumed that there is no diffusion of atoms on the crystal surface and the substrate. This assumption is justified, for example, with distinctive surface roughness of the crystal or when migration is suppressed by the presence of a sorbed impurity with a small diffusion coefficient. It is also assumed that the coefficients of condensation on the crystal surface a_c and the substrate a_s same. The surrounding space is filled with vapour atoms of the main substance and atoms of the residual gases in equilibrium with the temperature T_k. The substrate is sufficiently long compared to the size of the crystal.

The specific volumetric vapour flow incident on the element of unit area of the crystal surface can be written as [37]

$$j_f = nu\left[\exp\left(-S_\varphi^2\right)+\pi^{1/2}S_\varphi(1+\mathrm{erf}\,S_\varphi)\right]\Big/\left[M(\pi\chi)^{1/2}\right], \qquad (3.23)$$

where n is the partial density in the jet; φ is the angle between the velocity vector and the normal to the crystal surface; $S = u\sqrt{m/(2kT_f)} = M\sqrt{\chi/2}$ is the speed ratio; $S_\varphi = S\cos\varphi$; $M = u/a$, χ is the ratio of specific heats; $a = V\sqrt{\pi\chi/8}$ is the speed of sound; V is the arithmetic mean velocity of the thermal motion of atoms; T_f is the vapour temperature; $\mathrm{erf}\,S_\varphi = \dfrac{2}{\sqrt{\pi}}\displaystyle\int_0^{S_\varphi} e^{-\lambda^2}\,d\lambda$.

The vapour flow of atoms from the surrounding space is written as

$$j_k = n_k V_k / 4 = p_k (2\pi m k T_k)^{-1/2}, \qquad (3.24)$$

where n_k, p_k is the partial density and the partial vapour pressure in the surrounding area; V_k, m_k is the the arithmetic mean thermal velocity of atoms at a temperature of T_k, and their mass; k is the Boltzmann constant. It is assumed that the flux of the atoms reflected from the substrate and incident on the surface of the crystal is

$$j_r = nu(1-\alpha_S)\sin\varphi\left[\exp(-S^2)+\sqrt{\pi}\,S(1+\mathrm{erf}\,S)\right]\Big/\left(M\sqrt{2\pi\chi}\right). \qquad (3.25)$$

The evaporation flow of atoms from the surface of the crystal, taking into account the Thomson–Kelvin effect, can be written as

$$j_v = p_h \exp\left(-K\gamma\omega/kT_h\right)(2\pi mkT_h)^{-1/2}, \qquad (3.26)$$

where p_h is the saturation pressure of condensed matter at the crystal surface temperature T_h, $p_h = p_k/(1+\sigma_p)$; σ_p is vapour supersaturation; γ is surface energy; ω is the atomic volume in the solid phase; K is the local curvature of the crystal surface.

The evolution equation of the crystal surface takes the form

$$dr/dt = [\alpha_c\,(j_f + j_r + j_k) - j_v]\,\omega, \qquad (3.27)$$

where $dr = dx/\sin\varphi = dy/\cos\varphi$; x and y are the Cartesian coordinates longitudinal and transverse to the flow. Substitution of (3.23)–(3.26) into (3.27) and the introduction of the characteristic linear scale

$$r_* = 2^v\gamma\omega\Big/\Big\{kT_h\,\ln\Big[\alpha_c nu\Big/\Big(p_h\big(1+1/\big(M\sqrt{2\pi\chi}\big)+\alpha_c(1+\sigma_p)\big)\Big)\Big]\Big\},$$

obtained under the condition $dr/dt = 0$ as in [18] ($v = 0$ is the plane case, $v = 1$ the axisymmetric case), and the characteristic time scale t_0 give the equation (3.27) in the form

$$d\bar{r}/d\bar{t} = \alpha_c\left(\overline{nu}\right)\Big\{\exp\left(-S_\varphi^2\right)+\sqrt{\pi}\,S_\varphi(1+\operatorname{erf}S_\varphi)+(1-\alpha_c)\sin\varphi\exp(-S^2)+$$
$$+\sqrt{\pi}\,S(1+\operatorname{erf}S)]\Big\}\Big/\left(M\sqrt{2\pi\chi}\right)+\alpha_c\bar{p}_k - \bar{p}_h\exp\left(-\bar{r}_*\cdot K\right). \qquad (3.28)$$

In equation (3.28) the following notation are introduced:

$$\left(\overline{nu}\right) = \omega nut_0\big/r_*, \quad \bar{p}_k = \frac{p_k\omega t_0}{r_*(2\pi mkT_k)^{1/2}}, \quad \bar{p}_h = \frac{p_h\omega t_0}{r_*(2\pi mkT_h)^{1/2}}, \quad \bar{r}_* = \frac{\sigma\omega}{kT_h r_*},$$

$$\bar{K} = Kr_*, \quad \bar{r} = r/r_*, \quad \bar{t} = t/t_0.$$

Equation (3.28) can be solved numerically with a given initial form of crystallization at $t = 0$.

Under the assumption of no evaporation equation (3.28) was solved on a computer for values of the characteristic scale $r_* = 2.37\cdot10^{-9}$ m, $t_0 = 10^6$. Calculations were performed for the nucleus, which has initially the spherical shape. The nucleus is growing together with the planar condensate. The forms of the profiles of crystals that grew up to the same height $p_h = 0$, $p_k = 0$, $\alpha_c = \alpha_S = 1$; 0.7, are shown in Fig. 3.22 for different Mach numbers of the vapour flow. It is seen that with increasing Mach number of the directional vapour flow the shape of the crystals changes from conical to columnar (see Fig. 3.22 a). Moreover, the condition $\alpha_c < 1$ determines the appearance of the

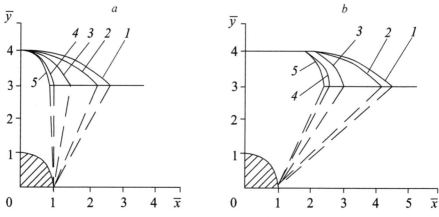

Fig. 3.22. The forms of crystals of the vacuum condensate. $a - \alpha_c = 1$, $b - \alpha_c = 0.7$. Curves 1–5 correspond to the Mach number M = 0.01, 0.1, 1, 10, 100.

flat top of the crystals (see Fig. 3.22 b). Comparison of the calculated results with experimental data (see Fig. 3.20) shows their qualitative agreement.

3.4.3. Partial analytical solution

It is shown in [36] that for $\varphi_0 = 0$ equation (3.28) has an analytic solution in the partial case under the assumption that the average mass and thermal motions of vapour atoms are additive and can be expressed as the sum of projections onto the normal to the surface $j_f = nu \cos \varphi + nV/4$ when neglecting the Thomson–Kelvin effect $e^{-\bar{r}_\kappa K} \to 1$, which is valid in the region where the curvature of the crystal surface at all points is relatively small: $K \ll 1/r_*$. For $\varphi \neq 0$ there is also an analytical solution. In this case, for the atoms reflected from the substrate $j_r = nu\left(1-\alpha_c\right)\sin\left(\varphi+(-1)^n \varphi_0\right)$, and the equation equivalent to (3.28) has the form

$$d\bar{r}/d\bar{t} = \alpha_c\left(\overline{nu}\right)\left[\cos\varphi + 1/\left(M\sqrt{2\pi\chi}\right) + (1-\alpha_S)\sin\left(\varphi+(-1)^n \varphi_0\right)\right] + \alpha_c \bar{P}_b - \bar{P}_h,$$

$$(3.29)$$

where $\eta = 0, 1$ for the left and right sides of the crystal shown in Fig. 3.23.

Equation (3.29) is self-similar in absolute magnitude, so the solution must be sought in the form of geometrically similar, time-varying surfaces. This hypothesis is in qualitative agreement with the experimental data described in section 3.4.1. The trajectory of any surface point in time when $y'_x = \text{const}$ is the isocline and can be used to search for such solutions. In this case, the projection of the crystal growth rate at a point located on the isocline, on the y axis can be expressed in terms of growth rate on the normal:

$$dy \, / \, dt = (dr \, / \, dt)\,(\cos \varphi + \sin \varphi \, \text{tg} \, \alpha), \qquad (3.30)$$

where α is the angle between the isocline and the y axis.

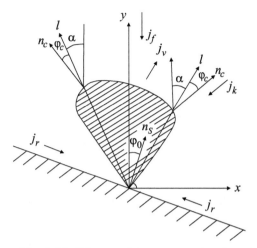

Fig. 3.23. Scheme of the computational model.

Substitution of (3.30) into (3.29) and time integration with φ = const, α = const lead to the equation of motion of the point of the condensation front along any particular isocline of the family of geometrically similar solutions:

$$\bar{y} = \frac{\alpha_c(\overline{nu})\left[\cos\varphi + 1/\left(M\sqrt{2\pi\chi}\right) + (1-\alpha_c)\sin(\varphi + (-1)^n\varphi_0)\right] + \alpha_c\bar{p}_k - \bar{p}_h}{\cos\varphi + \sin\varphi\,\mathrm{tg}\,\alpha} - t + c,$$

$$(3.31)$$

The integration constant c can be found from the initial conditions. For example, in the case of crystal growth from the point $\bar{y}|_{t=0} = 0$ $c = 0$. Equation (3.31) holds for any fixed φ and, consequently, for each point of the crystal.

Substitution of (3.31) $\cos\varphi = \left(1 + \bar{y}_x'^2\right)^{-1/2}$, $\sin\varphi = -\bar{y}_x'\left(1 + \bar{y}_x'^2\right)^{-1/2}$, $\mathrm{tg}\,\alpha = \bar{x}/\bar{y}$

gives a differential equation whose solution is the profile of the crystal at time t_0. In the new dimensionless coordinates $X = (\bar{x} - D)/R$, $Y = (\bar{y} - B)/R$, where

$$D = (-1)^{n-1}\alpha_c(\overline{nu})\bar{t}(1-\alpha_c)\sin\varphi_0,$$

$$B = \alpha_c(\overline{nu})\bar{t}\left[1 + (1-\alpha_c)(-1)^n\sin\varphi_0\right],$$

$$R = \left[\alpha_c(\overline{nu})/\left(M\sqrt{2\pi\chi}\right) + \alpha_c\bar{p}_k - \bar{p}_h\right]\bar{t},$$

there are two equal analytical solution of equation (3.28) in the form of

$$X^2 + Y^2 = 1, \qquad\qquad (3.32)$$

$$Y = c_1X + \left(c_1^2 + 1\right)^{1/2}. \qquad\qquad (3.33)$$

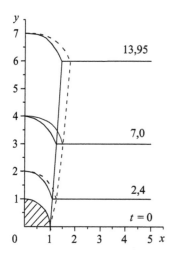

Fig. 3.24. Successive profiles of the crystals at different times, together with the planar condensate in the unoccupied surface of the substrate. The solid line – numerical calculation, dashed line – an analytical solution.

Expressions (3.32) and (3.33) represent a circle and a set of lines tangent to it. In the spatial coordinates x and y it is the family of circles with time-dependent radii with time and the position of the centre and the family of lines tangent to each of them. Successive profiles of the growing crystals at different times are shown in Fig. 3.24. The case where the initial shape of the nucleus does not contain parts of the surface facing the substrate is considered. The the degree of approximation of the analytical solutions (3.32) to the numerical solution of equation (3.28) is also indicated. Examples of graphic solutions of (3.32) and (3.33) for the planar case at $\bar{p}_h = \alpha_c \bar{p}_k$ are shown in Fig. 3.25. It is seen that for $\alpha_c = \alpha_s = 1$ the solution (3.32) represents forms of single crystals in the form of a cone with a hemispherical top. The growth rate of the radius of the centre of the spherical vertex is defined by the Mach number. The form of the crystal changes from whisker, directed to the vapour source, at $M = \infty$ to hemispherical when $M = 0$.

The angle of incidence of the vapour on the substrate has not effect on the shape of the crystals. The solution (3.33) for $c_1 = -\text{tg}\,\varphi$ describes the growth of the condensate film of uniform thickness. Figure 3.25 shows that the solution of the type (3.32) lies inside the region bounded by the solution of (3.33). Relevant examples of the growth of single crystals and a flat film of condensate are shown in Fig. 3.25 b for $\alpha_c = \alpha_s < 1$. We see that the cone-shaped crystallites in this case are the vertices describing part of the torus and a plane parallel to the substrate. The appearance of the flat top in the crystallites is due to the influence of the vapour, reflected from the substrate, on the formation of the lateral side of single crystallites.

Of special interest is the case when a heterogeneity in the form of a macroparticle of the condensed substance forms at some point on the flat substrate. This could be the nucleus, which formed in the initial stage of condensation or a macroparticle transferred from the vapour source. Development of the condensation front in this case can be analyzed using

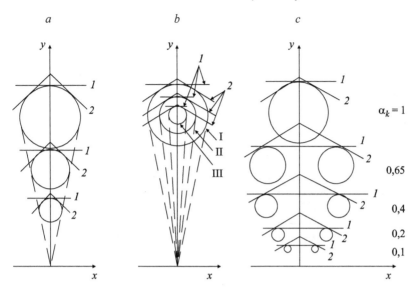

Fig. 3.25. Illustration of the solution of (3.32) and (3.33) when t, M, α_c vary. a – α_c = 1, M = 1.5; c_1 = 0 (1), 1 (2); b – α_c = 1; t = 2.75; M = 1.5 (I), 2.5 (II), 6 (III); c_1 = 0 (1), –1 (2); c – t = 2.75; M = 1.5; c_1 = 0 (1), –0.5 (2).

the analytical solution described above. For this purpose the form of the macroparticle trapped on the substrate should be compared with the solution (3.32) or (3.33) at the appropriate time. Analysis revealed the following characteristic features of the structures formed by further condensation.

1. In all cases of incidence of the macroparticle on the flat condensation front a cone-shaped crystals with a vertex on the particle forms.

2. The originally faceted particle gives the evolving faceted outlet on the outer surface of the condensate.

3. The macroparticle with no faceting forms a dome-shaped outlet on the outer surface at α_c = 1 and a domed-shaped outlet with a flat top at α_c <1 (Fig. 3.26).

4. The general form of the crystallite growing from the spherical macroparticle differs from the form of the cone, which is reflected in the experimental results (see Figs. 3.19 and 3.20). A gap forms at the boundary of the conical crystallite and the planar condensate due to mutual shading of the incident vapour flows in cases where the initial form of the macroparticles includes areas with a surface facing the substrate.

3.4.4. Columnar structure of the condensates. Orientation of the structure

Application of the model described in section 3.4.3 to describe the growth of an ensemble of crystallites, arranged relative to each other at distances of the order of their size, is complicated by the fact that, apparently, the flow of the atoms reflected from the substrate and moving to the crystal is affected by the presence, number and shape of the neighbours. Therefore, the model

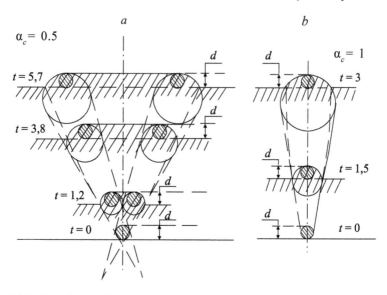

Fig. 3.26. The scheme of crystal growth with a domed top based on analytical solutions.

presented is applicable for the description of the ensemble only for $\alpha_c = \alpha_s = 1$. The model is also suitable, without limitation of α_c, α_s, for describing the collective growth of the crystallites, which grow simultaneously at all points of the substrate. In this case, the condensation front is described by the solution of (3.33) with $c_1 = -\text{tg }\varphi_0$, being both a set of points of the solution (3.32) with the property $\overline{y}_{\overline{x}}' = -\text{tg}\varphi_0$, i.e. lying on one of the isoclines (Fig. 3.27). The trajectories of points of the condensation front in time correspond to isoclines. This opens up the possibility to describe the internal structure of a flat vacuum condensate using the proposed mathematical model. If the surface of the condensate film has any local defects, preserved in the process of growth of the film thickness, it is natural to assume that their path will also correspond the isocline. The peculiarity of the case $\alpha_c = \alpha_s < 1$ it that there is a scatter of the characteristic angles of orientation of the structure, as shown in Fig. 3.27 b.

The angle between the normal to the condensation front and the isocline corresponding to the condition $\overline{y}_{\overline{x}}' = -\text{tg}\varphi_0$, can be determined by means of simple transformations from analytical solutions (3.32), (3.33) presented in the coordinates x, y:

$$\varphi_c = \varphi_0 - \text{arctg}\left(\frac{D + R\sin\varphi_0}{B + R\cos\varphi_0}\right). \qquad (3.34)$$

In the partial case at $\alpha_c = \alpha_s = 1$, $M = \infty$, $p_h = 0$, we can consider the scattering effect of the residual gases. This case corresponds to the growth of

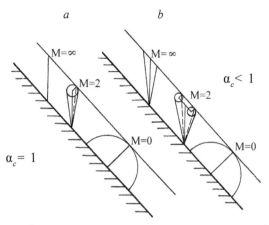

Fig. 3.27. Graphical interpretation of the solutions of (3.32) and (3.33).

condensed films in vacuum from a molecular beam. If the flux of atoms in a molecular beam at the output of the source is $(nu)_0$, then at a distance x_i [37]

$$\overline{nu} = (nu)_0 e^{-x_i/\lambda}. \tag{3.35}$$

Here λ is the mean free path of the vapour atoms between collisions with molecules of the residual gas. The value of λ/x_i is the Knudsen number Kn_x. Molecular beam scattering on residual gas is the background vapour flow

$$P_k = (\overline{nu})_0 (1 - e^{-x_i/\lambda}), \tag{3.36}$$

From (3.35) and (3.36) with (3.34)

$$\varphi_c = \varphi_0 - \text{arctg} \{ \text{tg } \varphi_0/[1/(\exp (1/Kn_x) - 1)] + 1 \}. \tag{3.37}$$

Using (3.37), Figs. 3.28 and 3.29 show a qualitative comparison of the results of calculation of the angle between the normal to the substrate and the direction of growth of columnar crystals with the experimental results. Figure 3.28 shows a comparison with the experimental data for condensation of magnesium by the method described in section 2.3.1, Fig. 3.29 – for the spraying of iron (curve 1) [33]. Comparison with the experimental data takes into account the presence of random motion of the vapour atoms in the molecular beam associated with the geometry of the vapour source. In analyzing Figs. 3.28 and 3.29, it can be concluded that the model described is adequate with respect to the experimental data in the angular range $\varphi_0 = 0 \div 70$ °, and the numbers $Kn_x = 0.3-10^2$.

To double-check the adequacy of the model, experiments were carried out with the condensation of the magnesium flow on a transverse copper wire of 1 mm diameter at different temperatures. The results of measurements of the relative rate of condensation at various points in the contour of the wire, corresponding to different angles φ_0, are shown in Fig. 3.30 compared with

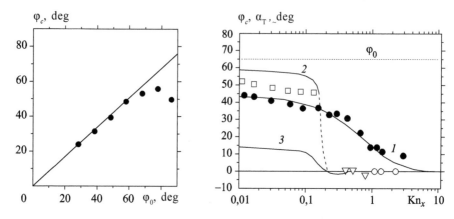

Fig. 3.28. The deviation of the direction of the structure from the normal to the substrate depending on the angle of incidence of the vapour flow to the normal to the substrate; solid line – calculation by formula (3.34), points – experimental data.

Fig. 3.29. (right) The deviation of the growth direction of columnar crystals and crystallographic directions α_T from the normal to the substrate, depending on the Knudsen number. 1 – calculation by formula (3.34), black dots - experiment [33], 2, 3 – α_T calculated for the [111] and [100] directions, open circles – experiment [33].

the calculations by formula (3.29) with $\varphi = \varphi_0$, $p_k = p_h = 0$. We see that when $\varphi_0 = 0$–$70°$ the experimental points correlate with curve 5, obtained for M = 0.1. This characterizes the vapour flow used in the experiments.

The discrepancy with the experimental estimates at $\varphi_0 = 70$–$90°$ can be explained by the shadowing effect of the substrate. Characteristic angle φ_0 of the spatial velocity diagrams of the vapour atoms with directed and random motion with velocities u and V is determined from the relation tg $\varphi_0 = V/u$. When $\pi/2-\varphi_0 < \varphi_0$ some of the vapour atoms do not reach the substrate, as their absolute velocity is directed away from it. When $\varphi_0 = \pi/2$

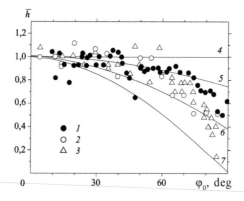

Fig. 3.30. The dependence of the thickness of the condensate on the angle of incidence of the floe to the normal to the substrate. 1–3 – an experiment at a substrate temperature $T_S = 381$, 530, 582 K; 4–7 – calculation for M = 0, 0.1, 0.5, respectively.

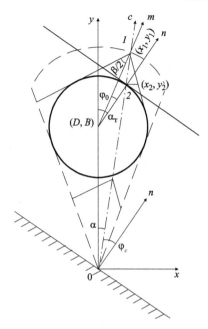

Fig. 3.31. Evolution of a faceted crystal in the process of condensation and the scheme for calculating the orientation of the structure and texture.

half of the atoms in the velocity distribution is directed from the substrate, which should correspond to less than half of $\overline{dr}/dt$ in the experiment compared with the calculation. This is also seen in Fig. 3.30.

3.4.5. Effect of faceting of crystals

When nucleus with faceting forms on the substrate the growth of faceted crystals from it can be described using the above model. The crystal faces correspond to the solution (3.33) in the selection of the appropriate value of the constant c_1 in the initial time. Obviously, the faceted crystals retain geometric similarity during growth. The evolution of a faceted single crystal is illustrated in Fig. 3.31 for $a_c = 1$. The characteristic element of the crystal is the joint of faces which moves on one of the isoclines during growth.

When the columnar structure of the condensate forms due to, for example, poor vacuum conditions, the formation of defects and the influence of capillary condensation of impurities, the faceting can occur only cut on the top of each column of the crystallite. If we assume that the faceting of the columns represents an intersection of the planes at angle β, characteristic of the given crystal structure, then the shape of the top of the single crystallite must be described by the locus of the vertices of faceted crystals whose faces are tangent to the circle described by (3.32). Such a locus is part of a circle with radius $R/\sin(\beta/2)$, as shown by the dashed line in Fig. 3.31. The resultant form of the crystallite with faceting of the columnar structure is in qualitative agreement with experiment.

If a single faceted crystal grows among the solid condensate with a flat condensation front, parallel to the substrate, then only part of the crystal protrudes above the condensation front. The height of the protruding part of the crystal increases in the condensation process. The superposition of such faceted crystals, growing in different parts of the substrate, can give a well-known stepped condensation front. If we imagine that these crystallites grow at all points of the substrate, then a condensate with a flat front, having faceted microroughness, forms at the last point.

This chain of logic leads to the assumption that the condensation model described above can, in addition to describing the structure, explain the known experimental observations of the orientation of the texture. Taking the axis of orientation of the texture as the median angle between the faces, by simple geometric calculations we can determine the angle of orientation of the texture:

$$\alpha_T = \text{arctg}\left[\left(\frac{y_1 - B}{x_1 - D} - \text{ctg}\,\varphi_0\right)\middle/\left(\frac{y_1 - B}{x_1 - D}\text{ctg}\,\varphi_0 + 1\right)\right],$$

where $x_1 = \sin \alpha\{(D\,\text{tg}\,\alpha + B) + [(D\,\text{tg}\,\alpha + B)^2 - (D^2 + B^2 - R^2/\sin^2 (\beta/2))/\cos \alpha]^{1/2}\}$, $y_1 = x_1/\text{tg}\,\alpha$. The results of calculations for $\varphi_0 = 65°$ and $\beta = \pi/2$, which corresponds to a cubic crystal structure, are shown in Fig. 3.32 as the dependence of φ_c (curve 1) and α_T (normal [110] – curve 2, the normal [100] – curve 4) on the Mach number of the vapour flow. Since the decrease in the Mach number of the vapour flow can occur, for example, in scattering of the by the atmosphere of residual gas in the chamber, it is possible to compare the results of calculations with experimental data [33] where the orientation texture was measured on the diffraction of reflected electrons. Such a comparison is shown in Fig. 3.29. The combination of curves 2 and 3 together with the corresponding points suggests a qualitative agreement of calculations with experimental results.

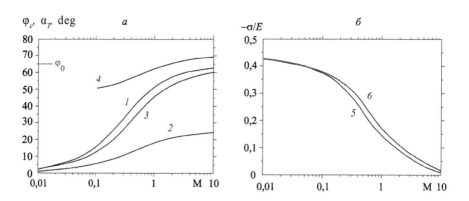

Fig. 3.32. The orientation of the structure and texture of columnar condensates with faceting (a) and the internal stresses in the condensate (b). Curves 1 – φ_c, 2 – α_T for the normal [110], 4 – α_T + 45° for the normal [100], 5 – $(-\sigma/E)$ – for the model of the orientation of the structure with respect to the isocline. Curves 3 – φ_c, 6 – $(-\sigma/E)$ model for the normal orientation of the texture. $\varphi_0 = 65°$, $\alpha_c = 1$.

For small Mach numbers of the vapour flow and when $\alpha_c < 1$ the directionality of motion of vapour to the condensate on the substrate decreases. In this case, all the more important may be the mutual segregation of acicular crystals whose vertices form the crystallization front. It is clear that preferential growth takes place in crystals whose tips are far away from the substrate and are the least shaded by neighbours. This would correspond to the orientation of the top of the faceting of the crystals normal to the substrate. According to Fig. 3.31, this corresponds to $\varphi_c = 0$. Then for the growth of columnar condensates with normally oriented faceting we obtain the appropriate formula for determining the orientation of the structure:

$$\varphi_c = \varphi_0 - \text{arctg}\left[\left(D + \frac{R\sin\varphi_0}{\sin(\beta/2)}\right)\middle/\left(B + \frac{R\cos\varphi_0}{\sin(\beta/2)}\right)\right].$$

The calculation results for this case are shown in Fig. 3.32, curve 3.

3.4.6. Internal stress and porosity of the condensates

The condensation model described in section 3.4.2 is based on the equation of mass transfer [38], which essentially represents the continuity equation of matter flow in the vapour and solid phases. However, when using it to describe the columnar faceted structure of a flat condensate on the substrate, attention is drawn to the fact that in this case, the total thickness of the condensate $l + \Delta l$ (distance 01 in Fig. 3.31) is greater than in the case of a continuous film growth (length 02). This is connected only with the gas-dynamic characteristics of the interaction of the vapour flow from the planes of different orientations. And if the kinetics of condensation within the adopted assumption of the absence of migration of adatoms enables the conservation of the faceting of the columns of the structure, then due to the need to satisfy the condition of continuity oriented microdeformation of the condensate $\Delta l/l$ may form in the condensate. This ratio of the microstrains can be interpreted as a cause of the formation of stresses in the condensate equal in the order of magnitude to $s = E \cdot \Delta l/l$, where E is Young's modulus.

If the resulting stresses exceed the tensile strength of the condensate, they may give rise to microcracks in the condensate, which is equivalent to the growth of a porous condensate. Under the assumption that the formation of a microcrack is accompanied by complete relaxation of internal stresses, the value of $\Delta l/l$ may be an estimate of the bulk porosity of the condensate.

For all cases of orientation of the structure considered in Section 3.4.5 we can derive equations for the calculation of $\Delta l/l$:
on the isocline –

$$\frac{\Delta l}{l} = \frac{[(D\,\text{tg}\,\alpha + B)^2 - (D^2 + B^2 - R^2)/\cos\alpha]^{1/2} - x_1}{(D\,\text{tg}\,\alpha + B) + [(D\,\text{tg}\,\alpha + B)^2 - (D^2 + B^2 - R^2)/\cos\alpha]^{1/2}};$$

in normal orientation –

$$\frac{\Delta l}{l} = \frac{1 - 1/\sin(\beta/2)}{B/R\cos\varphi_0 + D/R\sin\varphi_0 + 1}.$$

The corresponding estimates for the calculation of internal stresses are shown in Fig. 3.32. These data allow us to make a practically important conclusion that in order to reduce defects of film condensates associated with the appearance of internal stresses, it is necessary to increase the Mach number of the vapour flow.

3.4.7. Features of condensation of the vapour–droplet flow

It is interesting to consider a special case, when a heterogeneity forms on a flat substrate at any given time: the nucleus, which emerged in the early stages of condensation, or a macroparticle originating from the vapour source. Development of the condensation front in this case can be analyzed using the analytical solutions obtained above (3.32) (3.33). For this, the form of the macroparticle falling on the substrate should be compared with the solution of (3.32) or (3.33) at the appropriate time. This analysis revealed the following characteristic features of the structures formed by further condensation.

In all cases of the macroparticle falling on the flat condensation front a cone-shaped crystals with a vertex on the particle appears.

The originally faceted particle provides the evolving faceted outlet on the outer surface of the condensate.

The macroparticle which has no faceting gives at $\alpha_c = 1$ a dome-shaped outlet on the outer surface and at $\alpha_c < 1$ a domed-shaped outlet with a flat top.

The overall shape of the crystallite growing from the spherical macroparticle differs from the form of a cone, which is reflected in the experimental results (see Fig. 3.21). Owing to the mutual shading of the incident vapour flows where the initial form of the macroparticle includes areas with the surface facing the substrate, a conical gap forms at the border of the condensate.

3.4.8. Influence the efficiency of condensation on the structure formation of vacuum condensates

The formation of flat tops of individual crystals and crystals protruding above the flat condensation front at $\alpha_c = \alpha_s < 1$ was discussed previously. Equation (3.34) was used to analyze the effect of the condensation coefficient on the orientation of the columnar structure and internal stresses in the condensate. The results of calculations for $\varphi_0 = 65°$, $M = 1$ are shown in Fig. 3.33. A characteristic feature of the results is the ambiguity in φ_c at $\alpha_c < 1$ for both models (discussed in section 3.5) of textured condensates and σ/E for the model of the orientation of the structure of the isocline. The ambiguity is due to the fact that when $\alpha_c < 1$ the condensation front is formed by the top of the crystallites, described by two circles spaced-apart along the substrate with a common tangent. Formally, this reflects the equality of signs ± in the formulas for B and D. Apparently, the condensation process

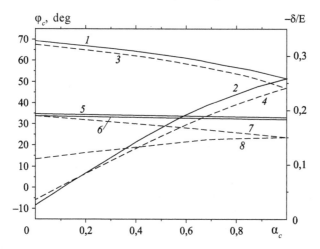

Fig. 3.33. The orientation of the structure (curves 1–4) and the internal stresses (curves 5–8), depending on the condensation coefficient. Dashed line – orientation of the structure on the isocline, solid line – the normal orientation of the texture. $\varphi_0 = 65°$, M = 1.

may result in the formation of columnar structures with an orientation in the whole range between the calculated values of α_c. Characteristically, when $\alpha_c = \alpha_s < 0.1$ calculations predict the appearance of structures oriented normal to the substrate in a direction opposite the direction of the vapour flow that is sometimes observed in experiments [17]. Increase the range of possible variation of φ_c with decreasing α_c is also in qualitative agreement with numerous observations of the structure with the deterioration of the vacuum conditions in the experiments, which is always accompanied by a decrease in α_c. The dependence of the magnitude of internal stresses in the condensate on the value of α_c is weak.

Conclusions

1. Processes of heterogeneous condensation of metal vapours on solid surfaces, intensively studied during the last three decades in the direction of obtaining single crystals, epitaxial thin films for microelectronics needs in high vacuum and absence of impurities, can be used for the formation of disperse systems with definable characteristics, if these processes are controlled with the help of impurities significantly affecting the structure of the condensates.

2. In the experiments on heterogeneous condensation of magnesium it was revealed that the known critical temperature of condensation, corresponding to the global changes in the condensation coefficient, depend on the presence and amount of non-condensable gas impurities.

3. It has been established that the condensation in the vicinity of the temperature region of 'collapse' leads to the formation of dispersed condensates with a predominantly columnar structure and that an active role in the formation of the latter is played by capillary condensation of foreign gas and vapour impurities.

4. A mathematical model of sorption processes of multicomponent gaseous mixtures taking into account the chemical reactions, condensation, capillary condensation, qualitatively explains the reasons for the existence of three critical temperatures of heterogeneous condensation and describes the results of experimental investigation of heterogeneous condensation near the region of 'collapse'.

5. Experiments show sequential filling of a hard surface by the dispersed magnesium vapour condensate. The model of the sorption processes of two-component gas mixtures, taking into account surface migration, was used to show the possibility of the chain mechanism of heterogeneous nucleation, when the formation of the first nucleus and capillary condensation of impurity components stimulate the emergence of new nuclei near the existing ones.

6. The geometric similarity of microformations, known as 'hillocks' in the processes of heterogeneous condensation, has been established by exeriments.

7. A mathematical model of growth of single crystals and polycrystals and of the formation of columnar structures of condensates, taking into account the gas dynamic parameters of the vapour flow and the geometry of the condensation process, was proposed. If the Kelvin–Thomson effect is disregarded, in addition to the similarity of the dimensions, the model shows the geometrical similarity of forms of all microformations, observed in heterogeneous condensation experiments. It allows to calculate the orientation of the columnar structure and texture of columnar crystals at various angles of incidence of the vapour flow. The calculations are consistent with literature and experimental data.

8. It seems that the most significant and unexpected result of this work is the one which shows that the model that takes into account mainly the structure of the vapour flow was able to explain many facts, known from the experimental study of the process of formation of condensates. Part of the findings is in qualitative agreement with the available experimental results. Another part of the calculations requires for verification special experiments, in which the parameters of the vapour flow are measured together with the characteristics of vacuum condensates.

List of Symbols

D - reduced the surface diffusion coefficient;

d - particle size, m;

G - vapour flow rate, kg / s;

h - thickness of the condensate, m;

J - the probability of nucleation, $m^{-2} \cdot s^{-1}$;

j - specific density of the particle flux, $m^{-2} \cdot s^{-1}$;

K - the reduced rate constant, the local curvature, m^{-1};

k_q and k_b - the rate constants of reactions in the Q-decomposition and synthesis;

k - Boltzmann constant, J;

l - distance, m;

M - the dimensionless mass;

M - Mach number;

m - mass of the particle, kg, molecular weight, at. units.;

n_a - Job density of adsorption, m^{-2};

n_c - the local normal to the surface of the crystal;

n_S - normal to the substrate;

P - flux of particles, $m^{-2} \cdot s^{-1}$;

p - pressure, Pa;

$p_{h\infty}$ - saturation pressure at a radius of curvature infinity, Pa;

r - radius of curvature, the normal coordinate, m;

r^* - radius of the critical nucleus, m;

S - speed ratio;

T - temperature, K;

t - dimensionless time;

U_D - the activation energy of surface diffusion, J;

u - velocity, m / s;

W - heat flow, J / $(m^2 \cdot s)$;

x, y - Cartesian coordinates, m;

$X,$ - the distance from the source;

α - angle, degrees;

α_c, α_S - efficiency ratio of condensation on the condensate and the substrate;

α_{li} - the probability of desorption of particles of i-th component of L different energy impacts;

α_{12}, α_{21} - the probability of desorption of atoms by bombarding the other component;

γ - specific surface energy, $J \cdot m^{-2}$;

φ - angle, rad, deg;

ε - the binding energy, J;

η - the probability of flow of particles in the trap;

Θ - the degree of filling of vacancies of adsorption;

μ - molecular weight, AU;

ν - frequency of normal and tangential vibrations, c^{-1};

ξ - surface concentration of particles of the adsorbate, m^{-2};

$\dot{\xi}$ - Rate of change of the surface concentration of adsorbate, $m^{-2} \cdot s^{-1}$;

ξ_{nm} - the concentration of adsorbed molecules of the reaction products of synthesis;

ρ - density kg/m^3;

σ - the supersaturation;

τ - time, s;

τ_i - reispareniya time, s;

τ_0 - the average time reispareniya, c;

χ - the ratio of specific heats;

ω - the volume of an atom, molecule in the condensed state, m^3.

Indices:

to - the critical temperature

cr - critical section;

t - texture;

and - an atom, adsorption;

b - chamber, a vacuum;

c - the condensate;

f - the stream;

h - saturation;

i - number of components;

k - the surrounding area;

Literature

1.   Knudsen M., Ann. Phys., 1916, Bd. 50, 472.
2.   Langmuir I., Phys. Rev., 1916, 149.
3.   Wood R.W., Phil. Mag., 1916, V. 32, 364.
4.   Palatnik L.S., Komnik Yu.F., Dokl. AN SSSR, 1959, V. 124, No. 4, 808–811.
5.   Palatnik L.S., Komnik Yu.F., Fiz. Met. Metalloved., 1960, V. 10, No. 4, 632-633.
6.   Palatnik L.S., Gladkikh N.T., Fiz. Tverd. Tela, 1962, V. 4, No. 2, 424–428.
7.   Palatnik L.S., Gladkikh N.T., Dokl. AN SSSR, 1961, V. 140, No. 3, 567–570.
8.   Palatnik L.S., et al., Fiz. Tverd. Tela, 1962, V. IV, No. 1, 202–206.
9.   Palatnik L.S., Gladkikh N.T., Fiz. Tverd. Tela, 1962, V. 4, No. 8, 2227–2332.
10.   Komnik Yu.F., Measurement of thermal effects in the condensation of thin films. Moscow, Atomizdat, 1979.
11.   Sircar S., Sur. Sci., 1985, V. 164, 393–402.
12.   Movchan B.A., Demchishin A., Fiz. Met. Metalloved., 1969, V. 28, No. 4, 653–660.
13.   Zaberin A.G., et al., in: Metallization in vacuum, Riga, Avots, 1965, 25–34.
14.   Roikh I.L., et al., Protective coating in a vacuum, Moscow, Mashinostroenie, 1976.
15.   Physics of Thin Films, Ed. G. Hass, R.E. Thun, Springer-Verlag, 1967, V. 2, 64.
16.   Zeilya R.I., Pinkevich I.A., in: Metallization in vacuum, Riga, Avots, 1968, 15–18.

17. Protod'yakonov I.O., Siparov S.V., The mechanics of the adsorption process in the gas–solid system, Leningrad, Nauka, 1985.

18. Geguzin Ya.E., Kaganovsky Yu.S., Diffusion processes on the surface of the crystal, Moscow, Energoatomizdat, 1984. 124.

19. Palatnik L.S., et al., Fiz. Met. Metalloved., 1963, V. 15, No. 3, 371–378.

20. Komnik Yu.F., Physics of metal films, Moscow, Atomizdat, 1979.

21. Langmuir I., J. Amer. Chem. Soc., 1916, V. 38, 2211–2295.

22. Brunauer S., et al., J. Amer. Chem. Soc., 1938, V. 60, 309–319.

23. Greg S., Singh K., Adsorption, surface area, porosity. Springer-Verlag, 1984.

24. Slyusarenko Yu.V., Koval' A.G., Poverkhnost'. Fizika, Khimiya, Mekhanika, 1986, No. 2, 32.

25. Bochkarev A.A., Polyakova V., in: 2nd Proc. Conf. Modeling the growth of crystals, Abstracts, Reports, Riga, 1987, V. 1, 90–91.

26. Chukanov V.N., Kuligin A.P., Teplofiz. Vysokikh Temper., 1987, V. 25, No. 1, 70–77.

27. Bochkarev A.A., Polyakova V., in: Thermophysics of crystallization and high-temperature processing of materials, Proc., Academy of Sciences, Institute of Thermophysics, Novosibirsk, 1991, 89–108.

28. Bochkarev A.A., Polyakova V., 2nd Proc. Conf. Modeling the growth of crystals, Abstracts, Reports, Riga, 1987, V. 1, 84–86.

29. Modern Crystallography, A.A. Chernov, et al., Moscow, Nauka, 1980, V. 3, 400.

30. Trofimov V.I., Poverkhnost'. Fizika, Khimiya, Mekhanika, 1983, No. 10, 22–25.

31. Aleksandrov L.N., Entin I.A., in: Synthesis and growth of perfect crystals and films of semiconductors, Novosibirsk, Nauka, 1985, 28–32.

32. Directory. The technology of thin films, V. 2, Sov. Radio, 1977, 768.

33. Okamoto K., et al., Thin solid films, 1985, V. 129, 229–307.

34. Hashimoto T., et al., Thin solid films, 1982, V. 91, 145–154.

35. Berdnikova V.V., et al., Thermophysics crystallization of substances and materials, Novosibirsk, Publishing House ITF SO AN SSSR, 1986, 122–132.

36. Berdnikova V.V., et al., Columnar structure of vacuum condensates of magnesium, *ibid*, 115–122.

37. Kogan M.N., The dynamics of rarefied gas. Kinetic theory, Moscow, Nauka, 1967.

38. Bochkarev A.A., Polyakova V.I.,Izv. SO AN SSSR, 1988, No. 5, No. 18.

4

The formation of disperse systems by homogeneous condensation with vapour jet expansion

The condensation of metal vapours by diffusion into the cooling medium (usually an inert gas) has been studied in [1] to develop the so-called gas-phase method for producing ultrafine powders. Unlike some isobaric processes, this section deals with the condensation of metal vapour with the rapid expansion into the vacuum. The advantage of this process relative to the isobaric process is the possibility of achieving, at adiabatic expansion, cooling rates of the vapour of the order of 10^{-7} K/s supersaturated by several orders of magnitude, by spontaneous nucleation in the pure vapour medium and produce a dispersed system as a stream of diluted metal vapour containing solid or liquid particles of the metal. In this experiment we can produce metal particles with the size of an atom to a few microns. And because no foreign environment is involved here, the particles obtained are of particular interest. Condensation in a vacuum eliminates the ingress of impurity particles into the volume and adsorption of impurities on their surface. Such particles should show special physical and chemical activity. The physical activity of these particles can be used for practical purposes, such as for producing thin-film coatings of high quality – the so-called cluster deposition, chemical reactivity, for directional changes in their chemical composition in the process of producing the particles.

However, the high chemical and physical activity of clustered flows creates some specific features in using them. It is not difficult to use clustered flows to create clouds of ultrafine particles in the upper atmosphere layers, the targets for nuclear physics studies. Using the same clustered flows to obtain new materials dispersed in the form of powders is almost impossible due to the fact that the high activity of the clusters does not allow them to be separated from the flow intact without special measures. In this regard, in the studies described here, particular attention was paid to finding ways to stabilize the clusters obtained in the flow with a view to their subsequent use.

Apparently, the first notable work on the study of condensation of metal vapours in the process of rapid expansion is the article by Hill, et al [2]. The authors of this paper studied the problem of condensation in the expanding vapour jet divided into specific objectives:

1. The study of the differences between the theoretical equilibrium extension and expansion without condensation.

2. Determination of the start of condensation.

3. Definition of condensate fraction in the flow.

4. Calculation of the rate of growth of the condensate particles.

5. The influence of condensation on the dynamics of the flow.

For the general problem of condensation there is a huge number of papers dealing with the expansion of steam in turbines and nozzles, the thermodynamics of the formation and growth of droplets. Less represented in the literature are theoretical and experimental studies on the condensation of the rapidly expanding metallic vapours.

The first observations of supersaturation of vapour were made by Helmholtz [3], who noted the transparency of the initial part of the vapour jet, escaping from a hole. He also found the influence of an electric discharge on the formation of fog. Wilson [4] considered the supersaturation in a vessel with a piston. He noticed a strong influence of dust on the resultant supersaturation. After removal of dust from the vapour he received the maximum supersaturation [4]. Hirn and Casin [5] also noted the influence of dust particles on the condensation at low rates of adiabatic saturation. However, Stodola [6] observed a significant increase in supersaturation in expansion of vapour in the nozzles. Based on the experience of Hirn and Casin, Stodola concluded that the main influence is not the presence of dust and its amount but the characteristic time of the expansion process. He found that dust particles can serve as condensation nuclei only if the supersaturation process is relatively slow and vapour diffusion to the particles does not limit condensation.

In the absence of dust in vapour or in the case where the process of vapour expansion is faster by an order or two in comparison with the process in the Wilson chamber, the size of condensation nuclei can be determined by the Kelvin–Thomson equation

$$\ln\left(\frac{p}{p_\infty}\right) = \left(\frac{2\gamma_\infty \omega}{kTr_*}\right), \tag{4.1}$$

where p is the vapour pressure; p_∞ is its saturation pressure at local temperature; γ_∞ is the surface tension of the condensed phase in the vapour environment; r_* is the radius of the stable nucleus. The calculation of r_* by (4.1) at supersaturations up to 10 yields a value of 2–3 radii of the molecules. This circumstance cast doubts on the legitimacy of using in the calculations the surface tension determined on the macroscale. Tolmin derived an equation that gives the correction for surface tension, as

$$\gamma = \gamma_\infty / [1 + C_1 r_a / r_*], \tag{4.2}$$

where the constant C_1 is chosen from 0.2 to 0.6.

Stodola [6] showed that the equilibrium of the condensation nuclei, calculated by the formula (4.1), is unstable. If the particle size of the condensate is considerably greater than r_*, then the actual vapour pressure is supersaturated for this particle.

Experimental studies of the effect of condensation on the distribution of the vapour pressure in a nozzle were made in [7–13]. Keenan [14] showed that the increase in pressure in the nozzles due to the presence of condensate particles and their size. Analytical studies, successfully explaining most of the experimental data, are published by Oswatitsch [15]. The capillary theory of formation of nuclei was developed by Folmer [16], Becker and Doring [17], Frenkel' [18] and confirmed by experimental data [19]. Most of these studies were discussed in the monographs [20, 21]. A detailed review of studies on homogeneous nucleation is given by Sutugin and Lushnikov [22]. Attention was paid to the state of theoretical studies of the kinetics of formation of a new phase in a supersaturated vapour, growth and evaporation, the properties of small clusters, special features of nucleation in non-stationary and non-isothermal conditions, a review of methods for the experimental study of spontaneous condensation and analysis of the experimental results.

The main methods of experimental study of spontaneous condensation are:

1. The Wilson chamber, fitted in recent studied with isothermal walls, the programmed cycle of expansion and optical methods of diagnostics, allows to trace the time variation of the dispersion and concentration of particles.

2. The diffusion chamber, differs advantageously from the Wilson chamber by the stationarity of the condensation process.

3. Supersonic flows of vapour and vapour–gas mixtures through the nozzle. Research in this direction is stimulated by the opportunity to manage the process of formation of the dispersed phase for power engineering applications. Both conventional gas-dynamic methods of diagnosis and optical methods are used. The condensation in nozzles is characterized by heterogeneity in space, non-isothermal and non-stationary characteristics.

4. High intensity molecular beams. The process is characterized by rapid, strong supercooling and a substantial non-equilibrium flow. Diagnosis of beams and condensation processes is carried out by means of ionization detectors and mass spectrometers.

5. Rarefaction behind the shock wave in shock tubes. The process is characterized by small time scales and is highly non-stationary and non-equilibrium. Studying these processes requires the use of high-speed diagnostics.

The main experimental results reported in dozens of works in a very condensed form can be described as follows.

1. Most of the experimental results are compared with calculations based on the classical capillary theory. In general, all studies observed their qualitative agreement. The observed quantitative discrepancy between the experimental and calculated data in almost all cases can be explained by the peculiarities of specific experiments – the influence of non-stationary, non-isothermal processes, lack of homogeneous nucleation, the presence of external influences such as ionizing radiation, electric fields [23–32].

2. Experimental results obtained by condensation in supersonic flows are described by the classical isothermal theory only with the introduction of adjustable parameters in the magnitude of the kinetic coefficient and surface tension of the clusters [33–37].

3. The phenomenon of unsteady condensation shocks, interacting with the shock waves located above on a current of a vapour jet [38–40], is detected and studied.

4. An investigation of two-component vapour condensation when the homogeneous condensation of the less volatile component creates conditions for the heterogeneous condensation of the more volatile component [40].

5. The role of cluster coagulation processes in the early stages of condensation, when the coagulation process distorts the results of homogeneous condensation [41] was studied, and methods were investigated for non-coagulated condensates by diluting the vapour of the non-condensable component [42–44].

6. The active role of the non-condensing carrier component in the formation of clusters was observed [45–46].

7. The similarity of condensation processes in supersonic free jets, which is expressed in the constancy of the mean size of clusters under the conditions $p_0 T_0^{(\chi-1)\chi} = \text{const}$ and $p_0 d_*^q = \text{const}$, was studied [47, 48].

8. The interaction of cluster beams with a solid surface was studied, and it was found that the scattering of clusters after the reflection shows the maximum intensity along the solid surface [49].

The study of homogeneous condensation of metal vapour during expansion in nozzles started apparently by the work of Kearton [50, 51]. During the eighties, a considerable amount of experimental data, some of which are summarized in Table. 4.1, was accumulated. According to the table data, we can make several important conclusions.

1. Vapour sources in the form of the Knudsen cell were mainly used, without superheating the vapour above the line of saturation.

2. Condensation of vapour was carried out mainly in the medium of the carrier, usually an inert gas.

3. The clustered flows of metal vapour were diagnosed by the time-of-flight mass spectrometer, retarding potential probes, light scattering by clusters.

4. Flows of metal clusters in the size range $N = 2$–$2 \cdot 10^3$ atoms/cluster were produced. The size distribution functions of the clusters were, as a rule, 'smeared' over a broad range.

5. The fragmentation of metal clusters under electron impact was observed [52].

6. The experimental data of some studies are summarized in the coordinates $\psi = p_0 d_* (T_b/T_0)^{(\chi-1)\chi}$ (torr $\cdot$ mm), N/z, where T_b is the boiling point of the metal; N is the average cluster size; z is the multiplicity of ionization [53]. The results of the work of Japanese researchers [54–59], aimed at developing a method of ionized cluster beam deposition of thin films (ICB), are not included in this generalization.

Table 4.1. Summary of experimental data for spontaneous condensation of metal vapours

No.	Metal	p_0	T_0	Carrier gas	Cluster size, at/cluster	Vapour source	Diagnostics	p_k, torr	Reference
1	2	3	4	5	6	7	8	9	10
1	Na	$p_0 d = 135$ torr·mm; 15 torr·mm			1–10		Ionization 0–100 eV, time-of-flight mass spectrometer Cu/Be 0–100 eV, ion drift length 1.7 m		[60]
	I		480 K		1–48 Max. on I_2	Sonic nozzle $d_{cr} = 0.12$ mm		$5\cdot10^{-5}$ oil	
	Hg	760 torr	650 K		1–512 Maqx. on Hg			$2\cdot10^{-7}$	
2	Mg		500 K	He	In matrix 10–400 Å	Knudsen cell		10^{-6}	[61]
3	Ag		1295°C	Ar, N$_2$	1–5	Heated graphite tube $d_{cr} = 0.84$ mm Condensation in medium Ar, N$_2$	Time-of-flight	10^{-3}–10^{-4}	[62]
	Al		1465°C	Ar, N$_2$	1–13		»		
			1780°C	Ar, N$_2$	1–12		»		
			1483°C	Ar, N$_2$	1–12		»	10^{-7}	
			1476°C	Ar, N$_2$	1–2		Laser excitation		
4	Bi			He 310 K	1–5	Knudsen cell Condensation in inert gas	Time-of-flight mass spectrometer		[62] Dependence on ionization energy, fragmentation
	Bi			He 190 K	1–9				
	Bi			He 90 K	1–20				
	Sb			He	1–60				
					1–100				
					1–240				
	Pb			He	1–30				

1	2	3	4	5	6	7	8	9	10
5	Sb Bi Pb			He 10 torr H$_2$, N$_2$, Ar, Ne, Kr, Xe	1–100 max. on 4 1–21 1–120 Max. on 4	Knudsen cell	Time-of-flight mass spectrometer		[63]
6	Ag	670 torr 1150 torr	500°C 1250°C	Ne		Knudsen cells with glowing, sonic nozzle	Retarding potential probe		[53]
7	In Pb Bi	0.71–0.91 torr, $p_0 T_0 = (6–13) \cdot 10^2$ torr K 0.83–1 torr 0.83–0.55 torr	445–968 K 320–640 K 340–415 K	N$_2$	0–70 Å, $(1–6) \cdot 10^3$ $(1–2) \cdot 10^3$ 0–50 Å, $7 \cdot 10^3$ 0–50 Å, $7 \cdot 10^3$	Sonic nozzle with blowing, wire evaporator	Electron diffraction equipment		[64, 65]
8	Sb Bi Pb			He 0.5 torr, 15 torr	1–500 1–280 1–400	Knudsen cell with condensation zone	Mass spectrometer	10^{-7}	[66]
10	Ag	$2 \cdot 10^{-3}$–10 torr	450–1500 K		500–2000	Knudsen cell $d_{cr} = 0.3$–1 mm	Energy analyzer 127°	10^{-4}–10^{-7}	[54] [55] [56]
11	Pb Cu	$1 \cdot 10^{-2}$–10 tprr			500–1000	Knudsen cell $d_{cr} = 10$ µm	Energy analyzer 127°, retarding potential probe	10^{-7}	[57]
12	CdTe		1300°C		30–250 Å	Knudsen cell $d_{cp} = 100$ µm	100 eV ionizer, microscope, electron diffraction equipment	10^{-6}	[58]
13	Si					Electron heating of crucible	Retarding potential probe	10^{-5}–10^{-7}	[59]
14	Pb		1625–1923 K		1–10	Knudsen cell, sonic nozzle, supersonic, channel 0.3; 0.5; 0.7 mm	Time-of-flight mass spectrometer	10^{-5}–10^{-7}	[67]

As a result of this brief analysis of the published data we can formulate specific research objectives with the results presented in this chapter.

1. In connection with a complete lack of experimental data on homogeneous condensation during the expansion of superheated metal vapours it is necessary to formulate such research.

2. It is necessary to clarify the effect of carrier gas on homogeneous condensation of metal vapours.

3. It it important to identify the reasons for disagreement of the results of several works on the development of the ICB-method, consisting in the appearance of clusters with significantly lower levels of the stagnation pressure of vapour.

4. In connection with the experimentally observed wide size distribution functions of the clusters it is important to develop a methodology for finding ways to control the distribution function.

5. On reaching a certain clarity in the processes of formation of clustered flows it is necessary to find methods of using the disperse systems in practice.

4.1. Methods of calculating the start of homogeneous condensation of metal vapours

Calculation of the adiabatic expansion of the metal vapour from a sonic nozzle was carried out by the method described in [48] and built on the ability to describe the free jet by the flow from a spherical source. The coordinate x, measured from the nozzle outlet with a critical diameter d_{cr}, appears in the dimensionless form

$$\delta = x/d_{cr}. \tag{4.3}$$

The density of atoms at the stagnation parameters:

$$n_0 = p_0 / (kT_0). \tag{4.4}$$

The distribution of relative atomic density, temperature and pressure along the jet axis, assuming ideal vapour with the adiabatic index 5/3:

$$\begin{aligned} n/n_0 &= K_7\delta^{-2}, \\ p/p_0 &= K_8\delta^{-4/3}, \\ T/T_0 &= K_9\delta^{-10/3}. \end{aligned} \tag{4.5}$$

The speed ratio and the Mach number along the jet axis:

$$\begin{aligned} S &= K_{10}\delta^{2/3} \\ M &= S / (\chi / 2)^{1/2}. \end{aligned} \tag{4.6}$$

The local most probable thermal velocity of the metal atoms:

$$V = 129 \, (T/m_a)^{1/2}. \tag{4.7}$$

The mean free path at the stagnation parameters:

$$\lambda_0 = (\lambda_n) / n_0, \tag{4.8}$$

where $(\lambda_n) = 1 / (\sqrt{2} \cdot 4\pi r_a^2)$.

The Knudsen number for the stagnation and the diameter of the critical section of the nozzle:

$$Kn = \lambda_0 / d_{cr}. \tag{4.9}$$

The local Knudsen number, defined by the local mean free path and the reduced length of the jet:

$$Kn_j = Kn_0 K_{22} \delta^{5/3}. \tag{4.10}$$

The value $Kn_j \approx 0.04$ corresponds to the section of the jet, in which the disappearance of collisions between the atoms leads to the freezing of the translational degrees of freedom of the vapour. If the conditions for the start of homogeneous condensation do not form up to this section, homogeneous condensation cannot in general occur for this adiabate [48].

The conditions for the beginning of homogeneous condensation are determined by the magnitude of the local nucleation rate [51]:

$$J = \left(\frac{p}{kT}\right)^2 \omega \sqrt{\frac{2\gamma}{\pi m_a}} \exp\left(-\frac{4\pi\gamma r_*^2}{3kT}\right), \tag{4.11}$$

where the radius of the critical nucleus is calculated using the formula (4.1), and its surface tension by the Tolmin formula (4.2):

$$r_* = \frac{2\gamma_\infty \omega}{kT \ln S_n} - 0.6 r_a.$$

Supersaturation is defined as the ratio of the local vapour pressure to saturation pressure corresponding to the local temperature:

$$S_n = p/p_\infty.$$

For calculation of the expansion of a monatomic gas from the sonic nozzle equations (4.3)–(4.10) should use the values of the constants [48]: $K_7 = 0.15$; $K_8 = 0.282$; $K_9 = 0.042$; $K_{10} = 2.98$; $K_{17} = 0.75$; $K_{22} = 23.4$. The saturation curve required for the calculations was approximated by the formula of the form

$$p_\infty = K_1 T^{K2} \exp(K_3/T). \tag{4.12}$$

The values of the constants K_1, K_2 and K_3 for magnesium, silver and lead, obtained an approximation of data [68], are presented in Table 4.2. In the temperature range 300–2000 K the error of formula (4.12) does not exceed ±1.5%.

The results of calculations of the saturation curves for clusters containing different numbers of atoms and fixed frequencies of nucleation are shown in Fig. 4.1 for Mg, in Fig. 4.2 for Ag, and in Fig. 4.3 for Pb. Figures 4.2 and

Table 4.2. Values of constants K_1, K_2, and K_3 for approximation of the saturation curve, collision constant, the volume of the atom in the condensed phase, and surface tension

Metal	K_1, Pa	K_2	K_3, K	r_a, 10^{-10} m	(λ_n), m^{-2}	ω, 10^{-29} m^3	γ_∞, N·m^{-1}	m_a, 10^{-25} kg
Mg	$1.24288 \cdot 10^{20}$	-2.83944	-19462.0	1.60	$2.200 \cdot 10^{18}$	2.300	0.565	0.406
Pb	$1.84843 \cdot 10^{11}$	-0.43089	-22481.9	1.75	$1.837 \cdot 10^{18}$	3.145	0.442	3.460
Ag	$3.33682 \cdot 10^{17}$	-1.85960	-34881.7	1.44	$2.713 \cdot 10^{18}$	1.7I4	0.800	1.800

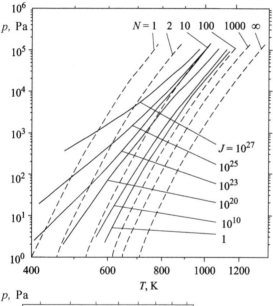

Fig. 4.1. p–T diagram of saturation for Mg clusters of particle size N and the nucleation rate J.

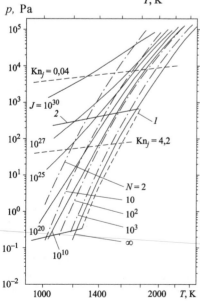

Fig. 4.2. Saturation curves for clusters of Ag, consisting of N particles, and the steady-state rate of nucleation. 1 – area of research [54–56], 2 – adiabate. Kn = 0.04; 4.2 correspond to the disappearance of collisions between atoms and atoms and clusters. $N = 800$, $d^* = 1$ mm.

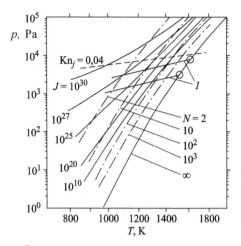

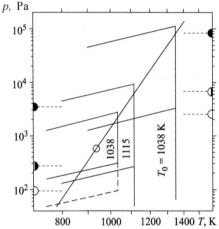

Fig. 4.3. Saturation curves for Pb clusters consisting of atoms, and fixed frequency of nucleation. 1 – adiabates investigated in [67], $d_* = 0.5$ mm.

Fig. 4.4. The area of investigated stagnation parameters. $d_* = 4 \cdot 10^{-4}$ m, Mg. The points on the axes of ordinates denote the stagnation pressure at three different temperatures at which saturation is reached: on the left – in a critical section of the nozzle, on the right – in the nozzle prechamber.

4.3 show the range of p_0, T_0, studied in [54–56] and [67], respectively. We see that the adiabates, corresponding to the modes studied in the cited works, are located below the line corresponding to $Kn_j = 0.04$. This means that the entire investigated range of stagnation parameters p_0, T_0 refers to the regime of free expansion of vapour from the source. In this regard, the origin of the clusters detected in [54–56] is unclear.

It is possible that the formation of clusters found in [54–56] and [67] is due to the influence of electron ionization in the vapour stream. The need to clarify this fact imposes additional requirements on the formulation of experimental studies of homogeneous condensation.

4.2. Features of the formulation of experimental research of clustered metal vapour flows

Experience in operating the vacuum system for studies of the possible homogeneous condensation of vapour flowing from the jet source identified a number of technological features.

1. Penetration of the ultrafine metal condensate into the backing pump of the pumping system is likely.

2. There is a change in the chemical composition of the oil in diffusion vapour–oil pumps related, apparently, to both the decomposition and polymerization of hydrocarbons in the presence of ultrafine metallic materials and reducing the service life of the pump without service.

3. The interior surfaces of the vacuum chamber are covered with a loose metal condensate worsening, due to sorption processes, the vacuum properties of the installation and requiring periodic cleaning of the chamber in compliance with the safety management of active ultrafine materials.

4. Condensation of metal vapours inside a vacuum chamber distorts the operation of pressure sensors and degrades the insulation of internal wiring lines.

5. The metal vapour jet in a vacuum chamber has an evacuation action that creates unsteady pressure, and possibly changes in the composition of residual gases.

With these features taken into account and considering the previously formulated tasks in studying homogeneous condensation, the system was equipped with the following methods of investigation:

1. Procedures associated with the ionization of the clustered flow and recording of ionic currents: mass spectrometer MX-7304, the energy analyzer with the beam rotated through 127°, the retarding potential probe.

2. Methods of measuring the enthalpy flow, as described in section 2.2.1.

3. Methods of measuring the steam flow rate and specific clustered flow rate using a Faraday type trap.

4. Methods of deposition of the clustered flow on a solid surface with subsequent research using electron and optical microscopes.

This combination of indirect but independent methods can give reliable information about the start of the process of homogeneous condensation and the properties of clustered flows. It is particularly important to obtain information on the occurrence of the process of homogeneous condensation in the vapour flow, without making intense disturbances in the flow that might cause condensation. Therefore, the techniques associated with the ionization of the flow or scanning the flow with a high-intensity laser beam are not suitable since there is direct evidence for the activation of condensation by high-energy particles and electromagnetic radiation [69]. In connection with this the determination of the conditions of homogeneous condensation in this research was conducted only with the use of methods of flow measurement, enthalpy flow, specific consumption of vapour in the flow and the local density of vapour atoms with a mass spectrometer.

As was already shown in Chapter 3, the metal vapour source constructed for the present study with adjustable superheating does not show any special features in the vapour flow in a wide stagnation pressure range. This means that if using these techniques which do not disrupt the flow record a non-monotonic variation of a parameter with a change of the stagnation pressure, it will point to the non-adiabaticity of the process of vapour expansion associated, apparently, with the appearance of condensation in the jet. Figure 4.4 shows the studied expansion modes of magnesium vapour from the sonic nozzle with a diameter of the critical cross-section d_{cr} = 0.4 mm at three different stagnation temperatures in comparison with the saturation curve on the flat surface of magnesium. Labels on the vertical axis: on the left – typical levels for the stagnation pressure of the corresponding stagnation temperatures, which in expansion of the vapour lead to saturation in the nozzle throat, right – the stagnation pressure levels for the respective stagnation temperatures at which saturated vapour is located in the nozzle prechamber. The need for a wide range of stagnation pressure changes is associated with the results of Japanese research, described in section 4.1, on the detection of formation of clusters in the free molecular expansion mode.

4.3. Results of experimental studies

Before presenting the results of measurements we should make several significant observations.

Since the start of spontaneous condensation may be due to the presence of impurities, after each new loading the vapour source with a fresh portion of magnesium the metal is degassed thoroughly for several hours at a temperature of 25–50 K above the melting point. All experiments were conducted using magnesium from the same ingot with a purity of 0.1.

Prolonged operation of the magnesium vapour source with a single nozzle leads to the erosion of its walls. This is particularly evident at the discharge modes p_0, T_0, close to saturation. With this in mind, the size and configuration of the nozzle were inspected before each experiment.

Due to the fact that the continuous gas-dynamic flow mode at all investigated modes p_0, T_0, occurs only near the nozzle at distances of no more than 2 mm, the local measurement of the parameters cannot be taken. Therefore, measuring sensors were located at distances of 75–240 mm, adjusted for changes in specific consumption of vapour along the jet as in the spherical expansion $\sim x^{-2}$.

For the convenience of diagnosis, a central section with an opening angle of 5–30° was 'cut out' from the vapour jet using a flat skimmer and a collimator. The skimmer and collimator temperatures were maintained at a level that provides the condensation coefficient near unity.

The measurements of the local specific flow rate were carried at the jet axis at a distance of 75 mm from the nozzle. For measurements of the specific flow rate the Faraday trap was equipped with additional apertures to eliminate the return flow and form flow tubes of the known diameter. The measurement results for T_0 = 1348 K are shown in Fig. 4.5 a. The increased scatter of experimental points for ρu in comparison with measurements of flow rate G can be associated with the possible capture (difficult to take in account) of the residual gases in the chamber during the condensation. The results of measurements of ρu in the range p_0 = $4 \cdot 10^2$–$2 \cdot 10^4$ Pa can be approximated as a function of $\rho u \approx p_0^C$, where C = 1.4. In the range $p_0 < 4 \cdot 10^2$ Pa, in spite of the increased scatter of experimental points, one can observe a deviation from this approximation.

The measurements of the enthalpy flux were taken by the method described in section 2.2.1. Before each measurement, the calorimeter was heated to a temperature of not less than 700 K to clean the surface. Under the assumption of complete thermal accommodation of vapour atoms incident on the calorimeter, the measured heat flux can be represented as the specific enthalpy flux $I = \rho u [C_P (T_0 - T_d) + L (1 - \alpha_c)]$, where c_p is the specific heat flow at the measured point, α_c is the possible condensate fraction in the flow, L is the specific latent heat of condensation. Since L is an order of magnitude higher than $c_p(T_0 - T_d)$ at the studied T_0, then for the qualitative interpretation of measurement results it is sufficient to use $I = \rho u L (1 - \alpha_c)$.

The results of measurements of I for three different T_0 are shown in Fig. 4.5 b. The markers on the x-axis correspond to the markers in Fig. 4.4. The experimental data for individual sites can be approximated by the function $I \sim p_0$. For measurements of I at different T_0 there is a correlation $I \sim T_0^{-0.5}$.

Measurement the density of a monatomic vapour in the jet were performed by a monopolar mass spectrometer MX-7304. To do this, the ionizer of the mass spectrometer was mounted on the nozzle axis at a distance of 240 mm. The measurement results for $T_0 = 1115$ and 1348 K are shown in Fig. 4.5 b. At $p_0 = 4 \cdot 10^2 - 2 \cdot 10^4$ Pa the experimental points can be approximated by the function $\rho_V \sim p_0^{1.15}$.

4.3.1. Discussion of the results of measurements by non-perturbing methods

Comparison of experimental data on the measurement of ρu, I, ρ_V for $T_0 = 1348$ K shows that in the range $p_0 = 3 \cdot 10^2 - 1 \cdot 10^3$ Pa non-adiabatic expansion of vapour in the jet takes place. The observed anomalies of the behaviour of these parameters in relation to p_0 can not be attributed to the influence of any non-stationary processes in the vapour source or in the nozzle. Non-equilibrium processes in evaporator and the viscous effects in the nozzle should be reflected in the dependence $G(p_0)$. The smooth shape of the curve suggests that the measurement results show the effect of condensation on the gas dynamic structure of the Mg vapour jet. The development of the condensation process can be seen clearly in the data corresponding to $T_0 = 1348$ K. When $p_0 = 3 \cdot 10^2$ Pa condensation apparently begins in the jet. Increase in p_0 in the range $3 \cdot 10^2 - 3 \cdot 10^3$ Pa probably corresponds to the movement of the condensation jump to the nozzle outlet. Heat dissipation due to the formation of clusters in the flow leads to a decrease in the polytropic coefficient of the expansion process and a more developed structure of the jet. Most sensitive to the condensation process are the density and specific enthalpy flux. In the range $p_0 = 3 \cdot 10^3 - 1 \cdot 10^5$ Pa the condensation jump is probably near the nozzle exit. The difference in the slopes of the curves $G(p_0)$ and $\rho_V (p_0)$, $G(p_0)$ and $I(p_0)$ shows the growth of the condensate fraction and the variability of the polytropic expansion coefficient.

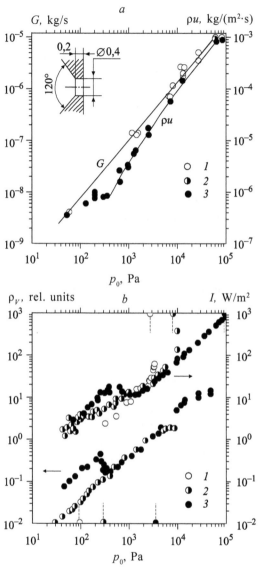

Fig. 4.5. Consumption, the specific consumption, heat content and density of magnesium, depending on the stagnation pressure. 1–3 correspond to $T_0 = 1038$, 1115, 1348 K.

The less pronounced form of the curves at $T_0 = 1038$ K and 1115 K can be explained by the fact that these modes of discharge of the vapour from the nozzle correspond to a higher degree of rarefaction of the flow. However, it is important to note the characteristic behaviour of the curves $I(p_0)$, obtained at $T_0 = 1038$ K and 1115 K, when the stagnation parameters transfer into the saturation region. There is a sharp increase in I, corresponding in the experiments to the discharge of liquid Mg on the nozzle edge. The formation of a liquid metal at the nozzle outlet is apparently due to the heterogeneous condensation of vapour on its walls and dripping on them downstream. The increase in the flow is due to intense evaporation of the metal, superheated relative to the pressure in the vacuum chamber. This phenomenon is important for correct interpretation of the results of investigations of clustered molecular beams conducted close to saturated parameters in the vapour source. Both supercooling of the vapour near the nozzle walls in the regime of heterogeneous condensation on them and the non-equilibrium evaporation in a vacuum of the melt flowing along the walls can give an uncontrolled contribution to the flow of clusters. However, in most publications dealing with the study of clustered beams there are no data for control of the temperature the nozzles.

The primary experimental data, shown in Fig. 4.5, give additional characteristics of the studied jet. The ratio $\rho u / I = A$ for fixed p_0, calculated from the experimental data, gives

$$\alpha_c = (AL - 1) / AL. \tag{4.13}$$

Similarly, from the measured $\rho u / \rho_v = B$, we have

$$u = D (1 - \alpha_c). \tag{4.14}$$

The results of processing the experimental data with the help of (4.13) (4.14) are shown in Fig. 4.6.

The curve for $\alpha_c(p_0)$ is qualitatively consistent with the results of [70] for condensation of a CO_2 jet. However, attention is drawn to an unusually high value of α_c at high p_0. Unusual is also the repeated reduction of velocity u with increasing p_0. The totality of these facts suggests that an important role in the condensation of metallic vapours in comparison with the condensation of gases is played some additional mechanism.

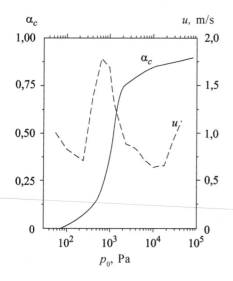

Fig. 4.6. The fraction of the condensate and average mass flow rate depending on stagnation pressure.

a   *b*

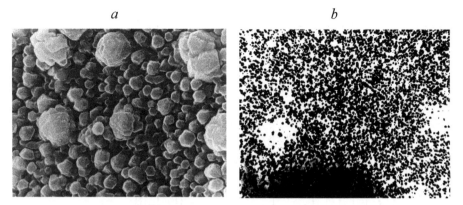

Fig. 4.7. The particles of the condensate, deposited on a solid surface. a – mica, $p_0 = 2.1 \cdot 10^3$ Pa, $T_0 = 1270$ K, deposition time 2 min, ×10⁴; b – carbon film, $p_0 = 3.5 \cdot 10^3$ Pa, $T_0 = 1153$ K, 3 s, ×2·10⁴.

4.3.2. Analysis of the fraction composition of the vapour jet by deposition on the substrate

Relatively high values of α_c at high p_0 indicate the possibility of diagnostics of the composition of the clustered jet by depositing it on a hard surface. The result of this experiment is shown in Fig. 4.7 a where it is shown that in the jet there are two factions of the condensate particles – single crystals with the typical dimensions of a micron and conglomerates of several microns.

The latter indicates, most likely, the important role of coalescence in the process of homogeneous condensation. An example of deposition on the carbon film with a short exposure, where relatively large conglomerates of particles have been removed, is shown in Fig. 4.7 b. Statistical analysis of electron micrographs makes it possible to determine the function of the particle size distribution (Fig. 4.8). We see that the size distribution function has two characteristic peaks – about $8 \cdot 10^{-8}$ and $3.2 \cdot 10^{-7}$ m.

So, using the diagnostics methods without introducing high-energy disturbances in the flow, indirect measurements cam be taken to find the area of the start of condensation in the vapour flow and identify some of its features.

4.3.3. Measurement of the distribution function of condensate particles by ionization methods

To further control the results obtained by indirect methods, measurement were made of the distribution function of clusters in a vapour jet through the retarding potential probe. The essence of the method and experimental setup (Fig. 4.9) are identical to those described in [54–56].

Part of the clustered metal jet is sent to the ionizer which is a radial electron gun, then to the analyzer of charged particle energy. The cumulative distribution function of charged particles is determined from the ion current at different retarding potentials U_z. Ideally, monoenergetic electrons once ionize

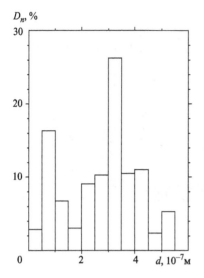

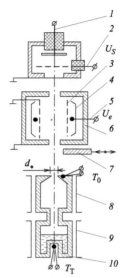

Fig. 4.8. The size distribution function of particles. Processing of photograph in Fig. 4.7 b.

Fig. 4.9. The scheme of measurements of the energy distribution function of clusters using the retarding potential probe. 1 – collector; 2 – control grid; 3 – screen; 4 – accelerating electrode; 5 – emitter; 6 – housing of the ionizer; 7 – flap, 8, 9 – heaters, 10 – the crucible with the melt.

vapour atoms and clusters. The results of measurements of the ion current at different retarding potentials can be converted to the mass distribution function of the charged particles at a known speed of their flight. The retarding potential probe can be used in the mode of cutting off the electron current of the ionizer by an additional grid with a negative potential and in the full current mode, when the difference of the ion currents with the diagnosed flux and without it is analyzed.

The following features that complicate the quantitative interpretation of the results of measurements were investigated in developing the method:

1. The uncertainty of the relative probability of ionization of atomic and cluster components of the jet, complicated by the possible non-monoenergetic nature of the ionizer electrons. The non-monoenergetic nature of the electrons may be associated with the generation of secondary electrons on the structural elements of the ionizer, the residual gas in the chamber and the vapour jet.

2. The process of ionization of atoms and clusters of the jet may be accompanied by the formation of negatively charged particles due to the absorption of electrons by clusters.

3. The presence of the space charge in the ionization zone involves the formation of a potential 'well' in the zone that distorts the energy spectrum of the analyzed particles.

4. In principle, the effect of electron impact may cause the collapse of the clusters, the collapse of charged clusters due to the Lippman effect which

reduces the binding energy of atoms in the cluster, the collapse of clusters in gradient electric fields of the probe, as well as the initiation of the coalescence of particles under the influence of ionization and charging.

5. Charged particles do not form in the ionization zone and instead form in the gradient field near the grid of the probe due to the auto-ionization of the vapor atoms excited in the ionization zone.

6. The secondary emission of electrons and ions from the collector under the influence of the photoelectric effect and bombardment of the collector with electrons, ions and charged clusters may take place.

7. The effect of polarization of the elements of the structure and of molecular layers of residual gases sorbed on them on the size and shape of the electric fields.

All these factors can distort the measured energy spectrum of the particles. At the present it is not possible to calculate the instrumental function of the measuring system, consisting of the ionizer and the retarding potential probe, taking into account all these factors. Experimental determination of the instrumental function is also not possible due to the lack of reliable data on clustered jets. Therefore, all results obtained using the retarding potential probe should be considered qualitative.

This section presents the results of measurements of the distribution function of negatively charged clusters of the metal vapour jet in the mode measurements of the total ionization current. The setup for determining the distribution function is shown in Fig. 4.10. Curve 1 denotes the ionization current fed to the collector when the retarding potential without the clustered flow changes, the curve 2 – with the flow. Comparing the curves, we can calculate the total current associated with the presence of the clustered flow in the ionizer. However, curve 2 should be corrected taking into account the absorption of electrons in the ionization zone, which takes place in the

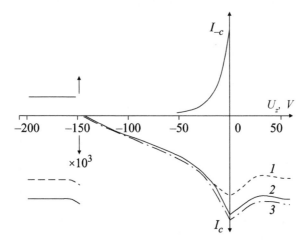

Fig. 4.10. Determination of the energy distribution function of the negatively charged clusters. The retarding potential probe, analysis of total current mode. 1 – no flow; 2 – with the flow; 3 – curve, corrected for the absorption of electrons.

analyzed flow. This can be done using parts of the curves 1 and 2, where their dependence on U_z is similar. The absorption coefficient of electrons, calculated at an accelerating voltage of the ionizer of 143 V, depends on the presence of the clustered flow. Its values are shown in Fig. 4.11, depending on the stagnation pressure at the vapour jet source of the jet of steam. Curve 3 in Fig. 4.10 is obtained taking into account the absorption of electrons.

The difference between the curves 1 and 3 can be interpreted as the current produced by the negatively charged clusters that reach the collector at different retarding potentials. This difference in the form of the curve I_{-c} is presented in the upper part of Fig. 4.10. It is worth noting two features of measurements carried out in this scheme. First, the curves 1, 2, 3, tend to the abscissa in the coordinate $U_z = -143$ V, which indicates the absence of a potential 'well' in the ionizer. Second, at $U_z < -143$ V a residual negative current is observed, which depends on the parameters of the analyzed flow and is independent of U_z in the range $-143...-200$ V. This can be interpreted as the presence in the flow of negatively charged clusters, whose energy exceeds 200 eV.

Curve I_{-c} (see Fig. 4.10) after normalization to its value at $U_z = 0$ can be interpreted as the cumulative energy distribution function of clusters in the range 0–143 eV. Figure 4.12 shows the results of measurements of the energy distribution function of the clusters for different accelerating voltages of the ionizer E_e at constant current of electron ionization I_e. In Fig. 4.13 similar measurements were carried out for different currents of ionization electrons at a constant $p_0 = 1.33 \cdot 10^4$ Pa. Figure 4.12 shows that in the range $E_e = 70-200$ eV the measurement results are practically unchanged. Figure 4.13 shows that at a current of ionizing electrons $I_e > 40$ mA the broadening of the measured distribution function is observed. This is in qualitative agreement with the data of [54–56] and can be interpreted as the effect of coalescence or disintegration of small clusters of fractions of more than 200 eV. In this regard, the diagnosis of the clustered flow was conducted using the $I_e = 40$ mA and $E_e = 143$ V.

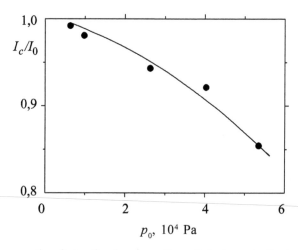

Fig. 4.11. The correction factor for the absorption of electrons in the ionization zone.

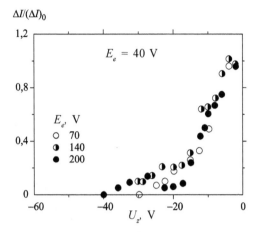

Fig. 4.12. The distribution function of the negative clusters at different energies of ionizing electrons.

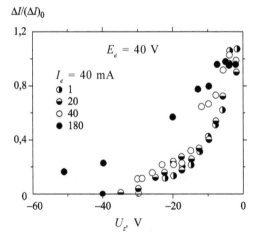

Fig. 4.13. The energy distribution function of the negative clusters, measured at different currents of the ionizer.

The results of measurements of the cumulative energy distribution function of clusters in the clustered metal vapour jet at T_0 = 1348 K and various p_0 are shown in Fig. 4.14. Data analysis showed that in the range p_0 = 4·10³–2.7·10⁴ Pa the distribution functions are almost identical. Measurements made at p_0 = 4·10⁴ Pa or more show the distribution function, indicating the appearance of smaller clusters in the flow.

Absolute values of ion currents, corresponding to the fraction of clusters with the energy below 50 eV (I_{-50}) and the fraction of clusters with an energy greater than 200 eV (I_{+200}) are shown in Fig. 4.15. It is evident that their dependence on p_0 significantly differs. One gets the impression that the clusters of the fraction less than 50 eV pass to the fraction of the size of more than 200

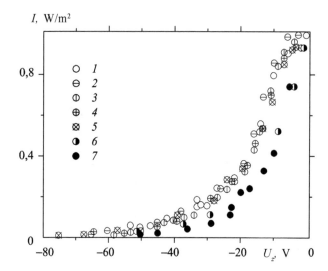

Fig. 4.14. Cumulative energy distribution function of the negatively charged clusters. The points 1–7 correspond to $p_0 = 4.655 \cdot 10^3$; $6.65 \cdot 10^3$; $9.975 \cdot 10^3$; $1.33 \cdot 10^4$; $2.66 \cdot 10^4$; $3.99 \cdot 10^4$; $5.32 \cdot 10^4$ Pa. $T_0 = 1348$ K.

eV. Moreover, this process takes place without the appearance of clusters with intermediate values of energy or mass. Such a process may be coagulation.

In the same range of pressures p_0 measurements were taken of the concentrations of Mg_2 dimers with a mass spectrometer. Measurements with a monopolar mass spectrometer in the range of relatively high p_0 are hampered by the fact that the high specific consumption of vapour causes intense metallizing of parts of the mass spectrometer in the area of the ionizer and analyzer. The results are shown in Fig. 4.15 in the form of the curve with black dots.

4.3.4. Start of spontaneous homogeneous condensation of magnesium

The results obtained in section 4.3 of the experimental results and the calculation procedure, outlined in Section 4.1, allow to draw conclusions about the beginning of the homogeneous condensation of magnesium in the investigated case of superheating the vapour and the sonic nozzle.

From Fig. 4.14 it is clear that at $T_0 = 1348$ K at pressures $p_0 = 4 \cdot 10^3 - 2.7 \cdot 10^4$ Pa the Mg vapour jet contains clusters are with energies up to 40 eV, or (at $T_0 = 1348$ K 1 eV corresponds for single ionization to approximately $N = 10$ atoms/cluster) with the size up to $N = 400$ atoms/cluster.

The first manifestation of the clusters in the flow is observed at $p_0 = 3.4 \cdot 10^2$ Pa in the measurements of ρu, I, ρ_v, shown in Fig. 4.5. If we assume that clusters are spherical, and we also assume that their radius is $r_c = r_a N^{1/3}$, the collision cross-section of clusters with the vapour atoms can be estimated as

$$\sigma_{c-a} = \sigma_{a-a} \, (N^{1/3} + 1)^2/4 \tag{4.15}$$

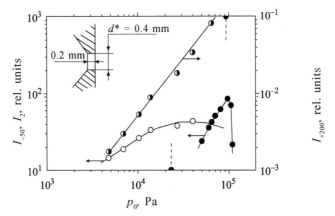

Fig. 4.15. The intensity of the flow of clusters with energy below 50 eV (I_{-50}), more than 200 eV (I_{-200}) and Mg_2^+ dimers of the stagnation pressure. $T_0 = 1348$ K.

where σ_{a-a} is the collision cross section of atoms with each other. For $N = 400$ $\sigma_{c-a}/\sigma_{a-a} \approx 17.5$. Figure 4.16 shows in the p–T coordinates the adiabatic expansion of the sonic nozzle, calculated by the method described in section 4.1, $T_0 = 1348$ K, $p_0 = 3.4 \cdot 10^{-2}$ Pa. It is seen that the adiabate intersects the saturation curve of the cluster, $N = 400$ at the value of the local number Kn $= 2.85$. Given the estimates made with the help of (4.15), for $N = 400$ the Knudsen number of collisions with atoms of the cluster is 0.163, which is plausible for the start of the first collisions of atoms with clusters, located in the state saturated with respect to the vapour. Apparently, for p_0, larger than $3.4 \cdot 10^2$ Pa, this will lead to condensation of magnesium vapour on these clusters, and then to their coalescence. These processes are reflected in the curves of $I_{-50}(p_0)$ and I_{+200} (p_0) in Fig. 4.15.

From the curve ρ_v (p_0), shown in Fig. 4.5, we see that for $p_0 = 2.3 \cdot 10^4$ Pa the density of the atomic component of the vapour in the jet greatly decreases. The corresponding adiabatic curve, calculated by the method described in section 4.1, is shown in Fig. 4.16 together with the line of the start of progressive 'freezing' $Kn_j = 0,04$. The point of intersection of these curves lies on the saturation line for clusters $N = 9$. This means that in the jet pair with a value of p_0 can be a cluster size of $N = 9$. Approximately the same pressure corresponds to the change of the measured energy distribution functions of clusters in Fig. 4.14 in the direction of smaller mean-sized diagnosed clusters, which may indicate the occurrence in the flow of a number of smaller particles in the condensate. The corresponding adiabatic point is marked at the bottom of the horizontal axis of Fig. 4.15. The appearance of dimers in the stream is actually registered at p_0 greater than $2.3 \cdot 10^4$ Pa. Unfortunately, due to the complexity of using the mass spectrometer for measurements in the flow of metal vapour and low sensitivity of the apparatus direct information about the cluster structure in this region p_0 could not be obtained.

The adiabate $T_0 = 1348$ K, $p_0 = 2.3 \cdot 10^4$ Pa, at which the measurements of ρ_v showed a decrease of density, indicating the possible beginning of

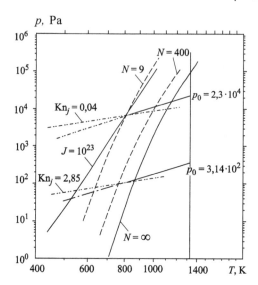

Fig. 4.16. The interpretation of experimental data at $T_0 = 1348$ K. Adiabates $p_0 = 2.3 \cdot 10^4$; $3.4 \cdot 10^2$ correspond to the beginning of homogeneous condensation and condensation on the artificial nuclei, $N = 400$. Lines $\text{Kn}_j = 0.04$; 2.85 correspond to the borders of extinction of collisions of atoms and clusters, $N = 400$.

homogeneous condensation in the flow with the formation of clusters, $N = 9$, and the clusters $N = 400$ detected by the retarding potential probe, can be qualitatively compared with the results of generalization [53] of the cluster sizes obtained by homogeneous condensation in the flow of gases, vapours and vapour mixtures of gases. Such a comparison in the coordinates N/z_j, Ψ, where z_j is the degree of ionization, $\Psi = p_0 d_{cr} (T_b/T_0)^{\chi(1-\chi)}$ torr·mm, T_b is the boiling point of substances used in the experiments, is given in Fig. 4.17. It is seen that for $d_{cr} = 0.4 \cdot 10^{-3}$ m, $T_b = 1376$ K for magnesium and $\chi = 5/3$ there is good agreement.

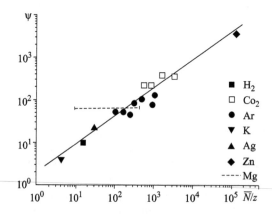

Fig. 4.17. Comparison of the data on the presence of clusters in the Mg vapour jet with the results of generalization [53] for other substances.

So, practically in the whole range of the investigated pressures p_0 at $T_0 = 1348$ K, including the region of free molecular expansion of vapour from the sonic nozzle, the flow was detected to contain clusters which condense when they are in conditions of saturation with respect to the vapour and there is a sufficient number of collisions with the vapour atoms. This result is in qualitative agreement with experiments [54–56] and [67]. Since most of the experiments were conducted without using techniques that perturb the flow, the origin of these clusters can not be explained by electrically stimulated condensation or coagulation. Apparently, there are other reasons for the appearance of clusters in the vapour jets not associated with homogeneous condensation. Section 4.4 will discuss one possible mechanism for the emergence of clusters.

Combination of experimental studies and calculations could indicate the conditions for the appearance of homogeneous condensation in the magnesium vapour je flows, despite their preliminary 'contamination' with clusters. The conditions for the beginning of homogeneous condensation of magnesium vapour correspond to similar results of studies of homogeneous condensation of gases and vapours in gas streams. The latter circumstance in connection with the existing guidance on the impact of carrier gas on the condensation in the flow [48] and on the condensation in the steady-state conditions [71] will be discussed in more detail in section 4.5.

4.4. Model of inhomogeneous origin of clusters

The results of experiments [72] set forth in section 3.1, showed that the interaction of the flow of vapour with the solid surface may be accompanied by the generation of heat corresponding to condensation but with no condensate. The results of experiments in section 4.3 also confirmed the presence of clusters in the vapour flow from a source in the conditions when the homogeneous condensation is not feasible. This information shows that it is important to consider the possible mechanisms for the origin of the free particles of matter in the condensed state.

4.4.1. Spontaneous withdrawal of solid condensate particles from the substrate

Suppose that on a solid surface there are solid particles of cubic shape with the side of the cube ma, where a is a cubic lattice constant. The cubic solid particle is isothermal with respect to the surface. The maximum Debye frequency of thermal oscillations in the particle

$$\omega_m = c\pi \left(\frac{6\pi^2 N}{V} \right)^{1/3}, \qquad (4.16)$$

where c is the speed of sound in the solid particle; V is its volume; N is the number of atoms in it. Assuming that $V = Na^3$, from (4.16) we obtain

$$\omega_m = \frac{c\pi}{a}(6\pi^2)^{1/3}.$$

If the particle is bound to the surface by the energy of desorption $m^2\varepsilon_s$, then the probability of the particles breaking off from the surface as a result of fluctuations will be determined by the Boltzmann formula for stochastic processes with overcoming the energy barrier:

$$F^{\cdot} = (c\pi/a)\ (6\pi^2)^{1/3}\ \exp\ (-m^2\varepsilon_s/(kT)). \tag{4.17}$$

The flux of particles leaving the unit surface area per unit time is given by

$$J^{\cdot} = (c\pi/a)\ (6\pi^2)^{1/3}n\ (N)\ \exp\ (-m^2\varepsilon_s/\ (kT)). \tag{4.18}$$

Here, $n(N)$ is the density of solid particles on the surface. It seems that of interest is the value $F^- \approx 1$. From (4.17) with this condition should be

$$m = \left(\frac{kT}{\varepsilon_S} \ln \frac{c\pi\left(6\pi^2\right)^{1/3}}{a} \right)^{1/2}.$$

Estimates for the oxides of magnesium on the magnesium surface at $T = 933$ K, $a = 2.84 \cdot 10^{-10}$ m, $c = 5 \cdot 10^3$ m/s, $\varepsilon_s = 0.1$, $\varepsilon_1 = 8.95 \cdot 10^{-21}$ J show that a particle having a size of $m = 6.9$ and containing $N = 330$ atoms leaves the surface in one second, with a probability equal to unity. If we assume that the entire magnesium surface is covered by separate particles of this size, the unit area per second emits into the surrounding space the particle flux $J^- = 2.6 \cdot 10^{17}$ m$^{-2} \cdot$s^{-1}.

The particle with the size approximately equal to the size of the atom in transfer to the gaseous medium should preserve the kinetic energy $1/2kT$. Condition $m_a Nu^2/2 = kT/2$ allows to estimate its speed in a gaseous environment: $u^2 = kT/(m_a N)$, where m_a is the mass of the atom. For magnesium at $T = 933$ K we have $u \approx 22$ m/s. The density of particles in the gas phase can also be assessed, $n_c = J^-/4 = 1.2 \cdot 10^{16}$ m^{-3}, and the total particle surface $f_c = 6$ $(m_a)^2\ n_c = 0.276$ m^2/m^3 contained in unit volume of the gaseous medium. At a temperature of 933 K the vapour pressure of magnesium is $p_{h\infty} = 4 \cdot 10^2$ Pa and the atomic density $n_0 = 3,1 \cdot 10^{20}$ m^{-3}. Despite the low relative concentration of solid particles in a gaseous medium, $n_0/n_c = 3.87 \cdot 10^{-5}$, per unit time in 1 m^3 of the vapour environment they collide with $\omega_C = f_C\ ph_\infty/(2\pi m_a kT)^{1/2} = 1.926 \cdot 10^{24}$ m$^{-3} \cdot$s^{-1} atoms. The characteristic time between collisions of a particle with any atom is $n_0/\omega_C = 1.61 \cdot 10^{-4}$ s. The characteristic time of flight of particles through the zone of the vapour jet in which condensation can take place is of the order of $10d_{cr}\ /u = 1.82 \cdot 10^{-4}$ s. The commensurability of the obtained characteristic times indicates the possibility in principle that particles of magnesium oxide at a temperature of 933 K, corresponding to $p_0 \approx 4 \cdot 10^2$ Pa, can be centres of condensation in the jet. At about this p_0 the experiments described in section 4.3 showed signs of condensation.

An indirect confirmation of the estimates given in this section is also the well-known fact that oxides are removed from any metal objects during heating. Equations (4.17) and (4.18) illustrate the strong temperature dependence. It is clear that the source of the particles, causing the condensation of supersaturated vapour in the experiments of section 4.3, may be various components of the installation, including the heat shields and heating elements.

4.4.2. The generation of clusters by the solid surface bordering with vapour

If the solid surface is in contact with a vapour medium under isothermal conditions then agglomerates can form on its surface from the adsorbed phase of the vapour atoms with a probability [73]

$$F^+ \sim \exp\left(-\delta\Phi\left(N\right) / \left(kT\right)\right),$$

where $\delta\Phi\left(N\right)$ is the energy barrier associated with the transition of N atoms from the adsorbed phase to the condensed state. If we assume the formation of faceted conglomerates with a square base ma and height of n_a, then the magnitude of the energy barrier is calculated as follows [73]:

$$\delta\Phi\left(N\right) = z_1 N \varepsilon_1/2 + m^2 \varepsilon_s - m^2 \varepsilon_1/2 - m^2 \varepsilon_1/2 - 4mn\varepsilon_1/2 - N\varepsilon_s. \quad (4.19)$$

The first two terms of (4.19) mean the energy compensated bonds of atoms within the agglomerate and the substrate, the third, fourth and fifth – the energy of uncompensated bonds of the agglomerate boundaries, the sixth – the energy of released adsorption vacancies. Here z_1 is the number of nearest neighbors in the crystal structure of the agglomerate.

The agglomerates with a minimum value $\delta\Phi\left(N\right)$ are most likely to form. With the condition $m^2 n = N = const$ minimization of $\delta\Phi\left(N\right)$ leads to the requirement of the dimensions m and n [73]:

$$m = N^{1/3}\left(1 - \varepsilon_s/\varepsilon_1\right)^{-1/3}, \ n = N^{1/3}\left(1 - \varepsilon_s/\varepsilon_1\right)^{2/3}. \quad (4.20)$$

From (4.20) it can be seen that if the binding energy with the substrate equals the binding energy of atoms in the agglomerate, then $m \to \infty$, $n \to 0$, i.e. flat nuclei of monatomic thickness are most likely to form.

Of greatest interest are the agglomerates that have reached the critical size, capable of further growth. The frequency of their formation per unit area of the substrate per unit time is given by [73]

$$J^+ = \zeta \left(\xi/n_a\right) vn\left(N\right) \exp\left[-u_D/(kT)\right], \quad (4.21)$$

where ζ is the proportion of peripheral adsorption vacancies facilitating the transport of adatoms to the nucleus; n_a is the density of adsorption vacancies; $n(N)$ is the density of agglomerates of the subcritical size; u_D is the activation energy of surface migration of adatoms $u_D \approx \beta\varepsilon_s$ ($\beta \approx 0.3$) [73]; ξ is the density of adatoms. The density of subcritical size agglomerates can be defined as [73]

$$n\left(N\right) = n_a \exp\left(-\delta\Phi\left(N\right)/(kT)\right).$$

Consider the situation where the formation of supercritical nuclei is accompanied by the process of departure of nuclei from the surface of the due to their thermal fluctuations and this prevents advanced heterogeneous condensation. The condition of equality of the rate of nucleation and recoil of the nuclei is the equality $J^+ = J^-$. The value of J^- is defined in section 4.4.1 as the probability of detachment of the particle from the surface due to thermal vibrations with overcoming the energy barrier $m^2\varepsilon_s$, representing the sum of the binding energies of the atoms of the base of the particle facing the solid surface. When considering the probability of detachment of the nucleus, spontaneously formed from the adsorbed phase, the energy barrier of separation $m^2\varepsilon_s$ should be corrected for the amount of energy released in the nucleus during its formation:

$$\delta\Phi^+ (N) = z_1 N \varepsilon_1/2 - m^2\varepsilon_1 - 2mn\varepsilon_1.$$

Here, as in (4.19), the first term reflects the energy of compensation of all the possible bonds of N atoms in the nucleus, the second term – a correction for uncompensated bonds of the base of the nucleus, the third – for uncompensated lateral faces. Depending on the mechanism of nucleation – instantaneous fluctuating concentration of N atoms in the nucleus or sequential addition of N atoms to it, the thermal effect of the nucleus may be partially reduced by thermal accommodation of the nucleus with the surface $\alpha_{tc}\delta\Phi^+ (N)$. The remainder of the thermal effect $(1 - \alpha_{tc}) \delta\Phi^+ (N)$ can be used to calculate the energy barrier for the separation of the nucleus from the surface:

$$\delta\Phi^-(N) = m^2\varepsilon_s - (1 - \alpha_{tc}) \delta\Phi^+ (N). \qquad (4.22)$$

In section 4.4.1 when considering the detachment of a relatively large particle from the surface in (4.17), the kinetic pre-exponential term was calculated with reference to the maximum Debye frequency. It was shown that the kinetic coefficient recorded in this way is independent of N. For small nuclei, with N of the order of several units, at $N \to 1$ the kinetic coefficient ω_m should apparently tend to $v_\perp$, as in the re-adsorption of a single atom [73]. Then, using (4.22) the probability of detachment of the nucleus from a few atoms can be written as

$$F^- = v_\perp \exp\left[-m^2\varepsilon_S + (1-\alpha_{tc})\delta\Phi^+ (N)\right].$$

The flow of nuclei detached from the unit surface area per unit time is determined by the formula

$$J^- = v_\perp n(N)\exp\left[-m^2\varepsilon_S + (1-\alpha_{tc})\delta\Phi^+ (N)\right]. \qquad (4.23)$$

The equiprobability condition of critical nucleus formation and its rebound from the surface can be written using (4.21) and (4.23):

$$\frac{J^-}{J^+} = \frac{n_a}{\varsigma\xi}\exp\left[\left(-m^2\varepsilon_S + (1-\alpha_{tc})\delta\Phi(N) + u_D\right)/kT\right]. \qquad (4.24)$$

The density of adatoms on the substrate surface without occupation of the vacancies by own adatoms is calculated as follows [73]:

$$\xi = p(2\pi m_a kT)^{-1/2}\nu_\perp^{-1}\exp(\varepsilon_S/kT), \qquad (4.25)$$

where p is the vapour pressure in the vapour environment; $\nu_\perp$ is the frequency of vibrations of the adatoms normal to the substrate. The density of adatoms on the substrate must be consistent with the critical nucleus with the aid of the Kelvin–Thomson effect and taking into account the size dependence of surface tension:

$$\xi = \xi_\infty \exp\{\gamma\omega/[kT\,(r_* + 0.6r_a)]\}. \qquad (4.26)$$

A comparison of (4.26) and (4.25) implies that a flow of atoms should fall on the surface of the substrate

$$p(2\pi m_a kT)^{-1/2} = \xi_\infty\nu_\perp\exp\{\gamma\omega/[(r^* + 0,6r_a) - \varepsilon_S]/kT\}. \qquad (4.27)$$

In (4.27) the equilibrium density of adatoms is defined as

$$\xi_\infty = p_\infty(2\pi m_a kT)^{-1/2}\nu_\perp^{-1}\exp(\varepsilon_1/kT). \qquad (4.28)$$

After substituting (4.28) into (4.27) and (4.25) we obtain

$$\xi = p_\infty(2\pi m_a kT)^{-1/2}\nu_\perp^{-1}\exp\{[\varepsilon_1 + \gamma\omega/(r^* + 0,6r_a)]/kT\}. \qquad (4.29)$$

From the condition that the perimeters of the circular and faceted nuclei are equal, the nucleus equivalent radius r_* in (4.29) can be replaced by $r_* = 2m_a/\pi$, where $a = \omega^{1/3}$ is the reduced lattice constant of the material of the nucleus. Finally, we have

$$\xi = p_\infty(2\pi m_a kT)^{-1/2}\nu_\perp^{-1}\exp\{[\varepsilon_1 + \gamma\omega\pi/(2a(m + 0,15\pi))]/kT\}. \qquad (4.30)$$

Substitution of (4.30) into (4.24) gives

$$\frac{J^-}{J^+} = \frac{n_a\nu_\perp}{\varsigma\,p_\infty(2\pi m_a kT)^{-1/2}}\exp\left[\left(-m^2\varepsilon_S + (1-\alpha_{tc})\delta\Phi^+(N) + \beta\varepsilon_S - \varepsilon_1 - \frac{\gamma\omega\pi}{2a(m+0,15\pi)}\right)\bigg/kT\right].$$

$$\qquad (4.31)$$

After taking the logarithm of (4.31) and introducing the notation

$$A = \ln \frac{\varsigma P_\infty (2\pi m_a kT)^{-1/2}}{n_a v_\perp}$$

with $N = m^3 (1-\varepsilon_s/\varepsilon_1)$, $n = m (1-\varepsilon_s/\varepsilon_1)$ according to (4.20) we obtain an equation for calculating m and N, corresponding to the probabilities of formation of supercritical nuclei from critical size agglomerates and their departure from the surface :

$$\beta\frac{\varepsilon_S}{\varepsilon_1} - m^2 \frac{\varepsilon_S}{\varepsilon_1} + (1-\alpha_{tc})\left[\frac{z_1}{2}m^3\left(1-\frac{\varepsilon_S}{\varepsilon_1}\right) - m^2 - 2m^2\left(1-\frac{\varepsilon_S}{\varepsilon_1}\right)\right] - \frac{\gamma\omega\pi}{2a(m+0,15\pi)\varepsilon_1} = \frac{kTA}{\varepsilon_1}+1.$$

$$(4.32)$$

It is seen that for the case of complete thermal accommodation of the nucleus on the surface of the substrate $\alpha_{tc} = 1$ from (4.32) we obtain a cubic equation for m with one change of the sign, which corresponds to a maximum of one positive root of m. For $\alpha_{tc} = 0$, which indicates a complete absence of thermal accommodation of the nucleus, equation (4.32) is transformed into an algebraic equation for m of the fourth degree with two changes of the sign, which corresponds to a maximum of two positive roots of m and N.

Figure 4.18 shows the results of the numerical solution of equation (4.32) for the condensation of magnesium on substrates with different binding energy for the Mg atoms. In the calculations the values of the parameters were: $\beta = 0.3$, $\varepsilon_1 = 8.95\cdot10^{-20}$ J, $z_1 = 6$, $\gamma = 0.568$ J/m², $\omega = 23\cdot10^{-30}$ m³, $a = 2.84\cdot 10^{-20}$ m, $T = 700$ K, $n_a = 1.24\cdot10^{19}$ m⁻², $v_\perp = 10^3$ s⁻¹; p_∞ was approximated by the formula (4.12). Figure 4.18 shows how with the decrease of α_{tc} the existence of a unique solution for m in the range $N = 1$–10 changes to a dual solution.

The physical meaning of the solutions (4.32), shown in Fig. 4.18, lies in the fact that each of the curves obtained at a certain ratio of thermal accommodation, divides the areas of the size of critical nuclei and hence areas of supersaturation on the surface of the substrate in which the rebound of nuclei of critical size or the beginning of development of heterogeneous condensation by sequential filling of the surface with supercritical nuclei predominates. The regions separated by the solutions of (4.32) can be identified by the analysis of signs of this equation. For a complete thermal accommodation $\alpha_{tc} = 1$ the main cubic term in equation (4.32) is negative, which means that for all m and N, larger than the solution, the rebound is less likely. Thus, heterogeneous condensation on the substrate may occur by filling the substrate only with islands larger than N, corresponding to the solution of (4.32). Figure 4.18 shows that condensation on substrates that exhibit low adsorption energy with respect to the condensing substance can occur only filling the surface with larger islands. Qualitatively, this coincides with the regime of droplet condensation, often studied experimentally on poorly wetted surfaces.

For the weak thermal accommodation of nuclei at the time when they reach their critical size there are two solutions for m and N between which is

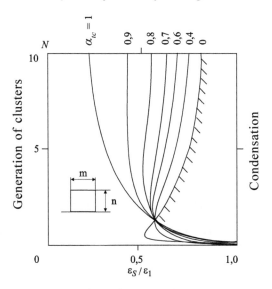

Fig. 4.18. The lines of equal probability of heterogeneous nucleation and maintenance of condensate from the substrate. Magnesium.

the area with greater probability of fixing the nuclei on the surface than the probability of the nuclei leaving from the surface. The positive sign of the quadratic term in m in equation (4.32) indicates that for all m and N, larger than the maximal solution, the likelihood of rebound of the nuclei is higher. This means that for weak thermal accommodation of nuclei the surface releases large nuclei and condensation takes place by filling the surface with nuclei of only the near-critical size. The behaviour of the curves corresponding to small α_{tc}, depending on ε_s shows that with decreasing binding energy with the substrate the surface of the substrate becomes an emitter of finer clusters.

Figure 4.19 shows the results of calculations of heterogeneous nucleation by the formula (4.32) for the above parameters for the extreme cases of thermal accommodation $\alpha_{tc} = 0$ and 1. For comparison, the results of calculations of the frequency of homogeneous nucleation in the vapour environment, which borders with the surface of the substrate, carried out using the formula (4.11) are presented. The increased value of heterogeneous nucleation for all α_{tc} indicates that the solid surface is always an emitter of particles of the condensed phase into the vapour region.

4.4.3. Experimental verification of models

The phenomenon of production of free clusters by the surface of a solid substrate, facing the flow of steam, first shown in experiments [72]. It was observed that the heat of condensation is supplied continuously to the surface of the substrate, but no condensate forms on the latter. The substrate, pre-contaminated with discrete nuclei of crystallization, after blocking the vapour flow was spontaneously cleaned within a few seconds.

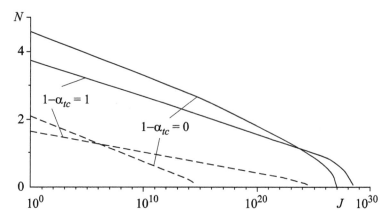

Fig. 4.19. The rates of heterogeneous (solid lines, $m^{-2} \cdot s^{-1}$) and homogeneous (dashed, $m^{-3} \cdot s^{-1}$) for the nucleation curves in Fig. 4.18.

Generation of aluminum clusters containing up to 20 atoms by the superheated nozzle surface of the evaporator was established in [74, 75]. The fluctuation mechanism of separation of extremely small particles from the condensation surface is described in [76, 77]. The authors of these studies also defined the conditions for removing the deposited condensate nuclei from the surface to the vapour phase, depending on their size and adsorption energy.

The authors of [78] found that another possible mechanism for the creation of microdust is re-evaporation during vacuum deposition. Special experiments have shown that this mechanism does exist [79]. The authors of this paper combined the accumulated results is a single study and examined the place and role of these two different mechanisms of generation of dust. Departure of the condensate from the substrate was observed in experiments carried out to adjust the method of measuring the energy distribution function of particles of a homogeneous condensate by the retarding potential probe as described in section 4.3.3. Ion current collector 1 in Fig. 4.9 was made of polished brass. During the experiments, the surface of the collector was covered with a Mg condensate produced by the vapour source. In servicing the retarding potential probe the collector was cleaned to remove the condensate and then again polishing. After each new polishing the collector in subsequent experiments, showed the appearance of pulsed electrical signals. When the collector surface was filled with the Mg condensate pulsed signals disappeared.

Figure 4.20 shows the results of measurements of such current pulses in the collector in the mode when the mesh 2 (see Fig. 4.9) adjacent to the collector received a positive potential of $U_z = 195$ V. It also graphically shows three successive measurements of the collector current when applying the flow of magnesium vapour for 3, 10 and 20 s. The letters 'o' and 'c' mark moments of opening and closing of the vapour flow by the valve. It is seen that after closing the valve after ≈3 s a group of pulse signals that are not resolved in time forms on the collector. After ≈40 s the pulses cease.

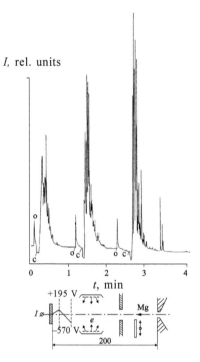

Fig. 4.20. Three cycles of deposition of the collector with magnesium with exposure times of 3, 10 and 20 s; o – opening the damper, c – closing the flap; p_0 = 1.46 · 10³ Pa; T_0 = 1348 K.

The results of these experiments can be interpreted as follows. When the vapour flow is applied nuclei of the condensed Mg phase form on the collector. After interrupting the vapour flow the magnesium nuclei are oxidized by oxygen and water vapour of the residual atmosphere. Particles of magnesium oxide with a low energy of adhesion to brass leave the latter by the mechanism described in section 4.4.1. The current in the collector comes from particles falling on the grid 2 (see Fig. 4.9) where they are charged and then returned to the collector with a positive potential. When the collector is cleaned the pulse signals disappear.

These pulses were also recorded when the flat, interrupting the Mg vapour flow was open, but with a different vapour flow rate. In these experiments, the grid nearest to the collector received a negative potential of U_z = –50 V. Figure 4.21 shows the results of comparative measurements of the collector current in the flap open (1) and closed (2). It is seen that the condensation of magnesium is accompanied by pulsed collector current peaks which can be explained by the withdrawal from the collector of condensate particles, recharging of the particles on the grid and their return to the collector.

Separate experiments were conducted to precipitate the particles of the condensate leaving the condensation surface. The experimental setup is shown in Fig. 4.22. A molecular beam of magnesium vapour was directed to the

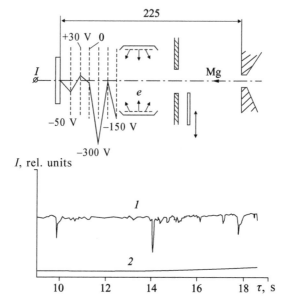

Fig. 4.21. The experimental setup and recording of ion current with time with the flap open (1) and closed (2). T_0 = 1348 K, p_0 = 6·10³ Pa, d_{cr} = 0.4 mm.

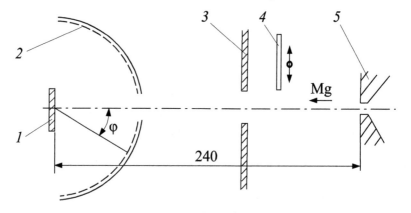

Fig. 4.22. Scheme of the experiment. 1 – target; 2 – mesh for the microscope; 3 – aperture; 4 – flap; 5 – vapour jet source.

target 1. In an arc around the target there were meshes with a carbon film for analysis on an electron microscope. In operation of the vapour source at p_0 = 266 Pa, T_0 = 1348 K with a seven-hour exposure and the use of polished copper as a target the mesh showed particles with the size of about 10 Å (Fig. 4.23). The surface of the target was not contaminated with magnesium dust. When the target was represented by the area of polished silicon no particles were detected and magnesium condensation took place on silicon.

Two fractions of ultrafine particles were produced in experiments with the vacuum deposition of zinc vapours [79]. The experimental setup is shown

Fig. 4.23. Detail of the electron micrograph of the mesh after 7 h of trapping particles. Magnification $1.5 \cdot 10^5$. Target – polished brass, $P_0 = 266$ Pa, $T_0 = 1348$ K, $\varphi = 34°$.

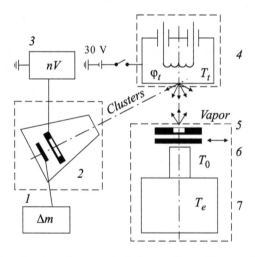

Fig. 4.24. Experiment setup. 1 – quartz balance; 2 – cell, revealing loose clusters; 3 – nanovoltmeter; 4 – glass–graphite target, emitting clusters; 5 – diaphragm; 6 – movable flap; 7 – evaporator with expanded graphite nozzle.

in Fig. 4.24. Zinc vapours from a closed heated graphite crucible 7 with an independently heated nozzle travelled to the polished surface of the target 4, made of glassy graphite. The temperature of the target was varied in the experiments using the built-in heater. A constant electric potential of +30 V was applied to the target. The vapour flow, reaching the target, was interrupted using the movable flap 6. To reduce the dispersion of vapours in the chamber the diaphragm 5 was installed in the vapour flow. On the side of the vapour source there was a cell for recording the clusters received from the target. The cell contains two probes: a ring collector for registration of the charge, which may be carried by the clusters from the target, and the sensor of the quartz resonance scales to measure small increments of weight. The cell was directed onto the target so that its sensors could received particles only from the target. The signal, generated by the particles travelling to the ring sensor was applied to the selective nanovoltmeter 3 through the frequency filter. The signal from the particles was recorded simultaneously with the change of mass in the sensor of the quartz balance and the appearance of the peak signals at the nanovoltmeter. The most reliable reproducible data were obtained with an electric potential on the target and with limiting the sensitivity of the nanovoltmeter by the frequency bandwidth.

Experiments were performed in a vacuum chamber with a residual pressure of $5 \cdot 10^{-5}$ Pa at a constant temperature of the evaporator $T_e = 687$–691 K and the nozzle $T_0 = 737$–741 K and the variable target temperature T_t. This corresponds to a change in the supersaturation of the target relative to the incident vapour flow of zinc. Measurements of the signal from the ring collector were taken with the open and closed valves. Measurements at $T_t = 288$–303 K showed the absence of zinc vapour deposition. Here the nanovoltmeter does not record the signal visible at the noise level. The absence of condensation on the target indicates that there is a phenomenon of 'collapse' of condensation found in [80]. The presence of the low-temperature zone of 'collapse' of condensation is due to the pollution of the adsorption of vacancies by impurities [81]. Stable condensation of zinc on the target takes place only within the range $T_t = 303$–523 K. The 'collapse' phenomenon is discussed in Chapter 3.

The results of measurements of the signal in the ring collector with the potential a the target of 30 V and a frequency bandwidth of 70–90 Hz are shown in Fig. 4.25 a by the points depending on the target temperature. The measurements were performed with an open flap, at a constant vapour source mode and periodic thermal cleaning of the target. The distribution of

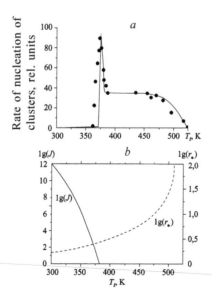

Fig. 4.25. The rate of nucleation of clusters (a), the nucleation rate J (m^{-2} s^{-1}) and the critical radius r_* (Å) of condensation nuclei (b) as a function of substrate temperature.

the experimental points shows both processes take part simultaneously in the process of particle nucleation at the target.

At T <525 K signals were recorded at the measuring instruments, reliably separated from noise, even in the case where the target has no electrical potential and the flap is closed. Pulsed signals, recorded in the nanovoltmeter, were accompanied by a weight gain recorded on the sensor of the quartz balance. The particles with a size of 0.1 μm, captured on the subject mesh in this temperature range, were analyzed with an electron microscope. The results of calculating the frequency of nucleation and the radius of critical zinc nuclei of the condensate on a graphite substrate with a wetting angle of $\pi/2$ under the experimental conditions are illustrated in Fig. 4.25 b.

The experimental data were interpreted using the model of nucleation of the clusters by the fluctuation mechanism [76], discussed in Section 4.2. The model is based on comparison of the probabilities of formation of condensation nuclei on the target and spontaneous rupture of their bonds with the target surface due to thermal vibrations. The model takes into account the dependence of the formation of solid or liquid crystallization nuclei with a spherical surface on the conditions. Formally, this corresponds to the two mechanisms of heterogeneous condensation: vapour–crystal, vapour–liquid. For clarity, we repeat briefly the main points of the model.

If the solid substrate is in contact with the vapour medium under isothermal conditions, its surface may show the formation of agglomerates from adsorbed atoms. Because of the thermal vibrations the agglomerates have an opportunity to break away from the surface. The probability of these processes will be determined by the Boltzmann formula for stochastic processes: $F' = \exp(-\delta\Phi'(N)/kT)$ and $F^- = \exp(-\delta\Phi^-(N)/kT)$ in accordance with the overcoming of the energy–efficient barrier $\delta\Phi'(N)$, associated with the transition of the N atoms from the adsorbed phase to the condensed state, and the energy-efficient barrier $\delta\Phi^-(N)$, for breaking up the bond of the agglomerate with to the substrate. The calculation of the equality of these probabilities in the heterogeneous classic nuclei in the theory of crystals gives an equation relating the number of atoms in the nitrogen nucleus, the binding energy of adatoms with the substrate, the binding energy of two neighbouring atoms in the agglomerate and the coefficient of thermal accommodation α_c. The results of the numerical solution of this equation for the condensation of magnesium on substrates with different binding energies for the magnesium atoms are shown in Fig. 4.18.

The fluctuation model of formation of free surface of clusters helps to explain some experimental results. Figure 4.25 shows that the appearance of a strong signal from the microparticle falling on the sensor of the measuring devices roughly coincides with an increase in nucleation frequency to a value of about 100 $m^{-2}\cdot s^{-1}$. Taking into account that the measurements were made with a bandwidth of 70–90 Hz, we can conclude that the sharp rise of the pilot signal takes place in the range T_r, where the frequency of microparticles emitted by the target coincides with the estimates of particulate nucleation frequency at the target in the conditions of the line of equal probability of nucleation and particle escape from the target. It can be considered established that the peak in the experimental data in Fig. 4.25 a at about 375 K is due to the generation of the cluster by the fluctuation mechanism.

The behaviour of the signal at a temperature greater than at the sharp rise when nucleation is virtually absent, indicates that its origin is not directly related to the nucleation in the classical view. In the range T_t = 370–520 K, where the frequency of formation of nuclei of critical size is $J < 1$ m^{-2}·s^{-1}, nucleation of Zn nuclei from the vapour may occur due to capillary effects. Zinc condensation nuclei with the surface curvature less than that of the critical nuclei may occur in the microscopic shells and closed pores of the graphite target. Such nuclei are able to grow due to vapour condensation on the surface under conditions where the saturation on the target surface is small or absent. The actual curvature of the surface of such condensate nuclei must correspond with the local saturation determined by their temperature and characteristics of the vapour flow on the target. With the growth of these nuclei a heterogeneity of the degree of supersaturation on their surface and, consequently, the anisotropy of their growth may appear. They may depart from the substrate by the evaporation of their side not facing the vapour source. This re-evaporation generation mechanism of free clusters by the condensation surface is discussed in detail in section 5.4 and in [79].

Since the surface concentration of such condensate nuclei is proportional to the concentration of capillary defects in the substrate, i.e. is a constant for this substrate, is proportional to the rate of growth of condensate nuclei and their lifetime on the substrate is proportional to the transverse dimension, divided by the rate of evaporation, the flow of microparticles emitted in this region of T_t, is given by $J_x = J_v (J_f - J_v) / R$. Here J_f and J_v are the incident and evaporating vapour flows. Assuming that each microparticle transfers from the substrate to the sensor of the measuring device a charge proportional to its capacity, i.e. size, to describe the signal on the measuring device, we have the formula $I \approx J_v (J_f - J_v)$. The general form for the description of the signal over the entire range of temperature is $I = JJ_v + \eta J_v (J_f - J_v)$, where the proportionality factor can be found by comparison with experiment. The result of evaluating I at the limited frequency of 100 Hz is shown in Fig. 4.25 a by the linear graph. There is a convincing agreement between the calculations and experiment.

So, it was found that when the vapour hits the substrate at supersaturation providing spontaneous nucleation emission of near-critical clusters from the substrate by the fluctuation mechanism. At supersaturation below the critical value larger clusters are emitted by the re-evaporation mechanism.

The generation of free microparticles by the two described mechanisms may be a significant component in the formation of smoke. In addition to the homogeneous formation mechanism of smoke particles the effect of the fluctuation and re-evaporation mechanisms provides an additional source of condensation nuclei in smoke streams. These results are important in vacuum technology. They explain why the vacuum systems, working with the processes of evaporation, are subject to intense dusting.

4.5. Effect of non-condensable impurities on homogeneous nucleation

In section 4.3 it was shown that the beginning of homogeneous condensation of magnesium corresponds to calculations by the classical theory. At the same time, study [48] showed a strong influence of the presence of He and Ne on the particle size of the condensate of Ar, CO_2, N_2. In [71] the authors presented experimental results on the nucleation of water in the presence of helium in the steady-state conditions in which there was an increase of the critical supersaturation of water vapour with increasing helium concentration. The authors of [71], considering helium as an effective coolant, expected the opposite effect – reduction of the critical supersaturation. In this regard, there is a need to find out the reasons for this contradiction. This section attempts to explain the experimental facts by the effect of the processes of sorption-desorption of impurities in the homogeneous nucleation process.

Unique experimental data [71] on the kinetics of nucleation in a mixture of water vapour and carrier gas helium, obtained by the thermal diffusion chamber method, showed a linear dependence of the critical supersaturation corresponding to the frequency of nucleation of 10^6 $m^{-3} \cdot s^{-1}$, on the partial pressure of the carrier gas. The authors [71] observed the coincidence of the extrapolation of the experimental data for the nucleation of pure water vapour to calculations by the classical capillary nucleation theory. To reconcile the results of measurements of the nucleation rate with the calculations by of the classical theory in the presence of significant partial pressure of helium the authors of [71] had to introduce an amendment from four to nine orders of magnitude. The mechanism of the influence of an inert carrier gas on the kinetics of homogeneous nucleation has remained unclear. In this book we attempt to resolve the above discrepancy between theory and experiment by taking into account the adsorption of atoms of the carrier gas on the nuclei of the condensate of the supersaturated component.

The classical representation of the fluctuation mechanism of the formation of aggregates of atoms or molecules in a vapour environment is accepted [73]. Spontaneously arising aggregates exchange material with the vapour phase over the adsorbed phase on their surface. The aggregates which reached the critical size become nuclei of the condensed phase and gain the opportunity for growth. To account for the effect of the presence of impurities in the vapour phase on the formation of critical nuclei, a model of sorption–desorption of the two-component gas mixture on the surface of aggregates is considered with the following assumptions:

1. Adhesion of the adsorbed particles takes place only on the part of the surface of the aggregate free from the adatoms with a constant coefficient of adhesion. Thermal accommodation of particles on the surface of the aggregated is instantaneous.

2. The particle adsorbed on the surface of the aggregate can be desorbed by thermal fluctuations of oscillations of the surface atoms of the aggregate. The desorbed particle travels to the gaseous environment.

3. The particle adsorbed on the surface of the particle in the process of migration can be captured in the 'trap' to form additional bonds with the aggregate atoms, thus becoming part of the aggregate or absorbed into the aggregate. The density and the capacity of 'traps' for the adatoms are constant. The traps are located on the surface evenly with the characteristic spacing smaller than the mean free diffusion path of the adatoms on the surface of the aggregate.

4. An unstable aggregate of atoms or molecules which has reached the critical dimensions becomes the nuclei of the condensed phase in the case if, taking into account the Kelvin–Thomson effect, the vapour and adsorbed phases become supersaturated with respect to the aggregate.

5. The aggregates and nuclei of the condensed phase are isothermal with respect to the gaseous environment.

6. The presence of the adsorbate of a foreign component on the surface does not change its surface energy.

Under these assumptions, the dynamics of the concentration two-component particles of the gas phase adsorbed on the surface of the aggregate or nucleus can be described by equations similar to those outlined in section 3.2:

$$\dot{\xi}_1(\tau) = \alpha_1 P_1 \left(1 - \xi_1/n_1 - \xi_2/n_2\right) - \xi_1/\tau_1 + \eta_1(\xi_{h1} - \xi_1),$$

$$\dot{\xi}_2(\tau) = \alpha_2 P_2 \left(1 - \xi_1/n_1 - \xi_2/n_2\right) - \xi_2/\tau_2 + \eta_2(\xi_{h2} - \xi_2),$$

(4.33)

where α_i is the coefficient of adhesion of particles of i-th component of the gas phase on the surface of the nucleus; $P_i = p_i(2\pi m_i kT)$ is the flux of particles, p_i is partial pressure, m_i is the mass of the particles, k is the Boltzmann constant, T is temperature; n_i is the density of adsorption vacancies on the surface of the nucleus; $\tau = v_\perp \exp(\varepsilon_i/kT)$ s is the re-evaporation time, ε_i is the binding energy of particles in the i-th component adsorbed on the nucleus; $v_\perp$ is the frequency of normal vibrations of adsorbed particles; $\eta_i = \zeta v_\parallel \exp(\varepsilon_{Di}/kT)$ is the probability of an arbitrary particle of the adsorbate flowing to the centres of condensation or 'traps' on the surface of the nucleus as a result of surface migration, ζ is the fraction of the adsorption vacancies, bordering with condensation centres or 'traps' [73], $v_\parallel$ is the frequency of longitudinal vibrations of adatoms ($v_\perp \sim v_\parallel \sim 10^{12} \div 10^{13}$ s^{-1} [73]), ε_{Di} is the activation energy of surface diffusion of adsorbed particles of i-th component ($\varepsilon_{Di} \sim 0.3\ \varepsilon_i$ [73]). The value $\eta_i \xi_i$ denotes the flux of particles of the adsorbate to the condensation centres of the aggregate or the nucleus of the condensed phase. The reverse flow of particles of i-th component from the centres of condensation or 'traps' into the adsorbed phase is conveniently represented as $\eta_i \xi_{hi}$. The value ξ_{hi} corresponds to the surface concentration of particles of i-th component in the nucleus of the condensed phase of i-th component which is in stable equilibrium with the gaseous medium, when there is also an equilibrium between the adsorbate and the condensate. From (4.33) with $\xi_i(\tau) = 0$, $\xi_i = \xi_{hi}$ we have $\xi_{hi} = \alpha_i P_{hi} / (\alpha_i P_{hi} / n_i + 1/\tau_i)$, where $P_{hi} =$

$p_{\infty i}$ exp $(2\gamma_i\omega_i / kTr_i)$ $(2\pi m_i kT)^{-1/2}$, $p_{\infty i}$ is the saturation pressure; γ_i – surface energy; ω_i is the molecular volume for the i-th component; r_i is the radius of the nucleus. A feature of equations (4.33) in contrast to the commonly used kinetic equations of adsorption of one component [82] is a description of the exchange of the adsorbate with the substance of the surface as the last terms linear with respect to x and time-independent. In isothermal conditions, such a description is justified to account for the exchange processes with condensation centres and 'traps', located on the surface.

Making equations (4.33) dimensionless with $\Theta_i = \xi_i/n_i$, the mean lifetime of adsorbed particles $\tau_0 = (\tau_1 + \tau_2)/2$, $t = \tau/\tau_0$ and the introduction of $a_i = \tau_0/\tau_i$, $b_i = \alpha_i P_i \tau_0/n_i$, changes the equations (4.33) to a system of inhomogeneous linear differential equations:

$$\dot{\Theta}_1(t) = b_1 - (a_1 + b_1 + c_1)\Theta_1 - b_1\Theta_2 + c\Theta_{h1},$$

$$\dot{\Theta}_2(t) = b_2 - (a_2 + b_2 + c_2)\Theta_2 - b_2\Theta_1 + c\Theta_{h2}. \tag{4.34}$$

Here $\Theta_{hi} = b_{hi}/(b_{hi} + a_i)$, $b_{hi} = a_i Ph_i\tau_0/n_i$.

The roots of the characteristic equation of (4.34) are given by

$$\beta^{\pm} = (a_1 + b_1 + c_1 + a_2 + b_2 + c_2)/2 \pm [(a_1 + b_1 + c_1 - a_2 - b_2 - c_2)^2/4 + b_1 b_2]^{1/2}.$$

The solution of (4.34) has the form

$$\Theta_1(t) = G_1 \exp(-\beta^+ t) + G_2 \exp(-\beta^- t) + \Theta_1^S,$$

$$\Theta_2(t) = G_1\left[(\beta^+ - a_1 - b_1 - c_1)/b_1\right]\exp(-\beta^+ t) + G_2\left[(\beta^- - a_1 - b_1 - c_1)/b_1\right]\exp(-\beta^- t) + \Theta_2^S,$$

where

$$\Theta_1^S = \frac{b_1[a_2 + c_2(1 - \Theta_{h2})] + (a_2 + b_2 + c_2)c_1\Theta_{h1}}{b_1(a_1 + c_1) + (a_1 + c_1)(a_2 + b_2 + c_2)},$$

$$\Theta_2^S = \frac{b_2[a_1 + c_1(1 - \Theta_{h1})] + (a_1 + b_1 + c_1)c_2\Theta_{h2}}{b_1(a_1 + c_1) + (a_1 + c_1)(a_2 + b_2 + c_2)} \tag{4.35}$$

are the steady-state values of the degree of surface coverage by the adsorbed particles. Integration constants G_1, G_2 can be determined using the initial conditions $t = 0$, $\Theta_1 = \Theta_{01}$, $\Theta_2 = \Theta_{02}$:

$$G_1 = \left[(\beta^- - a_1 - b_1 - c_1)(\Theta_1^S - \Theta_{01}) - b_1(\Theta_2^S - \Theta_{02})\right]/(\beta^+ - \beta^-),$$

$$G_2 = -\left[(\beta^+ - a_1 - b_1 - c_1)(\Theta_1^S - \Theta_{01}) - b_1(\Theta_2^S - \Theta_{02})\right]/(\beta^+ - \beta^-).$$

In Fig. 4.26 shows the conventional phase portrait of system (4.34). The trajectories the solutions 2 reflect the dynamics of the relative concentrations of the adsorbate on the surface of the condensed phase instantly formed in the gas space. The different trajectories correspond to different states of the adsorbate at the formation of the surface. During the characteristic time $1/\beta^{\pm}$

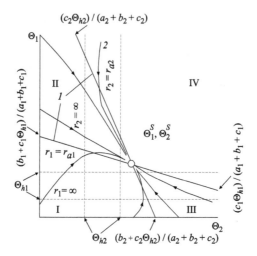

Fig. 4.26. The phase portrait of system (4.34). 1 – isoclines; 2 – trajectories of the solutions. Area I – the absence of condensation, II–III – nucleation of one component, IV – the area of joint nucleation.

the concentrations Θ_1^S, Θ_2^S are established on the surface of the nucleus. In the phase portrait the dashed lines mark the values of the surface concentrations of the adsorbate of the components on the surface of nuclei that are stable with respect to the gas phase. Indicated also are the extreme cases, corresponding to an infinitely large radius of the nucleus r_∞ and the minimum equal to the radius of an atom or molecule $r_i = r_{ai}$. Each intermediate value of r_i corresponds to a different line lying between the values on the chart.

The phase portrait of Fig. 4.26 allows a qualitative analysis of the conditions of nucleation. If a stable node with the (Θ_1^S, Θ_2^S) coordinates is in region I, stable are only nuclei with negative curvature nucleation to form a convex shape of nuclei is impossible. If the (Θ_1^S, Θ_2^S) node is in the region IV, the nuclei are stable at any size.

The areas II and III correspond to the nucleation of only the 1st and 2nd components. The areas enclosed between the dashed lines correspond to the joint nucleation with restrictions on the radius of stable nuclei of one of the components.

Such an analysis of the phase portrait can make an important qualitative conclusion: the presence in the gas phase of a condensing foreign component reduces the relative concentration of adsorbed homogeneous (in relation to the nucleus) particles on its surface. To ensure the stability of the nucleus of a given radius at a fixed partial pressure of the foreign component it is necessary to increase in the partial pressure of the component.

The characteristic times of establishment of the adsorbate concentration are determined by the characteristic time of re-evaporation of the adsorbed particles $\tau_i \sim \exp(\varepsilon_i/kT)$. The characteristic lifetime of the agglomerates of particles produced by fluctuations in the gas phase is given in [73]: $\tau_{Ai} \sim \exp(nN_{Ai}\varepsilon_i / kT)$, where n is the number of the nearest neighbours in the structure of the agglomerate, N_{Ai} is the number of particles in the agglomerate. Comparison of τ_i and τ_{Ai} shows that these times are equal only when $N_{Ai} = 2$. At $N_{Ai} > 2$ it is always $\tau_{Ai} > \tau_i$. This fact indicates that the formation of agglomerates in the gas phase, which reached the critical size and can become nuclei of the condensed phase, their surface has almost always a steady-state concentration of the adsorbate, defined by equation (4.35). These relations can be used to assess the impact of the presence of a foreign component on the stability of the nucleus. Putting in (4.35) $\Theta_1^s = \Theta_{h1}$ and $\Theta_2^s = \Theta_{h2}$ we can obtain the conditions for the stability of nuclei of the 1st or 2nd components:

$$b_1 = b_{h1}\left(1 + \frac{b_2 + c_2\Theta_{h2}/(1-\Theta_{h1})}{a_2 + c_2(1-\Theta_{h1}-\Theta_{h2})(1-\Theta_{h1})}\right),$$

$$b_2 = b_{h2}\left(1 + \frac{b_1 + c_1\Theta_{h1}/(1-\Theta_{h2})}{a_2 + \tilde{n}_2(1-\Theta_{h1}-\Theta_{h2})(1-\Theta_{h1})}\right). \tag{4.36}$$

The expressions (4.36) can be used for the analysis of experimental results [71] on the nucleation of vapours of tap water (component 1) in a helium atmosphere (component 2). Since helium cannot condense in the temperature range 100–400 K and its solubility in water at a temperature of 300 K and a pressure of 0.1 Pa is $7 \cdot 10^{-6}$ molar fractions, it is natural to set $c_2 = 0$. From the first of relations (4.36), after division by $b_{\infty 1} = \alpha_1 P_{\infty 1}\tau_0/n_1$, where $P_{\infty 1} = p_\infty(2\pi m_1 kT)^{-1/2}$, and substituting b_2, a_2 we have

$$s_1 = s_{h1}\left(1 + \frac{\alpha_2 p_2 \tau_2}{n_2\left(2\pi m_2 kT\right)^{1/2}}\right). \tag{4.37}$$

Here $s_{h1} = b_{h1}/b_{\infty 1}$ is the supersaturation required for the stability of water nuclei in the pure water vapour, calculated according to classical nucleation theory [71]; $s_1 = b_1/b_{\infty 1}$ is the corresponding supersaturation required to ensure the sustainability of the nuclei of the same size in the presence of helium. Note that the linearity of the type (4.37) was found in [71]. Since adsorption occurs at a polar liquid the surface molecules of which in the process of exchange with the surrounding vapour are in a continuous restructuring of their orientations, the coefficient of adhesion of helium should be proportional to the average lifetime of the surface molecules of water in the oriented state and have the temperature dependence of the type $\alpha_2 = \beta_2 \exp(\varepsilon_1^{ads}/kT)$ where β_2 is the coefficient of proportionality; ε_1^{ads} is the energy barrier of restructuring of the orientation, commensurate in value with ε_1. With this in mind and taking into account the temperature dependence of τ_2 from (4.37) we obtain

$$s_1 = s_{h1}\left(1 + \frac{\beta_2 p_2 \exp\left[\left(\varepsilon_1^{ads} + \varepsilon_2\right)/kT\right]}{n_2 v_\perp\left(2\pi m_2 kT\right)^{1/2}}\right). \tag{4.38}$$

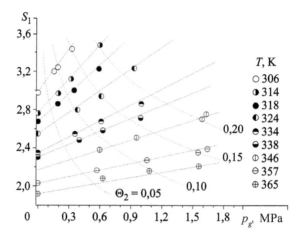

Fig. 4.27. The critical supersaturation for the nucleation rate of 10^6 m^{-3} s^{-1} of ordinary water based on the partial pressure of helium carrier gas at different temperatures. Points – experiment [71], solid lines – calculation by formula (4.38), dashed – lines of equal occupation of adsorption vacancies.

Figure 4.27 shows the experimental data [71] compared with the calculation by the formula

$$s_1 = s_{h1}\left[1+2,605\cdot10^{-10}\,p_2 T^{-1/2}\exp(3229/T)\right].\qquad(4.39)$$

The experimental data [71], obtained in the temperature range 306–365 K and pressures of helium of 0.16–1.6 MPa, show a maximum deviation from the calculation of ~2.5%. Figure 4.27 shows dashed lines, corresponding to a constant value of the occupation of the surface of water nuclei by the helium adsorbate using the formula $\Theta_2^S = b_2/(a_2 + b_2)$ from (4.35) when $b_1 \ll b_2$ and $\Theta_{h1} \ll 1$. Estimates $\Theta_{h1} \approx 1,5\cdot10^{-3}$ at $T = 360$ K show that the experimental data refer to the conditions when 5 to 25% of the nuclei are occupied by the adsorbate of a foreign non-condensing gas.

Figure 4.28 shows the results of generalization of experimental data [71] in the coordinates s_1/s_{h1}, Θ_2^S in comparison with the theoretical curve. The maximum deviation of points from the calculated line is about 2.5%, which can be considered quite satisfactory, taking into account the fact that in the calculations we have neglected the temperature dependence of the density of adsorption vacancies and the binding energy of the adsorbate with nuclei. Since $\varepsilon_1^{ads} \gg \varepsilon_2$ it is possible to normalize the value of β_2 in formula (4.38) in terms of $\alpha_2 = 1$ at $T = 273$ K. The obtained value of $\beta_2 = 7.3\cdot10^{-6}$ and comparison of (4.38) and (4.39) with $v_\perp = 10^{13}$ [72] gives $n_2 = 3.68\cdot10^5$ m^2 which is reasonable [82]. Comparing the exponents in these formulas gives the value of $\varepsilon_1^{ads} + \varepsilon_2 = 0.278$ eV, which is also consistent in the order of magnitude with the energy of the hydrogen bond.

Thus, consideration of sorption processes of foreign gases on the surface of the nuclei of the condensed phase allows us to describe the critical

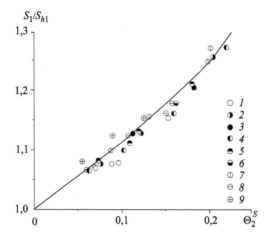

Fig. 4.28. The correction for the critical supersaturation of water in relation to the occupation of vacancies of nuclei by helium atoms. Solid line – calculations, points – experiment [71], 1–9 correspond to $T = 306, 314, 318, 324, 334, 338, 346, 357, 365$ K.

supersaturation of homogeneous nucleation by the classical capillary theory. A few comments should be made.

1. For a quantitative agreement of experimental data on the frequency of nucleation it is not necessary to introduce a factor A [71] which ranges from four to nine orders of magnitude. It is enough in (4.38) to correct the value of the supersaturation by the surface coverage of the nucleus by a foreign component. Physically, this means that during the formation of subcritical agglomerates of particles the vapour by thermal fluctuations the presence of a foreign gas reduces their growth rate due to occupation of parts of the bonds and reduction of the efficiency of collisions of particles with agglomerates.

2. At high degrees of occupation of the surface of subcritical agglomerates by a foreign component, especially the component with the binding energy very different from the binding energy of the condensing component, it is expected that the presence of the adsorbate on the agglomerates change their surface energy. In this case, it is not enough to consider the occupation of the adsorption vacancies in the adsorption process for the analysis of experiments. The magnitude of the energy barrier of critical nucleus formation in the calculation of the nucleation rate must be corrected for the defect of the surface energy associated with adsorption. This allows us to predict the need for further development of the classical capillary nucleation theory.

3. In the case of substantial solubility of the foreign gas component in the nuclei of the first component the assumption $c_2 = 0$, used to derive (4.37) and (4.38), is unjustified. The calculation of the real critical supersaturation at the known solubility, assuming equilibrium critical supersaturation, can be carried out using the expression $c_2 = \zeta \tau_0 v_\perp \exp(\varepsilon_{D2}/kT)$. In the absence of quantitative data on the solubility the critical supersaturation can be calculated from the expressions (4.36) as a system of equations for b_1 and Θ_{h2}.

4. An alternative model may be described the model of the processes of sorption–desorption disregarding the occupation of the adsorption vacancies by the particles homogeneous with respect to the nucleus. Formula (4.38) is also used in this case with the same assumptions. The adequacy of either model in respect of reality can be resolved by new experiments.

Conclusions

1. The concept of the need to combine the search processes of producing disperse systems by homogeneous condensation with finding ways to use them has been proposed. The need for such a combination is associated with high chemical and physical activity of the produced ultrafine particles.

2. The methodology of formulation of experimental studies of homogeneous condensation processes as a complex of studies using indirect methods not perturbing the flow, and methods of direct diagnosis of clustered jets has been developed.

3. The region of the beginning of condensation in the jet expansion of magnesium vapours, initiated by the presence of condensation nuclei of the inhomogeneous origin has been determined.

4. The region of the start of spontaneous homogeneous condensation in the jet expansion of magnesium vapour has been found. The region of the beginning of condensation can be predicted by calculations using the classical nucleation theory.

5. A theoretical model has been constructed to estimate the conditions for the appearance in the vapour flow of nucleation centres formed on solid surfaces by oxidation or heterogeneous nucleation.

6. The model of sorption–desorption of foreign matter on the nuclei of the condensed phase of homogeneous origin has been used to explain the experimental facts of the effect of foreign non-condensable impurities on the critical supersaturation of water vapour and the absence of such an effect in the condensation of metal vapours.

List of Symbols

a - lattice constant, m;

c - the speed of sound, m / s;

Wed - specific heat at constant pressure, J / (kg·K);

d - diameter, m;

E - energy, eV, J;

F - probability of the event, s^{-1};

G - flow rate, kg / c; the constant of integration;

I - enthalpy flux, W/m^2; current, A;

J - the frequency of nucleation, m^{-2}·s^{-1};

Kn - Knudsen number;

K_1, ..., K_{22} - constant;

k - Boltzmann constant, J / K;

L - heat of vapourization, Dzh/k^1;

M - Mach number;

m - mass, kg, the relative size of the cluster;

N - number of atoms in the cluster;

n - density of gas particles, m^{-3} density of particles on the surface, m^{-2}, the relative size of the cluster;

$n_{1,2}$ - adsorption density of vacancies, m^{-2};

P - flux of particles, m^{-2}·s^{-1};

p - pressure, Pa;

r - radius, m;

S - speed ratio, the supersaturation;

T - temperature, K;

t - dimensionless time;

U - potential, V;

u - velocity, m / s;

x - coordinate, m;

z_i - the degree of ionization;

α - the accommodation coefficient, the coefficient of adhesion;

α_c - fraction of the condensate;

γ - surface energy, J/m^2;

δ - dimensionless coordinate;

ε - the binding energy, J;

η - the probability of flow of adatoms at the center of condensation;

Θ - the degree of surface coverage by the adsorbate;

λ - the mean free path, m;

ξ - the density of adatoms on the surface, m^{-2};

ρ - density, kg/m^3;

σ - the collision cross section, m^2;

τ - the lifetime, s;

χ - ratio of specific heats;

ω - the amount of molecules in condensed phase, m^3; frequency, s^{-1}.

Indices

0 - braking;
1 - next-door neighbors;
∞ - smooth surface;
* - Critical;
+ - Stay;
- - Decrease;
cr - critical section;
and - the atom;
a - cluster;
e - the electron;
g - gas;
h - saturation;
i, 1, 2 - components;
m - the maximum;
s - surface;
z - the delay.

Literature

1. Fedorchenko N.M., Andrievskii R.A., Fundamentals of powder metallurgy, Kiev, Izd-vo AN USSR, 1963.
2. Hill M. et al., Teploperedacha, 1963, V. 85, No. 4, 17–34.
3. Helmholtz R.V., Ann. Phys. Chem., 1887, V. 32, 1–19.
4. Wilson C.T.R., Proc. Roy. Soc. London, 1897, V. 61, 240–243.
5. Hirn and Casin, Comptes Rendus de l'Acad. France, 1866, 63–66.
6. Stodola A., Steam and gas turbines, Mc-Graw-Hill Book Company, Inc., 1927, 117–128, 1034–1073.
7. Yellott J.I., Supersaturated steam, Trans. ASME, 1934, V. 56, 411–430.
8. Yellott J.I., et al., Trans. ASME, 1937, V. 59, 171–183.
9. Rettaliata J.T.. Trans. ASME, 1936, V. 58, 599.
10. Binnie A.M., Wood M.W., Proc. Inst. Mech. Engrs. London, 1938, No. 138, 229.
11. Binnie A.M., Green J.R., Proc. Roy. Soc. London, Bd. A 181, 1942, V. 181, 134/54.
12. Wegener P.P., Mack L.M., Condensation in supersonic and hypersonic wind tunnels, eds. H. Dryden, T. Karman, N. Y., Academic Press, 1958, V. 5, 307.
13. Stepchov A.A., J. Amer. Rocket Soc. Supplement, 1960, July, 645–699.
14. Keenan J., Thermodynamics (textbook), N.Y., John Wiley & Sons, 1941, 290.
15. Oswatitsch K., Z. Angew. Math. Mech., 1942, V. 22, 1–13.
16. Volmer M. Kinetik des Phasenbildung. Dresden, Leipzig, Th. Steikopf-Verlag, 1939.
17. Becker R., Daring W., Ann. Phys. (Leipzig), 1935, Bd. 24, 719–752.
18. Frenkel' Ya.L., Kinetic theory of liquids, Moscow, USSR Academy of Sciences, 1945.
19. Volmer M., Flood H., Z. Phys. Chem., Abteilung A, 1934, B. 170, H. 3 and 4, 273–285.
20. Deich M.E., Filippov G.A., Gas dynamics of two-phase media, Moscow, Ener-

giya, 1968.
21. Gorbunov V.N., et al., Non-equilibrium condensation in high-speed gas flows, Moscow, Mashinostroenie, 1984.
22. Lushnikov A.A., Sutugin A.G., Usp. Khimii, 1976, V. XIV, No. 3, 385–415.
23. Allard E., Kassner J., J. Chem. Phys., 1965, V. 42, No. 4, 1401–1405.
24. Allen E., Kassner J., J. Coll. Interface Sci., 1969, V. 30, 81–93.
25. Stahorska D., Acta Phys. Polonika, 1956, V. 15, 5.
26. Courtney W.G., J. Chem. Phys., 1963, V. 38, 1448.
27. Kortzeborn R.N., Abraham F., J. Chem. Phys., 1973, V. 58, 1529–1534.
28. Allen J., Kassner J.J. Proc. Intern. Conf. Weather Modification, Melbourne, 1973, Oxford, Oxford Univ. Press, 1974, 228.
29. Courtney W.G., J. Chem. Phys., 1962, V. 36, No. 8, 2018.
30. Lund L., Rivers J., J. Chem. Phys., 1966, V. 45, 4613.
31. Clark D., Noxon J., Science, 1971, V. 174, Issue 4012, 941–944.
32. Reist H., Heist G., J. Chem. Phys., 1973, V. 59, 665–671.
33. Baktar F., Campbell B., Proc. Inst. Mech. Engrs., 1970, V. 185, 345.
34. Kurshakov A.A., et al., PMTF, 1971, No. 5, 177.
35. Crane R., Moore H., Newton R., Proc. Inst. Mach. Engineering, 1970, V. 185, 345.
36. Barachdorf D., Proc. 3rd Int. Conf. Rain Erosion and Associated Phenomena Farnborough, 1970, Univ. Essex Press, 1971, 156.
37. Gyarmathy G., VDI Forschungsheft., 1, VDI-Verl. Düsseldorf, 1965, V. 1, 1.
38. Deich M.E., et al., Izv. AN SSSR, Ser. Energetika i transport, 1972, No. 17, 214.
39. Barschdorf D.V., Filippov, G.A, Izv. AN SSSR, Ser. Energetika i transport, 1970, No. 43, 252.
40. Merritt, S., Wezerstone, RTC, 1967, No. 5, 190.
41. Sutugin A., Fuchs N., PMTF, 1968, No. 3, 567.
42. Sutugin A., et al., J. Aerosol Sci., 1972, V. 2, 371.
43. Sutugin A., Fuchs, N., Interfacial liquid-gas boundary, Leningrad Univ. Press, 1972..
44. Fuchs N.A., Sutugin A.G., Fine sprays, Itogi nauki, Moscow, VINITI, 1969, P. 17.
45. Milne T.A., et al., J. Chem. Phys., 1970, V. 52, No. 3, 1552.
46. Langsdorf A., Rev. Sci. Instr., V. 10, No. 3, 91–103.
47. Hagena O., Obert W.J., J. Chem. Phys., 1972, V. 56, 1793–1802.
48. Hagena O., J. Surface Sci., 1981, V. 106, 101–116.
49. Becker E., et al., Rarefied Gas Dynamics, 6th Symp. N.Y., London, Acad. Press, 1969, P. 1449.
50. Kearton W.J., Proc. Inst. Mech. Eng., 1929, V. 11, 993.
51. Kearton W.J. in: Congr. Int. de Mecanique Generale, 1930.
52. Recknagel E., J. Phys. Chem., 1984, V. 88, 201–206.
53. Kuiper A.E.T., et al., J. Cryst. Growth, 1978, V. 45, 332–333.
54. Takagi T., et al., in: Proc. Int. Conf. on Low Energy Ion Beams, Salford, 1977 (2nd edn.), Inst. Phys. Cont. Ser., 1978, No. 38, Ch. 4. 142.
55. Takagi T., et al., ibid, 1978. No. 38, Ch. 5, 229–235.
56. Takagi T., et al., in: Proc. 7th. Int. Vac. Congr. and 3rd Int. Conf. Solid Surfaces, ed. by R. Dobrozemsky, F. Rüdenauer, F. P. Viehböck, and A. Breth, F. Berger and Söhne, Vienna (1977), 1603–1606.
57. Takagi T., et al., J. Vac. Sci. Technol., 1975, V. 12, No. 6, 1126–1134.
58. Theeten J.B., et al., J. Crystal Growth, 1977, V. 37, 317–328.

59. Takagi T., et al., J. Surface Sci., 1981, V. 106, 544–550.
60. Hoareau A., et al., Surface Sci., 1981, V. 106, No. 1–3, 195–203.
61. Millet J.L., Borel J.P., Surface Sci., 1981, V. 106, No. 1–3, 403–407.
62. Riley S.J., et al., J. Phys. Chem., 1982, V. 86, No. 20, 3911–3913.
63. Mühlbach J., et al., J. Surface Sci., 1981, V. 106, 18–26.
64. Yokozeki A., Stein G.D., J. Appl. Phys., 1978, V. 49 (4), 2224–2232.
65. Yokozeki A., J. Chem. Phys., 1978, V. 68 (8), 3766–3773.
66. Sattler K., et al., Phys. Rev. Lett., 1980, V. 45, No. 10, 821–824.
67. Saito Y., et al., in: Proc. of the 14th Intern. Symp. on Rarefied Gas Dynamics, July 16-20, 1984, Tsukuba Science City, Japan, V. 11, 839–846.
68. Roseberry F., Handbook of vacuum engineering and technology, Moscow, Energiya, 1972.
69. Biberman L.M., et al., Teplofiz. Vysokikh Temperatur, 1985, V. 23, No. 3.
70. Gaisky N.V., et al., Progr. in Astronautics and Aeronautics, 1976, No. 51, part 2.
71. Chukanov V.N., Kuligin A.P., Teplofiz. Vysokikh Temperatur, 1987, V. 25, No. 1, 70–77.
72. Bochkarev A.A., et al., Proceedings: Boiling and Condensation (Hydrodynamics and heat transfer), USSR Academy of Sciences, Institute of Thermophysics. Novosibirsk, 1986, 102–110.
73. Chernov A.A., ET AL., Modern crystallography, Moscow, Nauka, 1980, V. 3.
74. Knauer W., J. Appl. Phys., 1987, V. 62, No. 3, 841–851.
75. Hawley J.H., Ficalora P.J., J. Appl. Phys., 1988, V. 63, No. 8, 2884–2885.
76. Bochkarev A.A., Sib. Fiz.-Tekh. Zh., 1993, No. 2, 7–12.
77. Tanaka H., Kanayama T.J., Vacuum Sci. Technol., B, 1997, 15 (5), 1613–1617.
78. Bochkarev A.A., Pukhovoi M.V., PMTF, 1994, No. 3, 102–111.
79. Bochkarev A.A., et al., Vacuum, 1999, V. 53, 335–338.
80. Palatnik L.S., Komnik Yu.F., Dokl. AN SSSR, 1959, V. 124, No. 4, 808–811.
81. Bochkarev A.A., et al., in: Thermophysics of crystallization and high-temperature processing of materials, Novosibirsk, Publishing House, ITP SR RAS, 1990, 98–117.
82. Geguzin Ya.E., Kaganovskii Yu.S., Diffusion processes on the surface of the crystal, Moscow, Energoatomizdat, 1984.

5

'Genetics' and the evolution of the structure of vacuum condensates

It is known that the variety of structures of vacuum condensates is determined by many factors: the geometrical and crystallographic characteristics of the substrate [1], the kinematic and dynamic parameters of the vapour flow [2, 3], the presence of impurities in the vapour [4], substrate temperature, external energetic influences such as electromagnetic radiation. The impact of each of these factors has been studied and theories have been proposed. The influence of the crystallographic parameters of the substrate has been more thoroughly investigated during the growth of epitaxial films. They determine the size and orientation of the textures of the condensates. The direction of the vapour flow gives the orientation of the structures and the velocity distribution function of the atoms of the vapour determines the shape of the growing crystallites. With the development of individual crystallites we observed self-similar regimes that promote the growth of a large class of geometrically similar forms of the crystallites.

The parameters of the vapour flow are involved in competition in the growth of crystallites based on their mutual shading. The presence of impurities in the vapour flow results in a complex phenomenon of 'collapse' of condensation [5–7] determined by the occupation of the condensation surface by the adsorption vacancies. Another manifestation of the effect of impurities is their capillary condensation in areas of high negative curvature, which changes the nucleation process and contributes to the formation of columnar structures. The substrate temperature has a strong effect on the sorption processes, the state of the condensation surface, surface diffusion, and this all leads to changes of the condensation mode and the formed structures. Bombardment of the condensation surface by high-energy particles or irradiation with the electromagnetic field causes a change in the surface condition, activates a number of processes associated with overcoming the energy barriers, and changes the condensation mode.

The combination of these factors in real processes of condensation creates an extremely complicated physical picture which cannot be described by a single theory. The presence of such a theory would make it possible to find a connection between the conditions of the condensation process and the produced structure that is important in terms of technology. Research in this direction is also the aim of this chapter. The general concept is to collect the existing theoretical models in a single chain, the major links of which are sorption processes prior to nucleation, nucleation and evolution of nuclei of the formed crystallites.

5.1. Homologous matrix of structures at the initial stage of condensation

The fact that the above model of sorption kinetics gives results adequate to the experimental data for the study of processes related to nucleation, is the basis of using it for the analysis of conditions and the product of nucleation [8]. We note three principal provisions to be considered when creating an algorithm for the analysis of nucleation.

First, it is the strong effect of the multicomponent nature of the vapour mixture. Above it was shown that the capillary condensation of impurities radically changes the sorption pattern on the substrate and can cause a chain nucleation of condensation centres. Second, the surface migration as a mechanism for effective mass transfer plays an important role in nucleation. Therefore, the supersaturation determining the nucleation should be calculated from the concentration of components in the adsorbed layer. Third, for the uniqueness of a simple analysis of nucleation we should accept the condition of local 'round' vapour–adsorbate–condensate–vapour equilibrium. This condition allows us to prove an analogue of the Kelvin equation for the adsorbed phase, taking into account the local curvature.

We show an example of a simple two-component mixture of vapours of substances the condensed states of which can not form a homogeneous solution, and the construction of charts of possible structures in the initial stage of condensation. The main types of geometric surfaces, forming the shape of the nuclei of separate phases of the condensate are a plane, a cylinder, a sphere. Surfaces such as a cone, paraboloid, ellipsoid, have the local curvature and thus complicate the analysis, but do not make a fundamental difference in the course of further discussion.

Neglecting the anisotropy of the surface properties of the nuclei of individual phases, the stability of the existence of planar, cylindrical and spherical shapes can be traced on the basis of the Kelvin effect from the condition of their equilibrium with the adsorbed phase:

$$\xi_{hK} = \xi_{h\infty} \exp\left(\sigma\omega_a K/kT\right), \tag{5.1}$$

where $\xi_{h\infty}$, ξ_{hK} are the saturated concentration of adatoms on a flat surface and on a surface with curvature K; σ is the specific surface energy at the vapour–condensate interface; k is the Boltzmann constant; T is the temperature of the substrate; ω_a is the atomic volume in the condensed phase. Dimensionless saturation isotherms of the adsorbed phase as a function of the radius of curvature of spherical and cylindrical surfaces are qualitatively shown in Fig. 5.1. The x-axis marks the characteristic radii of curvature $\pm\beta r_a$, equal to several radii of the atoms or molecules of the component. Physically meaningful are only parts of the isotherms $r > \beta r_a$ and $r < -\beta r_a$. The region of the curves at $r_a = -\beta r_a \div \beta r_a$ has no physical meaning.

Any horizontal line in Fig. 5.1, implying a certain saturation in the adsorbed phase on the substrate, shows that the substrate can contain stable condensation nuclei only with the forms that lie below the horizontal line. All nuclei with smaller radii of curvature, lying above the horizontal line, need more saturation for sustainable existence. This circumstance makes it possible to build for a two-component mixture in the phase space (Θ_1, Θ_2) a chart of structures in the initial stage of condensation. Kinetics of sorption of the components by the substrate surface for the two-component mixture is described by two equations of type (3.8). Neglecting the chemical reactions it is a system of two linear homogeneous equations:

$$\dot{\Theta}_1(t) = b_1 - (a_1 + b_1 + c_1)\Theta_1 - b_1\Theta_2 + c_1\Theta_{h1},$$
$$\dot{\Theta}_2(t) = b_2 - (a_2 + b_2 + c_2)\Theta_2 - b_2\Theta_1 + c_2\Theta_{h2}.$$

(5.2)

Here $\Theta_{hi} = b_{hi}/(b_{hi} + a_i)$ is the partial occupation of the adsorption vacancies; $b_{hi} = a_i P_{hi}\tau_0/n_i$ is the dimensionless flux of the component atoms on a substrate from a saturated vapour phase. According to [9], the coefficients in (5.2) can be calculated as

$$a_i = \tau_0 v_\perp \exp\left(-\varepsilon_i/kT\right)$$

– the probability of thermal desorption;

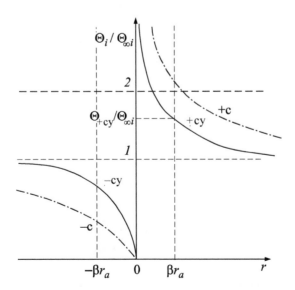

Fig. 5.1. Saturation curves for the crystallites with cylindrical and spherical surfaces. ±c – convex and concave hemispheres; ±cy - the same for a cylinder.

$$b_{hi} = \tau_0 \left(v_{\perp} n_a^2 (2\pi m_i kT)^{1/2} / \kappa_i \right) \exp\left(-3\varepsilon_i / kT\right);$$

$$c_i = 2\left(\tau_0 n_{4i} v_{\parallel} / \pi n_a^2\right)\exp\left(-u_{Di}/kT\right)$$

– The probability of flow of adatoms into the 'trap' of the substrate; $v_{\perp}$, $v_{\parallel}$ are the frequencies of thermal vibrations in the transverse and longitudinal directions in relation to the substrate; ε_i is the energy of one bond of the adatoms; κ_i are the effective coefficients of condensation of vapour atoms on the surface of the condensed phase; n_{4i} is the concentration of condensation vacancies. The vapour flow to the substrate $b_i = P_i \tau_0 / n_a$ is made dimensionless by the average time of re-evaporation of the adatoms $\tau_0 = (\tau_1 + \tau_2)/2$ and the density of adsorption vacancies n_a.

There is a common analytical solution of equations (5.2) as a sum of exponential terms and the constants in the form of coordinates of the stationary point:

$$\Theta_1^S = \frac{b_1[a_2 + c_2(1 - \Theta_{h2})] + (a_2 + b_2 + c_2)c_1\Theta_{h1}}{b_1(a_2 + c_2) + (a_1 + c_1)(a_2 + b_2 + c_2)},$$

$$\Theta_2^S = \frac{b_2[a_1 + c_1(1 - \Theta_{h1})] + (a_1 + b_1 + c_1)c_2\Theta_{h2}}{b_1(a_2 + c_2) + (a_1 + c_1)(a_2 + b_2 + c_2)}.$$

The phase portrait of system (5.2) is shown in Fig. 5.2. If the stationary (Θ_1^S, Θ_2^S) point is in region I, only nuclei with negative curvature are stable, nucleation to form convex nuclei is impossible. If the (Θ_1^S, Θ_2^S) node is in the region IV, convex nuclei of any size are stable. Dashed lines Θ_{hri} were calculated from (5.1) for the curvature of the atomic size of nuclei. Areas II and III correspond to the nucleation of convex nuclei of the 1st or 2nd components. The area enclosed between the dashed lines correspond to the

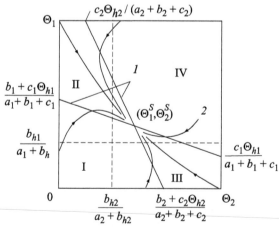

Fig. 5.2. The phase portrait of system (5.2). 1 – isoclines; 2 – trajectories of solutions. Region I – no condensation; II, III – nucleation of one of the components; IV – joint nucleation region

joint nucleation with restrictions on the curvature and size of stable nuclei of the components.

The above representation is sufficient to construct a qualitative picture of the distribution of combinations of forms of condensate nuclei of the components in the coordinates (Θ_1, Θ_2). Figure 5.1 shows that each particular combination $\Theta_i/\Theta_{\infty i}$ at a substrate temperature corresponds to a combination of stable cylindrical and spherical nuclei of the components. This fact can be used to map the generated structures (Fig. 5.3). The division of the coordinates by the critical values Θ_i is performed conventionally for the convenience of imaging. The chart is a matrix with diagonal asymmetry. Each element of the matrix corresponds to a particular structure, combining the forms of the crystallites of the individual components. The element of the matrix, adjacent to the origin, corresponds to the vapour phase of both components when condensation is impossible even with the formation of concave shapes. The element opposite to the diagonal corresponds to the condensation of both components in the amorphous state. The top row and the last column correspond to the amorphous state of one component, the region near $\Theta_{h\infty i}$ – to the condensation of single crystal flat films. Note that each subsequent element

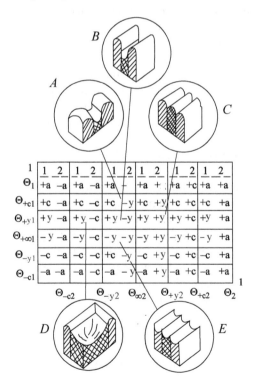

Fig. 5.3. Types of structures: A – columnar, B, C, D – flat cellular, E – volume-cellular, and the chart of the modes of condensation of the binary vapour mixture of mutually insoluble substances with the formation of dispersed crystallites of components. +a – amorphous state; –a – vapour; $\pm c$, $\pm y$ – crystallites with convex or concave spherical and cylindrical surfaces, respectively.

– up the column or right on the line – should include all forms available in the lower or left elements. For simplicity, only the codes specific to the form element are given.

Thus, a map of possible structures of the dispersed condensates of the vapour mixture at the initial stage is constructed. The implementation of a structure is determined by solving the problem of the kinetics of sorption processes on the substrate and of entering of the kinetic equations in a given region of phase space (Θ_1, Θ_2). A quantitative refinement of the map can be carried out by calculation the rate of nucleation of crystallite forms belonging to one element of the matrix. For the frequency of heterogeneous nucleation we can use a formula from [9] $J = \zeta \Theta v_{\parallel} \, n(N) \exp(-u_D/kT)$, where ζ is the fraction of the adsorption vacancies involved in the transport of adatoms to the nucleus; $n(N) = n_a \exp(-\delta\Phi(N)/kT)$ is the surface density of the subcritical agglomerates; u_D is the activation energy of surface migration of adatoms ($u_D \cong 0.3\varepsilon$) [9]. The forms of crystallites having a maximum rate of nucleation, J, appear.

Figure 5.3 shows schematic representations of the structures of condensates in random matrix elements. Options A, B, D are observed in experiments. The transition from structure A to structure D is checked by the joint condensation of oil vapours and magnesium vapours of magnesium. This transition occurs at increasing the flow of oil vapours onto a substrate at fixed other parameters.

The matrix predicts the existence of structures D which have not yet been discovered in the experiments. Unlikelihood of the existence of such structures is described by the contact of concave vapour – crystal-1 – crystal-2 phase boundary. In the presence of the third component this difficulty is removed. These structures for the two-component mixture of vapours should be considered degenerate.

Three important observations should be made.

The above classification algorithm structure is easily extended to the condensation of vapour mixtures of n-components. In this case, we should build an n-dimensional matrix in the Θ_1^n space. The allowance for the anisotropy of the surface properties of the individual phases of the crystallites in the amendment of the Kelvin–Thomson equation can specify the orientation of the crystallographic texture of phases relative to the substrate.

Along with cylindrical and spherical surfaces, which describe the shape of nuclei of the condensed phase, there are other stable surfaces of constant curvature. For example, a family of surfaces of revolution with a meridian curve

$$x = \int \frac{y^2 + C_1}{(y^2 - (y^2 + C_1)^2)^{1/2}} \, dy + C_2. \tag{5.3}$$

The circular cylinder and the sphere are special cases of the family (5.3).

In addition to the type of surface, describing the shape formed on the substrate of the condensate nucleus, it is important to define specific parameters of the form. The possibility of such a calculation can be illustrated by a single spherical nucleus on a flat substrate. The question is to determine the height of the spherical segment, the corresponding to the form of the nucleus. This calculation can be performed by minimizing the energy barrier for nucleus formation from N-atoms:

$$\delta\Phi(N) = \frac{\pi m^2}{4a^2} + \left(\varepsilon_s - \varepsilon_1/2\right) + ZN\varepsilon_1/2 - N\varepsilon_s - S\varepsilon_1/2a^2. \tag{5.4}$$

Here m is the diameter of the base segment; ε_1, ε_s are the binding energies of adatoms with the substrate and the surface of the nucleus; Z, a is the number of the nearest neighbours and the lattice parameter of the nucleus. The volume, radius, diameter of the base and the free surface area are calculated by the formulas using the contact angle and the size of the nucleus:

$$V = \pi R^3 \left(2/3 - \cos \varphi + \cos^3 \varphi/3\right) \cong Na^3, \; R = a \left(N/[\pi / \left(2/3 - \cos \varphi + \cos^3 \varphi/3\right)]\right)^{1/3},$$

$$S = 2\pi^{1/3}N^{2/3}a^2 \left(1 - \cos \varphi\right) \left(2/3 - \cos \varphi + \cos^3 \varphi/3\right)^{-2/3}.$$

Substituting these formulas into (5.4) and minimizing over φ gives an unambiguous relationship, similar to the well-known Young formula:

$$\varepsilon_s / \varepsilon_1 = (1 + \cos\varphi)/2, \tag{5.5}$$

The result (5.5) means that the equilibrium shape of nuclei with a spherical free surface has a maximum probability for nucleation. In calculating the nucleation rate of such nuclei it is necessary to know only the binding energy of adatoms with the substrate or the wetting angle of the substrate by the condensate material. A similar analysis can be done for other forms of nuclei. It is also possible to calculate the nucleation of heterogeneous components on a single substrate.

Thus, the above allows us to trace the unique relationship between the conditions on the substrate at the beginning of the process of adsorption of vapours of the mixture components and the result of nucleation of the condensate, when the substrate is filled with crystallites whose form and location can be calculated. This is followed by analysis of the evolution of the nuclei in the condensation process.

5.2. Heat and mass transfer on the surface of the crystallites in the condensation process

Evolution of the shape of condensate nuclei occurs due to a number of different factors: the convective and radiative heat exchange, incident and evaporated vapour flows, surface migration of adatoms. In condensation in vacuo with the formation of disperse condensates with crystallite sizes much smaller than the mean free path of the atoms in the vapour phase heat transfer calculations do not cause major difficulties. The proportion of convective heat coming to the front of condensation is determined by the excess heat energy of the vapour calculated from the temperature difference of the vapour and the substrate. The calculation of radiative heat transfer can be done with the knowledge of the temperature distribution and optical properties of the condensation front – the environment system. Calculation of the temperature of condensation front should take into account the thermal state of the substrate and the conductive properties of the condensate.

The vapour flow falling on the condensation front should be calculated from the initial state of the vapour flow, which is determined by the parameters of the vapour source and the conditions surrounding the substrate of the medium. The evaporation of the condensation front is determined by the local estimates: temperature and the curvature of the condensation front. Calculation of the surface migration is possible with the knowledge of the local temperature at the condensation front with definition of boundary conditions. Each of these processes has been well studied in isolation, there are known algorithms for formulating and solving problems. Difficulties in applying these algorithms to describe the evolution of the condensation nuclei at the front are, firstly, that all these processes are interdependent and require the simultaneous calculation of them with varying boundary conditions, and secondly, it is difficult to determine local physical properties of surfaces: the accommodation coefficients, the coefficients sticking coefficients of emissivity and emission. Therefore, at present can be delivered only to the problem of numerical simulation of the evolution of condensation front in several assumptions, namely:

1. All coefficients are: attachment, accommodation, emissivity and emission are equal to unity.

2. We neglect the surface migration due to the fact that we study the evolution of the crystallites, separated by foreign crystallites, condensation occurs in the presence of impurities, which reduce the length of the diffusion path of adatoms.

3. Conductive heat transfer from the condensation front to the substrate is calculated in a quasi approximation, without taking into account the temperature gradient along the substrate.

Under these assumptions the following equation can be proposed for the local specific heat flux on the element of the condensation surface

$$\delta q = \varepsilon_k \int_C \varepsilon_c \left(T_c^4 - T_k^4\right) dC + \varepsilon_k \int_S \varepsilon_s \left(T_s^4 - T_k^4\right) dS + \left(\alpha_k L + c_v (T_s - T_k)\right) \int_S J_s dS +$$

$$+\left(\alpha_k L + c_v (T_{ki} - T_k)\right) \int_K J_{vi} dK + c_m (T_m - T_k) \int_C J_m dC - L J_v,$$

(5.6)

where ε_k, ε_c, ε_s are the emissivities of the crystallite, the vacuum chamber and the vapour source, respectively; T_k, T_c, T_s are their temperatures; L, c_v is the specific latent heat of condensation of vapour and heat capacity; c_m, T_m, J_m

are heat capacity, temperature and specific flow of impurity vapours; J_v is the evaporation flux; α_k is the local condensation coefficient.

Individual members of the equation (5.6) take into account:

– The first and second terms of the equation – the radiation heat flux from the chamber walls and from the vapour source;

– The third term of the equation – the heat of condensation and convective heat flux from the vapour incident on the substrate and coming from the vapour source;

– The fourth – from the neighbouring crystallites;

– The fifth member – convective heat exchange with impurity vapour components;

– Sixth term – the heat loss by local evaporation of the crystallite.

In the quasi-one dimensional quasi-stationary approximation the integral of the specific heat flux on the crystal from the top to the given section determines the temperature gradient over the height of the crystallite:

$$\int_{S_k} \delta q dS_k = \lambda_k F_k(y)\left(dT_k/dy\right).$$
(5.7)

Replacing the left side of (5.7) by (5.6) gives the integral-differential equation for the local surface temperature of the crystallite. For the local surface normal rate of increment of the crystallite we can write the equation of local mass transfer:

$$dr/dt = \omega_a\left(\alpha_k \int_S J_s dS + \alpha_k \int_K J_{ki} dK + J_k\right).$$
(5.8)

The system of equations (5.7), (5.8) relates the evolution and the thermal state of the crystal of one of the components, developing by condensation from the vapour mixtures. Similar equations hold for each of the components present in the mixture.

For the calculation of local coefficients of condensation we can use the equations of the kinetics of sorption processes. For example, for a two-component mixture of a system of kinetic equations (5.2) the coefficients of condensation of the components are defined as $\alpha_{ki} = c_i(\Theta_i - \Theta_{hi})$.

Since there is no analytical solution of the above equations, we can recommend the procedure of numerical solution. The initial approximation is given by solving the system of the kinetics of sorption processes and the choice of the structure of a system of nuclei formed on a substrate, as described in the previous section. For this configuration of the nuclei we compute local vapour fluxes of all components using artificially defined condensation coefficients. The temperature distribution on the surface of the crystallites is calculated. We solve the system of kinetic equations of sorption with the definition of the condensation coefficients of condensation and clarification of the re-evaporation flows. If the calculated coefficients of condensation do not match the required values, the iterative refinement of the calculations is performed. After reaching a satisfactory accuracy the local increment of the surface of the crystallites in a given time interval is calculated. It is clear that the proposed algorithm of calculations is associated with a large amount of computation that is apparently the reason why there have been only a small number of attempts to solve these problems.

5.3. Evolution of two-dimensional crystallites in the condensation process

It is assumed that the vapour source is an infinite strip of width D, located at a distance L from the point at which we are to calculate the vapour flow (Figure 5.4). The convenience of this representation of the vapour source is the ability to vary the width of the strip, thereby simulating a variety of different sources: from the filamentous to the half-space without changing the formulas.

On the site of the condensation surface, where the entire surface of the source is fully visible, there is no shadowing and the vapour flow to the area normal to the axis of the source can be defined by the formula

$$J_s = (J_0(T_s)/2)(\sin \psi_2 - \sin \psi_1), \qquad (5.9)$$

where J_0 is the specific vapour flow diffusively emitted from the surface of the source; ψ_1 and ψ_2 are the extreme angles of a spherical wedge of visibility of the source, measured from the normal to the site. The specific vapour flow on section of the condensation surface from the source with the intensity variable in the angle is defined by the formula

$$J_s = (1/2) \int_{\psi_1}^{\psi_2} J(T_s, \psi) \cos \psi \, d\psi. \qquad (5.10)$$

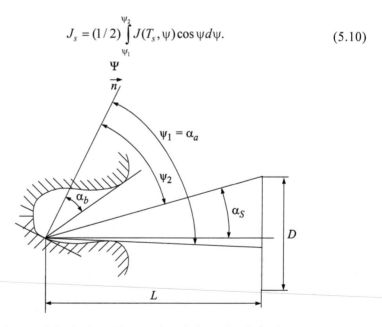

Fig. 5.4. The scheme of the limits of integration of flows for the strip vapour source.

The expression (5.9) and (5.10) are sufficient for the numerical simulation of the evolution of the set of crystallites in condensation of the vapour of the source. In the simulation of the set of the crystallites there can be a situation where part of the vapour flow from the source can be shaded by the adjacent crystallites. Since the exact calculation of the shading requires knowledge of the position and shape of neighbouring crystallites and such a spatial calculation is extremely cumbersome, it is assumed that the surrounding neighbours have the same shape as the analyzed crystallite and they are located in the middle at the same distance. To account for shading by the neighbours in the case of two-dimensional crystallites, it is convenient to assume in (5.9), (5.10) $\psi_1 = \alpha_a$, $\psi_2 = \alpha_b$ as shown in Fig. 5.4.

The sections of the body of the adjacent crystallites, visible from the analyzed point of the condensation surface, also contribute to the local vapour flow:

$$J_b = (1/2) \int_{\alpha_b}^{\pi/2} J_b(T_b, \psi) \cos \psi d\psi,$$

$$J_a = (1/2) \int_{\pi/2}^{\alpha_a} J_a(T_a, \psi) \cos \psi d\psi.$$

The specific vapour flows, emitted from the surface of adjacent crystallites, can be defined as

$$J_b(T_b, \psi) = p_h(T_b)(2\pi mkT_b)^{-1/2} \exp(K_b(\psi)\sigma\omega_a/kT_b),$$
$$J_a(T_a, \psi) = p_h(T_a)(2\pi mkT_a)^{-1/2} \exp(K_a(\psi)\sigma\omega_a/kT_a),$$

where T_b, T_a are the local surface temperatures of the crystallites, which determine the saturation pressure p_h; m, k is the mass of the atom of the vapour and the Boltzmann constant; σ, ω_a are the surface energy at the vapour–crystal interface and the atomic volume in the condensed phase, $K_i(\psi)$ is the local curvature of the surface of adjacent crystallites.

Calculations of heat flows to the surface of the crystallite can be conducted using the same integrals (5.9) and (5.10), but in the integrands the flows of the vapour molecules J should be substituted by the corresponding thermal values.

To illustrate the possibilities of the proposed algorithm for calculating the evolution of the crystallites, specific calculations were carried out for the condensation of magnesium on a flat substrate with nuclei located on it in the form of half-cylinders. It is assumed that portions of the substrate between the nuclei are filled with the adsorbate of some organic impurity, preventing the development of the nucleus along the substrate. Contact angle $\pi/2$ corresponds approximately to the condensation of magnesium on the carbon substrate at its room temperature. The space around the substrate is filled with saturated magnesium vapours at a temperature of 750 K. The surface of the nucleus, in general, receives magnesium atoms from the half-space, from the neighbouring nuclei and the free sites of the substrate. To reduce the amount of computation the condensation coefficient of magnesium in the nucleus is assumed to be 1 and on the substrate 0. These initial conditions for the analysis of the homologous matrix (see Fig. 5.3) correspond to the matrix element +cyl –cyl, if we assume that the magnesium is component 1.

The results are presented in Fig. 5.5 in the form of changes in the magnesium crystallite profiles with time at three different substrate temperatures. We see that the result of condensation of magnesium at substrate temperatures less than 729 K is a columnar structure. At the initial stage of their growth the columns evolve both along the substrate and height. Then, the development along the substrate slows down and the growth continues only at the tops of columns with an increase in their transverse dimension. Similar calculations for different initial distances between the columns showed that the profiles of the vertices of the columns made dimensionless with respect to the distance between their axes do not have any geometric similarity.

As the temperature of the substrate increases the adsorbate of t he organic impurity is desorbed from the substrate, and the re-evaporation of the columns becomes significant. Starting from a temperature of 730 K the development of nuclei along the substrate leads to the closure of their profiles on the substrate and then levelling of the condensation surface into a plane. For the homologous matrix (see Fig. 5.3) the increase in the temperature of the substrate corresponds to a change $\Theta_1 \Rightarrow \Theta_{\infty 1}$, which is qualitatively confirmed by the calculations.

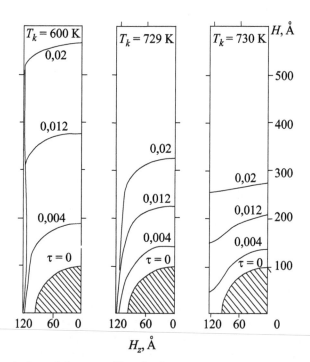

Fig. 5.5. The evolution of the crystallites in time at different substrate temperatures. $T_S = 750$ K, $H_z = 250$ Å, $R_{z0} = 100$ Å, $\alpha_S = \pi/2$.

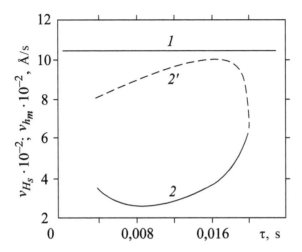

Fig. 5.6. Changing the growth rate of the peaks and troughs between the nuclei. $T_s = 750$ K, $H_z = 250$ Å, $R_{z0} = 100$ Å, solid lines – v_{max}, dashed – v_{min}. $T_v = 729$ (1); 740 (2, 2′).

The growth rate of crystallites peaks and troughs between them depends on time (Fig. 5.6). At a substrate temperature of 729 K the growth rate of the peak (curve 1) is constant, the growth rate of the trough is zero. The zero growth rate in the trough is associated with shading of the substrate between the crystallites and with the presence of the joint vertices of crystallites, which reduces the flow of vapour into the basin. Profiles of the columns (see Fig. 5.5.) indicate the arrest of the growth of the trough already after 0.12 s. The curves 2 and 2′ in Fig. 5.6 were obtained at a temperature at which the re-evaporation of the crystallite surface becomes important and the vapour is transported in the cavity by successive re-evaporation between the adjacent crystallites. In the beginning of condensation the growth rate of the trough significantly exceeds the rate at the tip which due to the influence of curvature of the tip and the Kelvin–Thomson effect.

The results of calculations allow to draw important conclusions. Despite the fact that the calculations have not incorporated the mechanism of surface migration, there is a redistribution of matter between different parts of the condensation front qualitatively similar to the action of surface diffusion. This means that there is a fundamental question about the role of each of these processes in the formation of condensates which is beyond the purpose of this work. Note also that the re-evaporation mass transfer mechanism is accompanied by significant convective energy exchange associated with the transfer of latent heat of phase transformation. This suggests that in the formation of complex non-dense structures in the regimes where re-evaporation is significant we should expect the state of the condensate close to equilibrium. This predicts the existence in such condensates of equilibrium shapes of the crystallites. Thus, the action described in section 2 of the homologous matrix can also be extended to the stage of evolutionary growth of the crystallites.

An important issue is the critical temperature of the transition of the structure of the condensate from columnar to continuous. Figure 5.7 presents the results of calculating the height of the columns, the height of the condensation front in the troughs and the relative depth of the gap between the columns, depending on the substrate temperature at two different distances between the nuclei:

$$\bar{h} = (H_s - h_m)/H_{s0}. \qquad (5.11)$$

Here H_s and H_{s0} is the height of the tip of the crystallite above the substrate in the current and the initial moment, respectively; h_m is the current height of the junction of the crystallites above the substrate. According to these parameters we can determine the critical temperature of a radical change in the structure of the condensate T_c^*. The value $\bar{h} = 1$ is typical for the growth of crystallites without deepening the gap between them and must meet the critical temperature of the transition of the structures.

Figure 5.7 a shows that the nuclei at the starting time are close to each other (H_z = 200 Å), the height of the condensate increases also in the troughs between the nuclei. However, with increasing height of the tops of the columns their shading effect prevails, the growth in the troughs of the condensate slows down and substrate temperatures less than 728 K results in the formation of the columnar structure of the condensate. The height of the condensate in the trough increases with increase of the substrate temperature. At a substrate temperature of T_k = 728 K we have the mode in which the growth of the columns is not accompanied by increase of the depth of the trough between the columns, at T_k> 728 K the growth rate of the condensate in the troughs is higher than the growth rate of the tip of the nucleus, which tends to equalize the condensation front and form a continuous condensate film.

This definition of the critical temperature allows some clarification of the homologous matrix of the structures at low supersaturation, set out in section

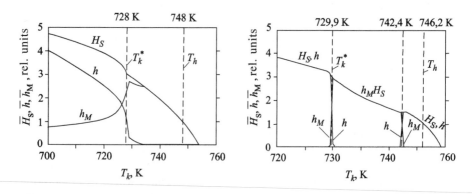

Fig. 5.7. The dependence of the height of peaks H_s and troughs h_m and their relative depth h on the substrate temperature T_k for the source when $\alpha_s = \tau/2$. T_s = 750 K, R_{z0}= 100 Å, τ = 0.02 s; H_z = 200 (a); 250 Å (b).

2. The transition from structures with the code +cyl to the condensate in the form of a flat film is not a smooth increase in the radius of curvature of the surface of the nucleus, but a radical change in the evolution of the nuclei.

The substrate temperature 748 K (see Fig. 5.7 a, dashed line) corresponds to the saturation of the nuclei with an initial radius of 100 Å. The H_s and $h\square$ curves show that the substrate temperatures higher than the saturation temperature for nucleation result in the formation of a condensate film.This means that the merger and alignment of the condensation front are faster than the time available to the nucleus evaporate. Condensation does not occur only at a substrate temperature above the saturation temperature for a flat surface.

The calculations for the distance between the nuclei of 250 Å (see Fig. 5.7, b) revealed some special features. We see that the zero height of the condensate between the nuclei rises sharply and reaches the height of peaks of nuclei at a substrate temperature higher than 730 K. The critical temperature, corresponding to the change of the structure of the condensate, is 729.9 K. However, in the film condensation region there is another characteristic temperature, 742.4 K, at which there is no contact between the nuclei during the evolution period a there is a slow change in the height of individual nuclei. In the range of substrate temperatures of 742.4–746.2 K the evolution of the condensation front into a plane occurs over a longer period of time with the formation of large-amplitude irregularities. This phenomenon is due to the fact that with the approach of the substrate temperature to saturation temperature the commensurability of the rates of evaporation and condensation significantly reduces the development of the nucleus along the substrate due to shading of its base by the vertex. When $T_k > 746.2$–750 K the nuclei evaporate with preservation of the curvature of the surface during the evaporation time.

Comparison of the curves $\bar{h}\,(T_k)$ for different initial distances between the nuclei shows a change of T_k^*. It is therefore important to study the dependence of T_k^* on the critical parameters such as the initial contact angle, the presence of impurities in the vapour flow, the dynamic parameters of the vapour. The calculated values of T_k^* lie in the range found experimentally [10] of 0.3–0.6 of the melting point of metals, the condensation of which has been studied.

Figure 5.8 presents the results of calculating the thickness of the condensate layer on tops of columns and in the troughs between them, as well as the depth of the troughs, depending on the substrate temperature at a small width of the strip evaporator $D/2L = 0.1$. In comparison with the previous results obtained for the condensation of the vapour coming from the half-space here there are a number of features. At substrate temperatures less than 622 K the developed structure is columnar, the columns are closing together. The value $\bar{h} > 1$ means the development of irregularities on the external condensation front. If $T_k > 622$ K $\bar{h} < 1$ which corresponds to the alignment of the condensation front. Between the columns there are local contact areas, forming a porous condensate. With increasing substrate temperature the porosity of the condensate decreases, the last pores disappear by condensation on a substrate with $T_k = 710$ K. There is joining of nuclei on the substrate. The full alignment of the condensation front is achieved in the temperature

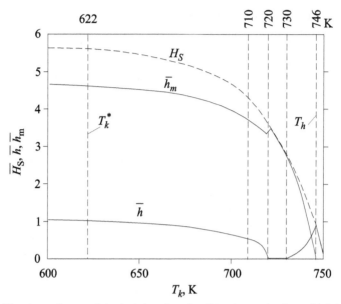

Fig. 5.8. The dependence of the height of peaks H_s and troughs h_m and of their relative depth $h = (H_s - h_m)/H_{s0}$ on the substrate temperature T_k for the source at $\alpha_s = 0.05$ rad. $T_s = 858.8$ K, $R_{z0} = 100$ Å, $\tau = 0.02$ s; $H_z = 250$ Å.

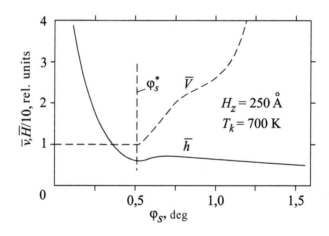

Fig. 5.9. The ratio of growth rates on the top $\bar{v} - v_{H_s}/v_{hm}$ and the relative depth of the troughs $\bar{h}$, depending on the angle of wetting of the nucleus with the substrate. $T_s = 750$ K, $T_k = 750$ K, $\alpha_s = 0.05$ rad.

range 720–730 K. When the substrate temperature approaches the saturation temperature for the initial size of nuclei observed partial alignment is obtained for the produced condensate film, then there is no contact amongst the nuclei and they evaporate.

An important parameter influencing the process of structure formation is the contact wetting angle of the substrate by the material of the nucleus.

Figure 5.9 shows the calculated $\bar{h}$ and the ratios of growth rates at the top and in the trough as a function of contact angle. Curves $\bar{h}\,(\varphi_s)$ lie above unity, which means that at a given temperature, source and process time the initial roughness – nuclei on the substrate – increases in the process of condensation. However, the curves $\bar{v}\,(\varphi_s) = v_{max}/v_{min}$ show that there exists a critical contact angle φ_s^* separating the two characteristic regions. At $\varphi_s < \varphi_s^*$ $\bar{v} = 1$ which corresponds to the alignment of the condensation front. In this area, further condensation leads to the formation of a planar monolithic film. At $\varphi_s > \varphi_s^*$ the columnar structure is formed with increasing depth of the troughs between the columns.

5.4. The evolution of axially symmetric condensation nuclei

It is assumed that the vapour source is a spherical segment of diameter D with the centre at the point where we want to calculate the specific vapour flow. The convenience of this representation of the vapour source is the ability to vary the angle of the segment, thereby simulating a variety of sources – from a point to half-space without changing the formulas. The scheme for the calculation of the characteristic angles is shown in Fig. 5.10.

In the section of the condensation surface where the entire surface of the source is fully visible there is no shadowing and the vapour flow to the area normal to the axis of the source can be defined by the formula

$$J = J_0 \sin^2 \omega,$$

where J_0 is the specific vapour flow emitted from the surface of the source; ω is the half-angle of the spherical sector of visibility of the source. If the normal of the area of the condensation surface form an angle ω with the axis of the source, the specific vapour flow on the site is

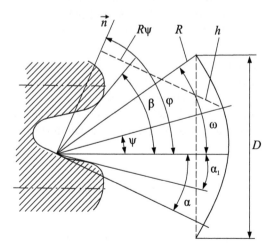

Fig. 5.10. Scheme of the integration limits of the flows for axisymmetric vapour sources and the crystallites.

$$J_k = J_0 \sin^2 \omega \cos \varphi. \qquad (5.12)$$

In the simulation of a group of the crystallites there can be a situation where part of the vapour flow from the source can be shaded by the adjacent crystallites. Since the exact calculation of the shading requires knowledge of the position and shape of neighbouring crystallites and a spatial calculation is extremely cumbersome, it is assumed that the surrounding neighbours have the same shape as the analyzed crystallite and they are located in the middle at the same distance. In this case, to calculate the shading by the neighbours it is convenient to assume that the contours of the adjacent crystallites produce a circular shadow of the point from the source to the solid half-angle β. Then the flow on the area of the crystallite can be calculated by the formula

$$J_k = J_0 \sin^2 \beta \cos \varphi. \qquad (5.13)$$

The meaning of this formula is that the shading by the neighbours reduces the diameter of the source visible from the analyzed point. It should be noted that the calculation using (5.13) gives some error if the neighbours are closely spaced, even if they are symmetric with respect to the calculated crystallite, but the analyzed point is not situated on the axis of the crystallite.

In the case of shading the source by the body of the analyzed crystallite it is assumed that from the observation point we can see the part of the source limited by a section of large diameter, tangential to the body of the crystallite. For convex crystallites it is the plane tangent to the surface of the crystallite at the observation point forming the angle α with the source axis. The integration of the vapour flow to the analyzed ares of the surface of the crystallite is conveniently performed on sections of the visible surface of the source by the planes parallel to the plane tangent to the body of the crystallite at the observation point. If ψ, Θ are the azimuthal angles and meridian points of the evaporator, visible from the observation point, we can then calculate:
  – element of the source surface $dF = Rd\psi R_\psi d\Theta$;
  – radius of rotation of an integrable element of the source surface $R_\psi = R \cos \psi$;
  – elementary vapour flow coming from the source $dI = J_0 R^2 d\Theta \cos \psi$;
  – element of the normal projection of the vapour flow to the surface area of the crystallite $dI_k = J_0 R^2 d\Theta \sin \psi \cos \psi d\psi$.

Since the integration over Θ is performed by the rotation around the normal to the tangent of the surface at a constant ψ, we can define the normal vapour flow

$$I_k = J_0 R^2 \int_{\psi_1}^{\psi_2} \Theta_a \sin \psi \cos \psi d\psi \qquad (5.14)$$

and the specific normal vapour flow to a local region of the surface of the crystallite

$$J_k = \left(J_0 / \pi \right) \int_{\psi_1}^{\psi_2} \Theta_a \sin \psi \cos \psi \, d\psi.$$

Here $\Theta_a = 4 \, \mathrm{arctg} \, (2h/a)$ is the length of the arc (in radians), described by the integrable part of the surface during the rotation around the normal, it is calculated using height

$$h = R \left(\cos (\psi + \alpha) - \cos \beta_1 / \cos \alpha \right) \qquad (5.15)$$

and the chords of the segment of the vapour source in the integration plane

$$a = 2(2hR \cos \psi - h^2)^{1/2}.$$

The limits of integration in (5.14) for convex shapes of the crystallites for the different cases of shading are computed: if $\beta > \omega$ and $\omega - \alpha < \pi/2$, then $\psi_1 = 0$, $\psi_2 = \omega - \alpha$, and in (5.15) $\beta_1 = \omega$; if $\beta < \omega$ and $\beta - \alpha < \pi/2$, then $\psi_1 = 0$, $\psi_2 = \beta - \alpha$, and in (5.15) $\beta_1 = \beta$. When $\omega > \pi/4$ there are two other possible cases, where part of the integral (5.14) can be taken in quadratures

$$J_k = J_0(1 - \sin^2 \psi_2) + \left(J_0 / \pi \right) \int_{\psi_1}^{\psi_2} \Theta_a \sin \psi \cos \psi \, d\psi, \qquad (5.16)$$

if $\beta > \omega$ and $\omega - \alpha > \pi/2$, then $\psi_1 = 0$, $\psi_2 = \pi - \omega - \alpha$, and in (5.15) $\beta_1 = \omega$; if $\beta < \omega$ and $\beta - \alpha > \pi/2$, then $\psi_1 = 0$, $\psi_2 = \pi - \beta - \alpha$, and in (5.15) $\beta_1 = \beta$.

For the concave-convex forms of the crystallites when the shading is performed by the top of the analyzed crystallite, the lower limit in the integrals (5.14) and (5.16) changes $\psi_1 = \alpha - \alpha_1$, where α_1 is the angle between the source axis and the plane passing through the studied point and touching the surface of the top of the crystallite. Table 5.1 shows the different cases of the ratio of angles and the corresponding integration limits.

It should be noted that the assumed shading by the vertex as a plane slightly overestimated but this greatly simplifies the calculations.

Figure 5.11 shows the calculated evolution of hemispherical nuclei of the Mg condensate, located on a flat substrate, when magnesium vapour arrives from the half-space with a temperature of 750 K. The behaviour of the axisymmetric crystallite corresponds qualitatively to the behaviour of a two-dimensional crystallite. With time there is an increase of the transverse size to the formation of a gap with the neighbouring crystallites.

In the process of crystallite growth there is some difference in the forms of its vertices. At short condensation times the vertex is stretched, then the profile is converted to a form more flat in comparison with the two-dimensional crystallites presented in Section 5.3. When the neighboring crystallites approach each other their lateral surfaces start to oscillate. These oscillations sometimes lead to a merger of two adjacent crystallites with

Table 5.1 Formulas and parameters for calculating the specific vapour flow on the crystallite surface

Type of screening	Condition	Formula	Integration limit
No screening	$\alpha_1 < \alpha$; $\alpha < -\omega$; $\beta > \omega$	(5.12)	$\psi_1 = 0$; $\beta_1 = \omega$
Screening by neighbour	$\alpha_1 < \alpha$; $\alpha < -\omega$; $\beta < \omega$	(5.13)	$\psi_2 = \omega - \alpha$
Mutual screening	$\alpha_1 < \alpha$; $\alpha > -\omega$; $\beta > \omega$	(5.14)	$\psi_1 = 0$; $\beta_1 = \omega$
	$\omega - \alpha < \pi/2$	(5.16)	$\psi_2 = \pi - \omega + \alpha$
Mutual screening and	$\alpha_1 < \alpha$; $\alpha > -\omega$; $\beta > \omega$	(5.14)	$\psi_1 = 0$; $\beta_1 = \beta$
screening by neighbour	$\beta - \alpha > \pi/2$		$\psi_2 = \beta - \alpha$
	$\alpha_1 < \alpha$; $\alpha > -\omega$; $\beta < \omega$	(5.16)	$\psi_1 = 0$; $\beta_1 = \beta$
	$\beta - \alpha > \pi/2$		$\psi_2 = \pi/2 - \beta + \alpha$

No screening	$\alpha_1 > \alpha$; $\alpha < -\omega$; $\beta > \omega$		$\psi_1 = \alpha_1 - \alpha$; $\beta_1 = \omega$
Screening by neighbour	$\alpha_1 > \alpha$; $\alpha < -\omega$; $\beta < \omega$	(5.12)	$\psi_2 = \omega - \alpha$
Screening by own tip	$\alpha_1 < \alpha$; $\alpha > -\omega$; $\beta > \omega$	(5.13)	$\psi_1 = \alpha_1 - \alpha$; $\beta_1 = \omega$
	$\omega - \alpha < \pi/2$	(5.14)	$\psi_2 = \pi - \omega + \alpha$
	$\alpha_1 < \alpha$; $\alpha > -\omega$; $\beta > \omega$		
	$\omega - \alpha > \pi/2$	(5.16)	

Screening by own tip	$\alpha_1 > \alpha$; $\alpha < -\omega$; $\beta > \omega$	(5.14)	$\psi_1 = \alpha_1 - \alpha$; $\beta_1 = \beta$
and by neighbour	$\beta - \alpha < \pi/2$		$\psi_2 = \beta - \alpha$
	$\alpha_1 > \alpha$; $\alpha < -\omega$; $\beta > \omega$	(5.16)	$\psi_1 = \alpha_1 - \alpha$; $\beta_1 = \beta$
	$\beta - \alpha > \pi/2$		$\psi_2 = \pi - \beta + \alpha$

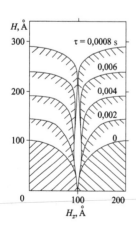

Fig. 5.11. The evolution of spherical nuclei. $T_S = 750$ K; $T_k = 300$ K; $H_z = 200$ Å; $R_{z0} = 100$ Å; $\omega = \pi/2$.

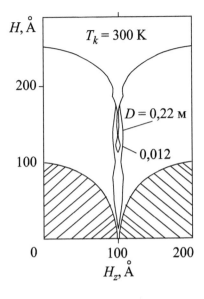

Fig. 5.12. The forms of columns for different dimensions of the vapour source. $H_z =$ 200 Å; R_{z0} = 100 Å. At D = 0.22 and 0.012 m T_s = 756.8 and 1106.7 K, respectively

subsequent separation of the profiles. The oscillations of the profiles in the process of calculations the profiles can be attributed to the instability of the calculation scheme in the conditions of a strong dependence of the arrival of the vapour of the source on the size of the gap.

Figure 5.12 shows the calculated growth of hemispherical nuclei for two different vapour sources, while maintaining the growth rate of the tops of the columns. It is seen that with decreasing diameter of the source, which is equivalent to approximation of the vapour flow to the atomic beam, the depth of the troughs and the gap between the columns decreases.

The results of such calculations for different D are summarized in Fig. 5.13. Two dashed lines show the zone of pulsed closing of the columns together in which a porous layer of the condensate forms. The pores, formed by periodic closing of the columns with subsequent divergence, are shown conventionally as ellipse-shaped points. It is evident that the porous layer of the condensate forms only at values $D/2L < 0.85$. At a higher $D/2L$ ratio there are gaps between the columns without no contact between them. In the condensation of vapour from a molecular beam, $D/2L = 0$, the porous layer is not formed and a monolithic film of the condensate appears.

The solid line in Fig. 5.13 denotes the size of the gap between the columns, averaged over the thickness of the condensate layer. The physical meaning of the quantity δ_z/R_z corresponds approximately to the bulk porosity of the produced condensate. This curve shows that the use of the atomic beam as a vapour source helps to ensure the formation of a monolithic non-porous condensate.

Figure 5.14 compares the profiles of the crystals formed by condensation of the vapour from half space on nuclei placed on a substrate at a distance of 500 Å from each other. The graph shows the profiles of approximately the same height, obtained at different substrate temperatures. The calculations

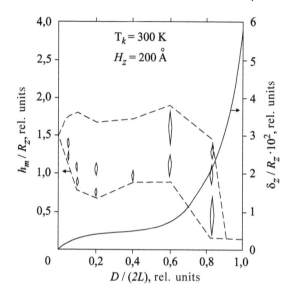

Fig. 5.13. The relative depth of the troughs between the colums h_m/R_{z0}, porosity of the condensate δ_z/R_{z0}.

do not take into account the vapour flows reflected from the substrate. These conditions correspond to the evolution of the nuclei formed on ledges, if the substrate has a micro-roughness. With increasing substrate temperature there-evaporation process reduces the size of the base of the crystal. At a substrate temperature of 730 K the crystal base completely evaporates and the crystal loses contact with the substrate.

These results lead to two important conclusions. First, the supersaturation, calculated from the parameters of the vapour flow, substrate temperature and the curvature of the nucleus is not a sufficient basis for the development of the nucleus. Portions of the surface of the nucleus, shaded from the vapour flow by its top, may be in the unsaturated conditions. This means that in the calculation of nucleation processes in determining the saturation it is more accurate to take into account the properties of the substrate. Second, these results clearly indicate the possible existence of the phenomenon of generation of free microparticles formed by nucleation on the surface and losing contact with it due to evaporation of the base. This phenomenon may be important in the mechanism of formation of fumes. Note that the mechanism of this phenomenon is fundamentally different from the previously discovered fluctuation generation of free clusters [11].

5.5. Some details of the evolution of the structures of the condensate

5.5.1. Steady-state profile of the top of the columns

The profiles of the developing columns continuously change with time, the gaps between the columns are reduced (see sections 5.3, 5.4). The fundamental

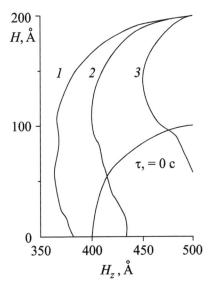

Fig. 5.14. The forms of the columns at different substrate temperatures. $T_s = 750$ K; $H_z = 500$ Å; $\underline{R_{z0}} = 100$ Å; $\omega = \pi/2$. $T_k = 300$ (1), 714 (2), 730 K (3); $\tau = 0.001$; 0.0015; 0.0023 s for the curves 1–3, respectively

question is: is there any stationary profile to which the dynamic profiles of the columns tends during a sufficiently large time of their development. This problem can be solved using equation (5.8). The projections of the normal increment on the x and y axes correspond to:

$$dx/dt = (dr/dt) \cos \alpha_a, \; dy/dt = (dr/dt) \sin \alpha_a.$$

In the steady profile all its points at which the tangents to the profile retain their direction, must move with the velocity components

$$dx_s/dt = (dr/dt) \cos \alpha_a = 0;$$
$$dy_s/dt = (dr/dt) \sin \alpha_a = \text{const.} \qquad (5.17)$$

Equation (5.17) together with (5.8) can be used for the numerical solution of the steady profile of the top of the column. The meaning of the constants in (5.17) – the velocity profile of the top of the column as a whole along the coordinate y. It is convenient to give the value of this constant at the highest point of the profile of the top of the column. For example, for a profile of the column with the tip smoothly convex, $\alpha_{a0} = \pi/2$, const $= (dr/dt)_0$, the peak tip grows only at the expense of arrival of the vapour from the source. It is important to note the special feature of the definition of the steady-state profile in the range of the crystallite size in which an important role is played by the Kelvin–Thomson effect. In this case, to calculate the integrals in (5.8) it is necessary to define the approximate profile of the crystallite and neighbours for the calculation of integrals, followed by iterative refinement. If the Kelvin–Thomson effect is not taken into account calculations do not require iterations.

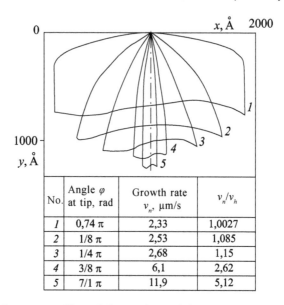

No.	Angle φ at tip, rad	Growth rate v_n, μm/s	v_n/v_h
1	0,74 π	2,33	1,0027
2	1/8 π	2,53	1,085
3	1/4 π	2,68	1,15
4	3/8 π	6,1	2,62
5	7/1 π	11,9	5,12

Fig. 5.15. Steady-state profiles of the vertices of the columns at different angles of taper $\alpha_{a0} = \pi/2 - \varphi$.

Calculations of steady-state profiles for the condensation of the vapour from the half-space were carried out for different values of the angle α_{a0} excluding the Kelvin–Thomson effect (Fig. 5.15). The resulting profiles are self-similar in terms of absolute size, so they are constructed in the coordinates assigned to the size between the columns. The meaning of various values of α_{a0} that faceting can appear on the tops of the columns in accordance with the preferred direction of growth of different crystallographic properties of the crystallites.

Analysis of Fig. 5.15 suggests the following conclusions.

1. The steady profile during the growth of crystallites with a smoothly rounded top is the plane; the section of the profiles of neighbouring crystallites degenerates into a straight line $x = \pm H_z$.

2. Sharpening the top of the crystallite $\alpha_{a0} < \pi/2$ causes the sharpening of the profile and increase of the rate of growth of the crystallite. This result agrees qualitatively with the known experimental fact of the abnormally high rates of whisker growth.

3. Coalescence of neighbouring crystals is possible only for $\alpha_{a0} = \pi/2$. For all $\alpha_{a0} < \pi/2$ at finite y there is a gap between adjacent crystallites that increases with the height of the crystal.

In accordance with the first finding all the columnar structures formed from nuclei with the spherical or cylindrical shape should degenerate during development into a flat condensation front. This is contrary to the majority of experimental data, which show no degeneration of the columns and instead show their enlargement in the transverse direction, while maintaining on average the round shape. This contradiction is explained in the next section.

The question of the existence of steady-state profiles of the evolution of nuclei for the condensation of the directional vapour flow has a certain special feature. When the vapour comes from a source of finite size, all parts of the condensation surface with the tangents to these parts forming with the source an angle greater than arcsin $(D/2L)$, receive a full flow of vapour without shading. The normal projection of this flow determines the normal rate of evolution. This case can be analyzed using the techniques proposed in [2, 3], where it is shown that solutions of such problems are geometrically similar profiles of crystals with the similarity ratio varying with time.

It should also be noted that for the condensation from an ideal atomic beam, where all the vapour atoms are unidirectional, the form of the condensation surface is indifferent and is determined only by evaporation and stochastic processes on the condensation surface.

5.5.2. Competition between the columns in the growth process

All the calculations in sections 5.3 and 5.4 were conducted assuming that the studied crystals have an identical shape. This significantly reduces the amount of necessary calculations. However, the abolition of the requirement of identity may give a new principal result. The calculation of the simultaneous evolution of three nuclei was carried out using (Fig. 5.16) the initial conditions in which the average size of the nucleus is half the size of neighbouring nuclei.

Figure 5.16 shows that the condensation process is accompanied by a reduction of the transverse dimensions of the smaller nucleus simultaneously with the increase of the size of its neighbours. Reasons for this are two: first, the higher nuclei receive more vapour from the source, and secondly, in mutual re-evaporation and recondensation the smaller nucleus loses in accordance with the Kelvin–Thomson effect more by evaporation than the amount of the vapour it receives from larger neighbours. This leads to slower growth of

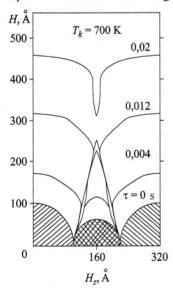

Fig. **5.16.** The evolution of three nuclei. T_s = 750 K; T_k = 700 K; H_z = 160 Å; R_{z1} = 100 Å; R_{z2} = 50 Å; α_s = $\pi/2$.

crystals of smaller size and even stops their growth. Neighbouring crystals are, therefore, able to develop in the transverse direction.

Given that the stochastic process of nucleation can not provide simultaneously a set of identical nuclei, it is clear that the evolution of real nuclei, not idealized in size, is a fierce competition, based on mutual shading of the vapour flows and inequality of the mutual re-evaporation, caused by the difference of curvature. The columns which for some reason got the advantage in size, displace and absorb smaller neighbours. This concept is consistent with the well-known experimental data: the size of the crystals of film condensates is proportional to the film thickness.

5.5.3. Clusters at the condensation front

After contact with the cluster with the condensation surface we have a situation similar in many respects to nucleation. All aspects of the described model are suitable for studying the evolution of the condensation surface after being hit by the clusters. Assume that the clusters trapped on the condensation surface are distributed uniformly on this surface with the average distance between H_z and that as a result of the interaction with the surface they form hemispherical projections. Under these assumptions, the algorithm for calculating the evolution of the nuclei should be amended in the integrals that describe the flow of vapour atoms arriving on the surface of the nucleus from the substrate. For the case study of the evolution of the cluster on the condensation surface it is expedient to admit that in the parts of the condensation front free from clusters, condensation occurs and the flow of vapour from these sites consists of two components: the evaporation flow and flow of the reflected atoms if the condensation coefficient differs from unity.

For the average distance between clusters of 500 Å (Fig. 5.17) we obtain parachute-like profiles which agree qualitatively with the shape of the crystallites, as calculated in [12] based on analytical solutions of the equation (5.8) with additional simplifications. The correspondence of these profiles presented in [12,13] with the experiments is better. The profiles in Fig. 5.17 shows deviations of the form from spherical shapes typical of real crystals.

To compare the evolution of clusters at the condensation front and of nuclei on the substrate, calculations were carried out with the parameters that are equivalent to those used in section 5.3. The calculation results are shown in Figs. 5.18–5.20. A characteristic feature of the profiles in Fig. 5.18 in comparison with Fig. 5.5 is that the formation of columnar structures in clusters does not depend on the need to fill the substrate with the condensate. Condensation fills the space between the clusters from the time they fall on the surface. Otherwise the profiles of the crystallites are similar to those shown in Fig. 5.5.

The dependences of the rate of condensation on the peaks and in troughs in between the clusters (see Fig. 5.19) show that the rate of condensation in the troughs is lower than the rate at the top at low temperatures of the condensation front and the inversion of the rate with increasing temperature is also visible. In accordance with the inversion of the velocity in condensation

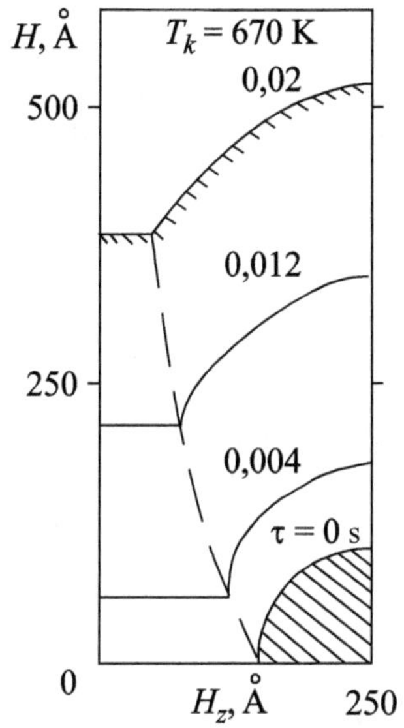

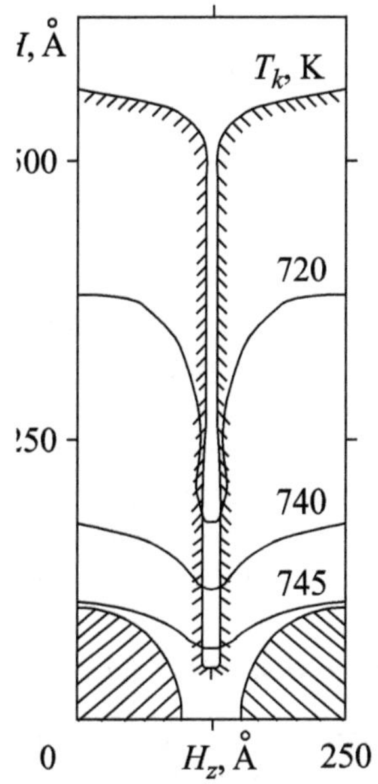

Fig. 5.17. The evolution of the condensation front after clusters falling on it. T_S = 750 K; T_k = 670 K; H_z = 500 Å; R_{z0} = 100 Å; $\alpha_s = \pi/2$.

Fig. 5.18. (right). Condensation front forms after being hit by clusters at different temperatures of the substrate with τ = 0.02 s, T_S = 750 K, H_Z = 250 Å, R_{z0} = 100 Å, $\alpha_s = \pi/2$.

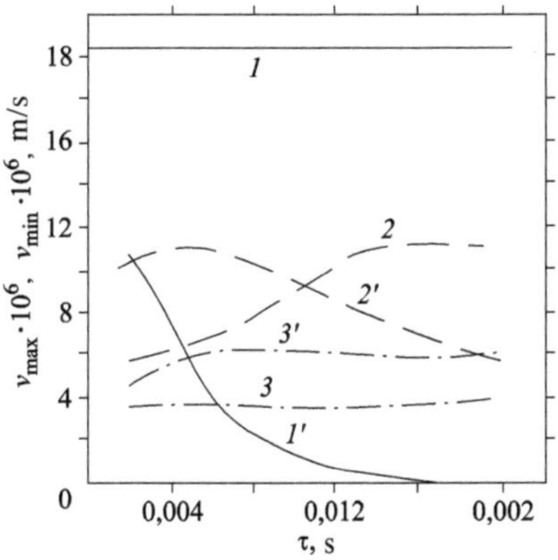

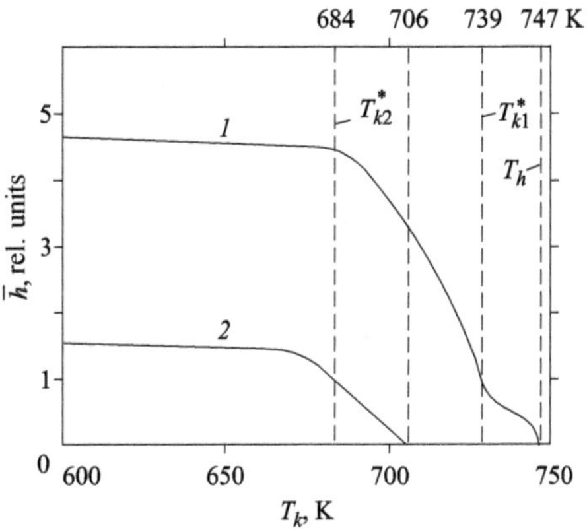

Fig. 5.19. Changes of the growth rates of peaks v_{max} and troughs v_{min} with time at different substrate temperatures T_k. T_s = 750 K; H_Z = 500 Å; R_Z0 = 100 Å; $\alpha_s = \pi/2$. For curves (1, 1'), (2, 2'), (3, 3') T_k = 300, 725, 740 K, respectively.

Fig. 5.20. (right) The relative depth of the depressions $\bar{h} = (H_S - h_m)/R_{Z0}$ depending on the substrate temperature for two different distances between clusters. T_s = 750 K; R_{Z0} = 100 Å; $\alpha_s = \pi/2$; τ = 0.02 s. H_z = 250 (1) and 500 Å (2), respectively.

the process of formation of the columnar structure gives way to the alignment of the condensation front.

The results of calculation of the dimensionless depth of the depressions between the clusters in constant time after the fall of the clusters on the surface as a function of temperature at the condensation front for two average distances between clusters (see Fig. 4.20) show that in contrast to the condensation on the nucleus there are only two characteristic regions of the structures, separated by a critical temperature T_k^*: the region of columnar structures and the region of the alignment of the condensation front into a plane. It should be noted that in contrast to the condensation on the nucleus on the substrate in condensation on the front with clusters the value T_k^* decreases with increasing distance between the clusters. As a result, there is a temperature range (for curve 2 T_k = 706–747 K), where the condensation front during τ = 0.02 s is aligned in a plane.

5.5.4. Discussion of the results

In conclusion, we give a qualitative description of the key moments of the formation of dispersed vacuum condensates.

At the stage of sorption of vapour molecules on a substrate competition takes place between the adatoms of the components of the vapour mixture on the same adsorption vacancies. Upon reaching the concentration of adatoms of supersaturation at which the nucleation probability exceeds a critical value, formation of nuclei of the condensate of one of the components starts. If the concentration of adatoms of other components remain unsaturated, they form condensate nuclei with concave surfaces around the nuclei of the supersaturated phase. This limits the transport of adatoms of the supersaturated component to the nucleus and stimulates the formation of next nuclei. If some other component becomes supersaturated, combined nucleation of the two components producing convex shape nuclei takes place. The possible combinations of the forms of the condensate nuclei are classified by the homologous matrix of the initial condensation structures.

The evolution of the resultant system of nuclei depends on the contact angles of components on the substrate, substrate temperature, its ideal state, the nature of the vapour source. There is a critical contact wetting angle of the substrate by the condensate; at smaller wetting angles nuclei of any form degenerate into a plane. There is also the critical temperature of the substrate, below which the columnar structure are realized, above – a flat condensation front. The critical temperature is determined by the wetting angle, the characteristics of the vapour source, the distance between the nuclei. Decrease of the contact angle, movement of the source to a point source and convergence of nuclei reduce the critical temperature of transition from columnar to planar.

The presence of the unsaturated vapour component determines the process of capillary condensation in the gaps between the crystals of the supersaturated components. Consequently, the columnar structure of the condensate is retained in the condensation process. The presence of additional supersaturated vapour components initiates the competitive growth of columns from substances of different components. This competition is based on the sorption of dissimilar adatoms on the same adsorption vacancies and on the mechanism of mutual shading of the crystals from the vapour sources.

The crystals of one component grow in a continuous competition with each other. The basis of competition is the mutual shading of the components from the vapour source. The result of competition is determined by random fluctuations of the shape and size. Larger crystals with a smaller curvature and with a sharp cut on the top have better growth conditions.

The thermal effects accompanying condensation cause desorption of the adatoms of the components with lower binding energy with the substrate, intensify the process of re-evaporation of the crystals and give rise to the appearance of temperature gradients. The latter generates an additional mechanism of mass transfer between the crystals which provides a high-quality contribution to surface migration, which leads to a trend to form equilibrium shapes of crystals in the structures of the condensates.

The presence of clusters in the vapour flow of vapour and their fall on the condensation front also create preconditions for the formation of columnar structures of the condensate. The principal difference from the condensation on the substrate with nuclei that have arisen as a result of nucleation, is the inverse dependence of the critical temperature of the section of the zone of the structures on the distance between the inhomogeneities of the condensation surface.

For the described qualitative picture of the formation of dispersed condensates this section of the book provides algorithms for quantitative calculations. The calculations correlate well with experimental data.

5.6. Some technological advice

The results presented here can be used to propose a number of technical recommendations for producing solid film or regular columnar structures in the process of vacuum vapour deposition. Some of these recommendations are well known in the existing technology, others are formulated for the first time. For the known technological methods we propose an interpretation of the effect achieved, which differs from the traditional one.

Known technological methods of producing single-crystal films. The use of molecular beams. In the single crystal effect, the molecular beams are minimally subjected to shading, which helps to fill with atoms the thin gaps between the irregularities formed on the substrate. With increasing energy of the atoms the molecular beam becomes single-speed???, with its susceptibility to shading decreased.

The temperature regime of the substrate. To obtain the maximum density of condensate nuclei with smaller sizes the substrate temperature in the initial stage of deposition should be low. Then, to improve the alignment of the inhomogeneities in the film, it should be increased. In addition to surface migration, the process of the condensation front includes re-evaporation of the sprayed substance that helps to fill defects–troughs, especially in the direction against the temperature gradient.

Vacuum hygiene. Removal of impurity vapours eliminates their capillary condensation in the areas having a large negative curvature. This reduces the density of resultant superatomic??? scale defects.

Cluster deposition. Impurity clusters in the vapour flow increase the density of filling of the substrate, thereby reducing the distance between the inhomogeneities of the condensation front and encouraging their more rapid merging into the monolith at the initial stage of condensation. During further deposition the clusters contribute to the closure of the crystallites with each other, if gaps form between them.

Spraying at oblique incidence. The effect of levelling the condensation surface is achieved at oblique incidence of the vapour atoms onto a substrate the large flow of matter is received by the sides of the inhomogeneities, which brings them closer together. However, at oblique incidence of the vapour flow the inhomogeneities which have come closed together cannot grow into each other due to strong shading effects at small distances between the inhomogeneities.

New production techniques. *The use of molecular beams.* The formation of the single-speed molecular beam is associated with the need for developing a special technique and with the loss of the efficiency the deposition process. It is therefore advisable to limit to the optimum extent the requirements imposed on the beam. It is enough if the indicator of the beam velocities lies within the minimum angle between the tangent to the profile of condensation and the direction to the beam source. This case eliminates shading effects. More accurate calculations of the parameters of the beam required for a particular technological problem can be made using the above models.

The temperature regime of the substrate. The calculations described in the sections 5.2 and 5.4 show that to produce a high-quality non-porous film the substrate temperature in the condensation stage should be higher than that at which the closing of the nuclei takes place by their development along the substrate and the first contact of the nuclei occurs on the surface of the substrate. At lower substrate temperatures the produced flat contours of the film do not mean that the structure is dense. Limitation of the upper substrate temperature is dictated by re-evaporation which is non-uniform along the contour of the nucleus. When approaching the saturation temperature the closing of the nuclei is delayed in time and the perturbation amplitude at the condensation front increases. It is in principle possible to calculate the optimum condensation temperature and time for producing dense films with minimal deviations of the surface from the plane.

Vacuum hygiene. At the nucleation stage the presence of unsaturated impurities in the vapour increases the packing density of the substrate by smaller nuclei. With further condensation the impurities must be removed from the working volume. Dense filling of the substrate with small nuclei ensures high quality of the films.

Cluster deposition. At the nucleation stage the impurity clusters in the vapour flow provide denser filling of the substrate, accelerating their closure and increasing the quality of the film. However, the cluster size of the vapour flow should be matched with the critical size of nuclei, formed on the substrate by heterogeneous nucleation. At the condensation stage the clusters contribute to the healing of joints of the crystals. However, the size of the clusters in the vapour flow in the condensation stage should match the scale of defects in the joints. Larger clusters and their high concentration in the vapour flow lead to an increase in the amplitude of perturbations of the condensation front and to the formation of non-dense structures in the film.

Spraying at oblique incidence. Oblique incidence of the vapour flow at the nucleation stage with transition to the stage of building a film should be combined with deposition with a normal molecular beam.

Technology for producing dispersed oriented films. Currently, there are no methods for obtaining film materials with an oriented disperse structure. However, due to the possibility of using these materials in electronics for compacted magnetic recording, in the chemistry of catalysts and filters, these technologies should be developed. A number of recommendations for the creation of such technologies have been made in this chapter.

1. *Structures with separate columns* correspond to the member of the homologous matrix with the code +c −cyl (see Fig. 5.3). They are realized in the joint condensation of vapour of the substance whose structure must be produced, and vapours of insoluble impurities. Adhesion to the substrate is determined by the supersaturation of vapour components with respect to vapour temperature. The size of the gap between the columns is calculated from the saturation in the process of growth of the condensate layer. To obtain coarse-crystalline columns, the substrate temperature should be maintained in the critical range T_k^*, but not above. To obtain polycrystalline and amorphous columns the substrate temperature should be reduced. Preservation of the transverse size of the columns in the condensation process is facilitated by the use of vapour flow in the form of a molecular beam.

2. *Disperse frame structures* are shown in the homologous matrix by members with the codes +c −cyl, −c+cyl, +c +cyl. The recommendations are similar to those for columnar structures. A special feature is the choice of the substrate temperature at all stages of the process below the melting point of the component for creating the frame. Structures with hollow pores in th frame structure are produced by the removal of the condensate of the impurity −c, +cyl.

3. *X-ray amorphous ultrafine composite materials* correspond to the members of the homologous matrix +c +c, +a +a. Modes of nucleation and condensation are calculated using the sorption kinetics (see section 2). These materials, due to their high metastable interfacial energy, have unique properties, such as the ability to produce film coatings by deposited of microparticles on the coated surface.

Conclusions

Thus, by mathematical modelling of the processes of sorption, nucleation and condensation we have traced 'genetic' and evolutionary aspects of the processes of structure formation of vacuum condensates. The 'genetic' factors may include factors which control the path of formation of the structures. These geometrical and crystallographic properties of the substrate, thermal and crystallographic properties of condensed matter, kinematic and dynamic properties of the vapour source. The specific combination of these factors include a specific set of physical processes responsible for the birth and evolution of the structures of vacuum condensates in three distinct stages: sorption, nucleation, development of nuclei. Apparent global changes in structures, such as moving from a monolithic film to the growth of columns occur in view of the criticality of processes involved in the evolution with respect to some parameters. For example, the strong dependence of the saturation curve on temperature with a slight change in the latter changes the balance between the rate of condensation and the rate of evaporation and this includes a mechanism for aligning the condensation front. It turns out that the fundamental puzzles have fairly simple solutions.

The authors do not claim that the studies described in this section are complete. However, the goal – to find a way to establish an unambiguous connection between the conditions and the result of condensation – has been fulfilled. Further prospects for refining and deepening research in this direction have been indicated.s.

List of Symbols

a - lattice parameter, m, chord segment sources nick a couple of meters;

a_i, b_i, with *i* - coefficients that reflect the probability of thermal desorptiontion, adsorption from the gas phase flow of the adsorbate in the center condensation;

b_{hi} - a dimensionless atomic flux components on the substrate;

C_v and s_m - specific heat of vapour and vapour impurity, J / (kg·K);

C_1 - the constant of integration;

D - diameter of the source, m;

F - cross-sectional area of the crystal on the target height in m²;

J - the frequency of nucleation, m⁻²·s⁻¹;

J_0 - flux vapour emitted from the surface of a diffusely vapour source, m⁻² ·s⁻¹;

J_s, J_m, J_v - flux from the source, impurities, and evaporation at the point of the crystal surface, m⁻²·s⁻¹;

J_{ki} - evaporation flux at the surface points visible out of a spot, m⁻²·s⁻¹;

N - number of different kinds of traps and sinks (condensation, capillary condensation in the pits, absorption);

H_z - the distance between the nuclei, Å;

$\bar{h} = (H_s - h_m)/H_{s0}$ - The relative depth of the gap between the columns;

h_m - the height of condensation front in the depressions between the posts, Å;

H_s - height of the columns in the condensate, Å;

K - curvature, m⁻¹;

K_q, K_b - rate constants of reactions of decomposition and synthesis;

k - Boltzmann constant, J / K;

L - the amount of energy impacts, specific latent heat condensation, J, the distance from the source to the substrate, m;

m - diameter at the base of the segment, m the atomic mass of vapour, kg;

n_a - adsorption density of vacancies, m⁻²;

n_{4i} - the concentration of vacancies condensation, m⁻²;

n (*N*) - the surface density of subcritical aggregates, m⁻²;

N - number of components;

Q - quantity of decomposition reactions and synthesis;

P_i - the specific density of the particle flux components on the substrate, m⁻²·s⁻¹;

R - radius of the nucleus, m;

R_{z0} - radius of the nucleus, Å;

S - area of the free surface of the nucleus, m²;

T - temperature, K;

T_a, T_b, - the local surface temperature of the crystallites, determine the saturation pressure p_h, K;

T_K, T_c, T_s - temperature of the crystallite, the vacuum chamber and vapour source, K;

T_k^* - the critical temperature of the radical changes in the structure condensate, K;

T_m - the temperature of the impurity vapour, K;

U_D - the activation energy of surface migration of adatoms, J;

V - volume of the nucleus, m³;

$v = v_{max}/v_{min}$ - the ratio of growth rates of the peaks and troughs of the condensate m / s;

x, y - Cartesian coordinates;

Z - number of nearest neighbors;

α_a, α_b - limiting the visibility angles source, rad;

α_k, α_{ki} - local rates of evaporation and condensation, respectively;

α_{li} - the probability of desorption of different energy L impact-viyami;

a_s - the angle, rad;

βr_a - the characteristic radius of curvature of the cluster, m;

$\varepsilon_s, \varepsilon_i$ - the energy of one bond of the adatoms with the substrate and with the surface of the nucleus, J;

$\varepsilon_k, \varepsilon_c, \varepsilon_s$ - the emissivity of the crystallite, the vacuum chamber and vapour source;

ζ - the vacancy rate of adsorption involved in the transportation-Adjustments of the adatoms to the fetus;

η_{hi} - the probability of interaction of adsorbed particles with traps and drains N of various kinds;

λ_k - thermal conductivity of the crystal, W / (m·K);

$v_\perp, v_\parallel$ - The frequency of thermal vibrations in the transverse and longitudinal directions to the substrate, c⁻¹;

$\delta\Phi(N)$ - the energy barrier of nucleation, J;

δ_q - local heat flux, J / (m²·s);

ξ_i - adatom concentration of i-th component, m⁻²;

ξ_{nm} - surface concentration of adsorbed molecules products of fusion reactions, m⁻²;

$\xi_{h\infty}, \xi_{hK}$ - saturated concentration of adatoms on a flat surface values and on a surface with curvature K, m⁻²;

σ - the specific surface energy, J/m²;

τ - time, s;

τ_i - the lifetime of the i-th component, c;

φ, φ_s - contact angle with the surface of the nucleus, rad;

φ_s^* - critical wetting angle at which the changing structure of the condensate, rad;

θ_i - dimensionless concentration of adatoms;

θ_{hi} - partial employment vacancies adsorption;

$\theta_{\infty i}$ - dimensionless concentration of adsorbed atoms to the plane the surface;

κ_i - the effective coefficients of condensation of vapour atoms on the surface of the condensed phase;

ψ_1, ψ_2 - extreme angles of a spherical wedge likely source measured from the normal to the surface area, rad;

ω_a - atomic volume in the condensed phase, m^3.

Indices:

a - atom;

c - the camera;

i - number of components;

h - saturation;

v - vapour;

m - an admixture;

l - the external action;

0 - initial;

s - the source of vapour;

k - a crystal, the substrate

z - nucleus.

Literature

1.   The technology of thin films, handbook, V. 2, ed. L. Mayssal, R. Gleng, Sov. Radio, 1977. 768 c.
2.   Bochkarev A.A., Polyakova V.I., Poverkhnost'. Fizika, khimiya, mekhanika, 1988, No. 3, 29–35.
3.   Bochkarev A.A., Polyakova V.I., in: Proc. IX Conf. on the dynamics of rarefied gases, Sverdlovsk, 1987, V. 116, 30.
4.   Bochkarev A.A., Polyakova V.I., in: Thermal processes in the crystallization of substances, Novosibirsk, 1988, 128–135.
5.   Palatnik L.S., Gladkikh N.T., Dokl. AN SSSR, 1961, V. 140, No. 3, 567–570.
6.   Palatnik L.S., Gladkikh N.T., Fiz. Tverd. Tela, 1962, V. 4, No. 2, 424–428.
7.   Palatnik L.S., Gladkikh N.T., Fiz. Tverd. Tela, 1962, V. 4,No. 8, 2227–2332.
8.   Polyakova V.I., in: New technologies and materials, Part 1, Novosibirsk, Nauka, 1992, 9–37.
9.   Modern crystallography, A.A. Chernov, et al., Moscow, Nauka, 1980, V. 3.
10.   Palatnik L.S., et al., Fiz. Met. Metalloved., 1963, V. 15, No. 3, 371–378.
11.   Berdnikova V.V., et al., in: Thermophysics of crystallization of substances and materials, Proceedings, Academy of Sciences of the USSR, Institute of Thermophysics, Novosibirsk, 1986, 122–132.
12.   Bochkarev A.A., Polyakova V.I., Izv. SO AN SSSR, Ser. tekh. nauk, 1988. No. 5, No. 18, 124–137.
13.   Berdnikova V.V., et al., in: Thermophysics of crystallization of substances and materials, Proceedings, Academy of Sciences of the USSR, Institute of Thermophysics, Novosibirsk, 1986, 115–122.

6

The formation of columnar structures in co-condensation of two components as a way of producing nanosized composites

In chapter 3 it was shown that the chain mechanism of nucleation is initiated by the influence of impurities on the nucleation stage, and then capillary condensation of impurities between the crystallites of the metal condensation leads to the formation of columnar structures in the condensate. In fact, the condensate with the columnar structure is a mixture of ultra-small particles of components insoluble in each other. This fact suggests a clear idea of developing a technology of nanosized materials by atom-by-atom folding of components in the condensate, thus obtaining a fundamentally new material which can not be obtained by macroscopic methods. The chemical and structural properties of such materials are determined by the parameters at which the process of joint condensation of the components is performed.

A general idea is developed in this chapter in a particular application area. Tests were carried out of the theoretical possibility of assembly of chemical power sources (CPS) by successive layer formation of the necessary components of their construction, including the housing, the insulator, the electrode, the separator and the electrolyte. The feasibility of this 'assembly' would fundamentally change the technology of production of simple electronic products required in large quantities. Instead of tedious manual assembly we can imagine an automated line of sealed chambers in which the whole product is 'grown' gradually during condensation or deposition. Moreover, the possibility of wide variation of the process parameters makes it possible to produce materials with the most useful properties for different components of structures. It is clear that such a general idea can be implemented in the short term and at low cost but it is important to explore in principle the possibility of its implementation. The focus of this work is dictated by the desire to find a way to use the already acquired knowledge on the formation of oriented condensates. In this regard, most of the effort was applied to the creation of an experimental setup in which the operating experience which could be easily transferred to the level of technology. This required the accommodation facilities to the new technology, as set out in the first section.

The second section will discuss the possibility of formation of various types of active anodes for the CPS in the process of condensation with varying chemical composition of the condensed material. In addition to the unusual chemical composition the condensation processes in principle can ensure the formation of the most useful structures of materials. For example, it is assumed that the columnar structure of the anode could provide a large reactive surface, which will increase the discharge current of the CPS.

The third section presents the results of testing the processes of evaporation and condensation of polypropylene in order to obtain porous films. This part of the work focuses on the future of manufacturing porous separators of the CPS together with the manufacture of electrodes and, possibly, impregnation with the electrolyte.

In the fourth section of this chapter we examine the feasibility of the idea of 'electric petrol', the essence of which is the creation of a composite anode active as a suspension of ultrafine lithium in propylene carbonate. It is expected that the 'liquid anode' will improve the CPS characteristics and make it able to be recharged many times.

The fifth section summarizes the results of a study of adsorption under the effect of light. This part of the work was carried out in accordance with the idea of finding ways to further control the processes involved in the formation of disperse systems by condensation. Interesting fundamental results that confirm the possibility of intensification of volume processes during adsorption and condensation were obtained.

Experience gained in the experiments in this chapter was used in further work to develop a production technology of nanosized metal–organic composites by joint condensation of the components. This is the subject of the next chapter.

6.1. Modernization of the experimental setup for work with lithium electrodes of chemical power sources

6.1.1. Modernization and basic requirements for equipment

The formation of columnar structures in the vacuum condensates is based on the growth of islands of one of the components with automatic filling of the gaps between them by the capillary condensate of impurities. Filling the gap with admixtures occurs even in the absence of saturation of the impurity vapour. This means that to produce a condensate with the given composition it is required to ensure an appropriate exact composition of the gas or vapour environment in which the process is carried out. Such conditions can only be created in vacuum. However, when the technological considerations require the transportation of goods through various processing chambers for performing various process steps, the use of vacuum causes additional difficulties. There is an engineering task of developing a simple process for rapid movement of goods in different processing chambers. Since we are talking about a specific technology, we can formulate a number of requirements to address this problem.

1. Production technology of the CPS is characterized by production of large numbers of products, which means that the displacement device should be rapidly acting.

2. Since it is necessary to transport articles from the atmosphere into the vacuum chamber and between the chambers, the gateway device should be sealed.

Other requirements for experimental and technological facilities derive from the specific process of vacuum deposition.

3. The implementation of joint or layer condensation from the vapour phase requires the presence in a single cell of several evaporators and the possibility of their simultaneous work in the regimes ensuring the production of the right structure of the condensate.

4. Results of vacuum condensation – the structure of the condensate and its adhesion to the substrate – are highly dependent on the temperature of the condensation front. This means that the installation must provide the necessary temperature regime of the workpieces.

5. The working chamber should be equipped with a device for cleaning the substrate before the deposition process to ensure the efficient adhesion of condensate.

The stated requirements were the basis of modification of the setup described in section 2.1.

6.1.2. The general scheme of the installation

In the present work, we used an experimental multipurpose metal–gas-dynamic installation designed for research into the processes of condensation of metal vapours in difficult conditions. The general diagram of the apparatus is presented in chapter 2 in Fig. 2.1. Because of the three-chamber configuration of the installation it can be used in different ways. The central vacuum chamber is equipped with a fixed evaporator with a power up to 60 kW. Ultimate vacuum in the chamber is $2.66 \cdot 10^{-4}$ Pa. In setting up experiments in the central chamber the side chambers can be used as lock chambers for the transportation of samples or diagnostic elements. The lock chambers can also act as a stand-alone pilot plant, as they have an autonomous pumping with ultimate vacuum of $1.33 \cdot 10^{-3}$ Pa. The lock chambers are equipped with manipulators, designed for installation on them of mobile working elements of the experiment.

The experiments were carried out in the lock chambers. One of them was equipped with evaporators of metals, a high-speed gateway for supplying substrates for deposition, and a box with an inert medium for the diagnosis of samples. The chamber was designed for joint and layer deposition of metals. The scheme of the chamber is shown in Fig. 6.1.

Another lock chamber was equipped to carry out experiments to investigate the joint condensation of vapours of metals and organics. It contained a metal evaporator, a collector to feed the vapour of organic matter, a capacitor, and a device for removing condensation products.

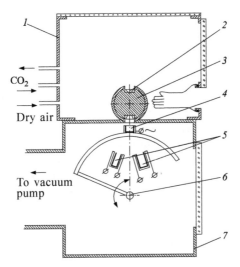

Fig. 6.1. The lock chamber for the deposition of metals. 1) casing of the bag box; 2) substrate; 3) high-speed gateway; 4) substrate heater; 5) metal evaporators; 6) swivel arm; 7) chamber housing.

6.1.3. Fast gateway device

A simple gateway device was developed, manufactured and tested in order to quickly move the substrate into the vacuum chamber for deposition. Its scheme is shown in Fig. 6.2. It is based on the well-known rotary plug valve circuit. The substrate is set in a depression in the plug and when the plug is rotated it appears in the workspace of vacuum chamber. Unlike similar devices, the casing

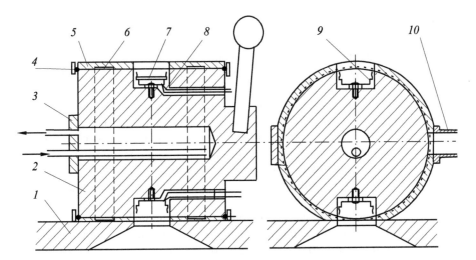

Fig. 6.2. The scheme of the high-speed gateway. 1) cover of the vacuum chamber; 2) plug; 3) cooling of the plug; 4) seal; 5) the body of the gateway; 6) vacuum channels; 7) substrate; 8) thermocouple; 9) substrate holder; 10) vacuum line.

of the gateway has channels with independent vacuum pumping, intended for preliminary evacuation of the depression of the plug during its rotation. This protects the vacuum chamber from penetration of the foreign gas.

The plug of the gateway provides channels for a thermally stabilizing and the input of thermocouples to monitor the temperature of the substrate. The thermocouples are mounted on the holders of the substrates.

Tests have shown high performance characteristics of the gateway device. When turning the plug the vacuum in the chamber is created by the vacuum tightness of the gateway. Several samples were deposited using the gateway. At the same time, a number of shortcomings of this design have been identified.

1. Placing the plug in the casing the gateway to the third class of accuracy requires vacuum grease for vacuum sealing. During operation of the gateway the vacuum grease tube enters the zone of 'visibility' of the evaporators. Radiation of the evaporators sublimes the vacuum grease in the chamber, pollutes the working environment, which affects the quality of samples. This shortcoming is amplified with the approach of the evaporators to the working window of the gateway. This structure can be improved by be more precise manufacturing of parts of the gateway.

2. The design of the chamber allows one to set the gateway on the chamber in such a manner that adjacent evaporators can be used only over limited distances. This model increases the loss of the evaporated metal and reduces the deposition rate. This disadvantage can be eliminated by changes in chamber design but this in not the subject of this chapter.

Identified weaknesses should be considered when creating a custom installation for the deposition of parts for CPS. In the present work, samples with satisfactory quality were obtained using the composition of the work area set out below.

6.1.4. The layout of the evaporators

The schematic of the lock chamber used for the study of deposition of metal coatings is shown in Fig. 6.3. The evaporators are secured in the chamber on the current supply. Sputtered substrates are mounted on a cartridge mounted on the manipulator. The evaporators are oriented in two possible options: either both are directed to the substrate for joint deposition, or set so that when the cartridge with the substrates is turned the desired component can be coated. For the option of simultaneous deposition of two components there is some difficulty due to the fact that the simultaneous orientation of the two evaporators for one substrate is possible only at oblique incidence of the vapour flow to the substrate.

Spraying experiments were carried out using the thermal method of cleaning used substrates to remove adsorbed impurities on them. To this end, the chamber contains a resistive heater. Before the beginning of deposition the substrate is placed using the manipulator near the heater for dynamic warm-up. The heating of the substrate to a temperature of 150°C was considered sufficient to obtain good adhesion of the deposited material.

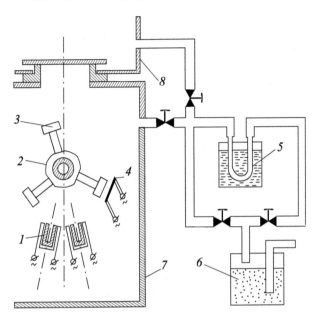

Fig. 6.3. Schematic of the chamber for the deposition of metals. 1) evaporators, 2) cartridge with samples, 3) sample holder, 4) the heater for cleaning, 5) liquid nitrogen air dryer, 6) silica gel dehumidifier, 7) the chamber body, 8) casing of an inert box.

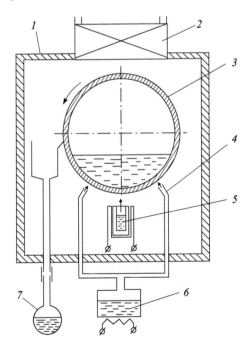

Fig. 6.4. Chamber for studying joint condensation of metal and organic matter. 1) casing, 2) gate, 3) capacitor, 4) nozzles for supplying organic matter vapours, 5) metal evaporator, 6) organic matter storage, 7) the collector of the condensate.

The schematic of the lock chamber used to study the joint condensation of vapours of metals and organics is shown in Fig. 6.4. Inside the chamber there is a rotating condenser, cooled by water or liquid nitrogen, depending on the requirements of the experiment. There is a metal evaporator oriented to the capacitor, close to the capacitor there are nozzles for supplying the vapour of organic matter. Organic matter vapours are generated by an individual evaporator located outside the chamber. A blade is used to remove the condensate from the capacitor, and the evacuation of the product from the chamber is carried out using a collector – a vacuum line with locking devices.

6.2. Formation of the active anode of the CPS in joint and stratified condensation of lithium and aluminum, lithium and organic matter on electrodes

The task of forming the anodes of the CPS requires a priori:

a) deposition of a given amount of metal corresponding to the electrical capacitance of the CPS. This condition can be satisfied by regulation of the time of deposition, if we have previously studied the deposition rate depending on the modes of evaporators;

b) the achievement of a given composition in the deposited electrode. In joint condensation this requirement may be satisfied by specifying the ratio of vapour flows of the different components on the substrate. If the condensation efficiency of the different components is different, then dosing of the vapour flows of the components is not sufficient to obtain the desired composition of the condensate. This requires control of the composition of the samples. Formation of the sprayed layer of a given composition by evaporation of an alloyed prepared in advance is extremely difficult due to the fact that the Raoult's law has been studied insufficiently for most mixtures and alloys, and even the well-known condensed state–vapour equilibrium diagrams are not satisfied in evaporation in vacuum.

These two requirements mean that to obtain the required samples certain methodical preparation is required prior to the experiments – study of the characteristics of evaporation and of the condensation process. In general, this methodical preparation is complicated and time-consuming. For the evaporator we must define: the indicator of the specific vapour flow, the velocity distribution function of the vapour atoms, the chemical composition of the vapour generated by the vaporizer, the presence and composition of the cluster component in the vapour flow, the intensity and spectrum of electromagnetic radiation. These parameters of the vapour flow and temperature and also the temperature and orientation of the substrate completely determine the properties of the resulting condensate. The formulation of these methodological studies within the framework of this study would not achieve the applied purposes. So here we confined ourselves to general tests of the aluminum evaporator.

Table 6.1. Results of tests of an aluminium evaporator

Mass of sprayed aluminium, mg	T_v, K	Mass of sprayed aluminium, mg	T_v, K
3.0	1553	4.6	1585
3.3	1572	5.0	1603
4.4	1576	8.0	1623

6.2.1. Methodical testing of evaporators

The values obtained in the evaporation of aluminum from an open cylindrical crucible 12 mm in diameter and deposition on a stainless steel substrate with a diameter of 18 mm at a temperature of 370 K at a distance of 30 mm are presented in Table 6.1. Spraying was carried out for 5 min. The amount of the metal deposited on the substrate was determined using an analytical balance.

The evaporator can be analyzed on the basis of the Langmuir representation of the evaporation process. In evaporation in high vacuum the atomic flux of evaporation is described by

$$J_v = \frac{p_h(T_v)}{(2\pi m k T_v)^{1/2}},$$ (6.1)

where p_h is the vapour pressure of aluminum at the temperature of evaporation T_v; m is the mass of the aluminum atom; $k = 1.38 \cdot 10^{-23}$ J/deg is the Boltzmann constant. The saturation curve is approximated by the formula of the form

$$p_h(T) = K_1 T^{K_2} \exp(K_3 / T),$$ (6.2)

where $K_1 = 1.04528 \cdot 10^{-9}$, $K_2 = 5.434572$, $K_3 = -27818.35$.

The calculation of the rate of increase of the mass of the condensate on the substrate under the assumption that the evaporation flow reaches in the unaltered form the substrate and condenses on it with the efficiency of condensation equal to unity, can be carried out by the formula

$$J_k = J_v m f_k,$$ (6.3)

in which f_k is the sprayed area on the substrate.

To calculate the linear velocity of condensation or evaporation we can use the equation

$$\dot{H} = J_v m / \rho_k.$$

Here ρ_k is density of the condensate.

The results of calculations by the formulas (6.1)–(6.3) are shown in Fig. 6.5 as the dependence of the rate of increase of mass on the substrate on the temperature of the evaporator. The graphs show in the form of the experimental points with the scaling factor of 60.2 the data from Table 6.1 after calculating the appropriate speed for them. It is seen that the experimental data repeat the nature of the calculated curve, indicating at least the normal operation

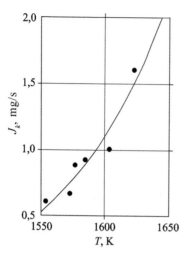

Fig. 6.5. Rate of increase of the mass of aluminum in the sample.

of the evaporator and the fact that the condensation of aluminum occurs at a constant efficiency. The need to introduce the scaling factor is due to the fact that the calculations do not take into account the geometrical parameters of the evaporator–substrate system and the vapour scattering from the source.

Thus, the calculation by the expressions (6.1)–(6.3) and the graph in Fig. 6.5 can serve as a guide for determining the parameters and the time of deposition of aluminum on the samples. However, it should be borne in mind that when changing the source–substrate configuration the scaling factor should be clarified by the experiment.

A similar analysis of the operation of the evaporators was also conducted for the evaporation of magnesium and lithium. The initial experimental data for magnesium are shown in Table 6.2.

The results of processing the data in Table 6.2 by (6.1)–(6.3) are shown in Fig. 6.6 for the required parameters: $K_1 = 5.483369 \cdot 10^{13}$, $K_2 = -1.029346$, $K_3 = -17330.55$; $\rho_k = 1.74 \cdot 10^3$ kg/m³, $\mu_{Mg} = 24.3$ and a scale factor $M_{Mg} = 26.7$.

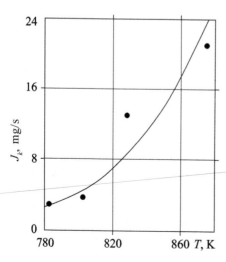

Fig. 6.6. Rate of increase of the mass of the sample during the deposition of magnesium.

Table 6.2. Mass of magnesium sprayed on substrates with a diameter of 18 mm for 5 min

T_v, K	Sprayed magnesium mass, mg	T_v, K	Sprayed magnesium mass, mg
783	33.5	829	46.4
803	42.3	875	225.7

The existing correlation between experiment and calculation shows that for magnesium we can use consideration for determining the mode of deposition.

At the highest condensation rates (see Fig. 6.6) the surface of the Mg condensate is covered with faceted crystallites. If the condensation rate is reduced to up to 9.5 μm/s and less the condensate is a set of inclined plates with a typical size of 3–5 μm. A micrograph of this sample is shown in Fig. 6.7.

Tests on lithium evaporators are difficult because of its chemical activity and the inability to weigh the samples on an analytical balance; therefore, in the case of lithium the thickness of the deposited layer was measured. The data are presented in Table 6.3.

The results of processing the data in Table 6.3 are presented in Fig. 6.8. The parameters used in calculations were: μ_{Li} = 6.94, ρ_{Li} = 0.534·10³ kg/m³, K_1 = 6.137384·10², K_2 = –5.799734, K_3 = –23894.5, M_{Li} = 7.48. In the figure the black dots indicate the values averaged over all measurements at a given temperature. It is seen that at 833 K the experimental points have a high spread. This may be due to the fact that when loading the evaporator Li is strongly oxidized and the oxides prevent evaporation. The data at 873 K were obtained by loading lithium, stored under a layer of vacuum oil. For these data, the scatter of the experimental results is smaller.

The structure of the surface of the vacuum condensate of lithium, as well as magnesium, depends on the condensation rate. At the condensation rate of about 2 μm/s the condensate surface is covered with plates. A micrograph of this sample is shown in Fig. 6.9 a. With increasing deposition rate of up to 4 μm/s the surface of the Li condensate shows growth of faceted crystallites

Fig. 6.7. Micrograph of a sprayed magnesium sample, ×1000.

Table 6.3. Duration and thickness of deposition of Li

$T_v = 833$ K									
Time, min	25	20	10	20	25	25	20	20	16	12
Thickness, mm	0.35	0.2	0.1	0.3	0.55	0.5	0.3	0.4	0.48	0.45

$T_v = 873$ K							$T_v = 898$ K		
Time, min	12	8	10	11	11	12	12	6	5
Thickness, mm	0.55	0.35	0.4	0.4	0.45	0.4	0.45	0.5	0.55

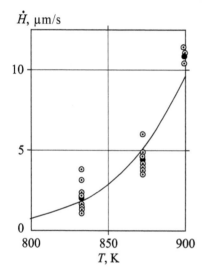

Fig. 6.8. Linear deposition rate of lithium.

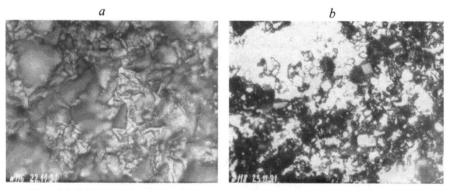

Fig. 6.9. Micrograph of the surface of a lithium sample; a) ×500; b) ×250.

(Fig. 6.9, b). In accordance with previous studies, these results indicate that for condensation of lithium with a porous structure is required to carefully select mode condensation. Judging from the data in Fig. 6.9, a and b, in the manufacture of CPS electrode samples it is necessary to reduce the rate of condensation.

6.2.2. Stratified condensation of lithium and aluminum

Layer-by-layer (stratified) deposition of lithium and aluminum on the cover of the casing for CPSs was conducted in the setup shown in Fig. 6.3. The evaporator temperature and deposition times required for deposition were determined using the calculated curves in Figs. 6.5 and 6.6. Spraying was conducted in the following order. The covers of the CPS placed in the cartridge were degassed under vacuum with heater 4 (see Fig. 6.3.) by heating to a temperature of 150–200°C. A lithium layer was then sprayed on all substrates in the cartridge. The power of the lithium evaporator was switched off and the aluminum evaporator was heated up. By rotating the cartridge the samples were oriented to the aluminum evaporator and an aluminum layer of the calculated thickness was deposited.

Some peculiarities were revealed in the process. Higher (in comparison with the Li evaporator) working temperatures of the aluminium evaporator heated up the sprayed specimen. This leads to melting of the lithium already existing on the sample to disruption to the resulting structure of the anode on it. The result of this deposition is the coating of the cover of the CPS with lithium in the form of droplets 2–4 mm in size. To avoid such detrimental consequences, aluminum was deposited produced in intermittent pulses with a duration of 5 s and the on/off ratio of 2 min. After coating all the samples in the cartridge the chamber was opened through an inert box, the cartridge was replaced and the samples were placed in transport containers.

Table 6.4. Anode samples sprayed for tests

Sample No.	1st Li layer	2nd Al layer		3rd Li layer	4th Al layer	
	mm	μm	mg	mm	μm	mg
7	0.3	0.2	0.2	–	–	–
14	0.5	0.15	0.1	–	–	–
15	0.5	0.3	0.25	–	–	–
18	0.4	0.3	0.25	–	–	–
19	0.4	0.15	0.1	–	–	–
20	0.4	0.2	0.2	–	–	–
24	0.45	0.35	0.3	–	–	–
27	0.3	0.07	0.06	0.3	0.2	0.2
28	0.25	0.07	0.06	0.25	0.15	0.1

Samples of anodes shown in Table 6.4 were coated and transferred to the Novosibirsk Plant of Chemical Capacitors (NCCP) for the installation in the CPS.

6.2.3. Joint condensation of metal vapours

The implementation of the process of the joint condensation of two different metals imposes additional requirements for the experiment. Firstly, there is a need to harmonize the regimes of condensation of two components for the same temperature on the substrate. Secondly, the mode selection of the evaporators of the different components must be made from the condition of the same duration of the process. Third, for the same structures of the condensate on the substrate as in the condensation of individual components, it is important to maintain the specific vapour flows to the substrate.

We show that these requirements can be satisfied on the whole. The specific vapour flows on a substrate from evaporators in a general form can be written as

$$J_1 = A_1 f_1(T_1) r_1^{-2}, \tag{6.4}$$

$$J_2 = A_2 f_2(T_2) r_2^{-2}. \tag{6.5}$$

Here A_1, A_2 are the scaling factors that reflect the real properties of the evaporators; f_1, f_2 are the functions of the type shown in Figs. 6.5 and 6.6; r_1, r_2 are the distances of the evaporators from the substrate.

The evaporators emit the following radiation fluxes on the substrate

$$Q_1 = B_1 T_1^4 r_1^{-2}, \tag{6.6}$$

$$Q_2 = B_2 T_2^4 r_2^{-2}, \tag{6.7}$$

where B_1, B_2 are the geometric form factors of the evaporators. Because the condensation mode and the structure of the condensate are defined by both the specific vapour flow and the radiation incident onto the substrate, for example, from (6.4) and (6.6) we see that for separate operation of the evaporators there is only one combination of T_i and r_i, corresponding to a particular condensation mode.

To preserve the condensation mode of the vapour components at joint operation of the evaporators, it is necessary to ensure constancy of J_1, J_2 and the sum $Q_1 + Q_2$. The total radiation flux of the jointly working evaporators Q_{i0} should remain the same as that which provides the structure of the condensate of the components. In this case, from (6.6) and (6.7) we obtain the equation

$$Q_{i0} = B_1 T_1^4 r^{-2} + B_2 T_2^4 r^{-2}. \tag{6.8}$$

The system (6.4)–(6.8) is not closed with respect to unknown T_1, T_2, r_1, r_2, which represent the set of its solutions – combination of T_1, T_2, r_1, r_2, suitable for the process. The choice of a suitable solution should be made on the basis of the design features of the equipment used.

The above desired formulation of the experiments to study joint condensation shows that the experiment require time-consuming preparatory work for the characterization of the evaporators of the type (6.4) and (6.5), but this is beyond the scope of this paper. Therefore, we confine ourselves

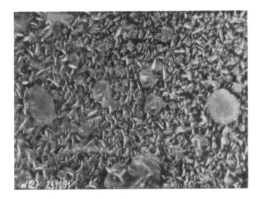

Fig. 6.10. Lamellar structure of the joint condensate of magnesium and aluminium. Deposition rate of magnesium and aluminum respectively, 8 and 1 mg/s.

to the joint condensation of magnesium vapours with a deposition rate of 8 mg/s and aluminum vapour with a deposition rate of 1 mg/s. The structure of the condensate at the joint condensation of magnesium and aluminum in this mode is shown in Fig. 6.10.

The figure shows clearly the lamellar structure of the condensate of aluminium and magnesium. Against the background of this structure there are larger polycrystalline aggregates whose origin is apparently not due to the heterogeneous condensation of aluminium and magnesium vapours. In all likelihood, the origin of these aggregates should be attributed to the formation of particles on the walls of open crucibles by the fluctuation mechanism described in chapter 4.

6.2.4. Magnesium with oil

Experiments with joint condensation were planned and conducted for the following reasons.

1. The lamellar structure of the condensate discovered in the previous experiment with joint vapour deposition of different metals can be practically useful. The lamellar structure of the condensate of the metal can serve as a power matrix for filling with the condensate of the liquid substance, such as electrolyte, in formation of the CPS.

2. Joint condensation of magnesium vapours and vacuum oil is the simplest case of joint condensation of metal and organic matter. It can be considered the first step to begin a more detailed study of the formation of ultrafine metal–organic composites, which is set out in the next chapter.

3. The magnesium–vacuum oil vapour simulates an inevitable process that takes place in a vacuum system vapour oil pumping which is used for the evaporation of the metal. The joint condensation of magnesium and oil simulates what happens on the walls of the vacuum elements. Knowledge of these processes is useful for the development of vacuum technology used for high-temperature applications.

a   *b*

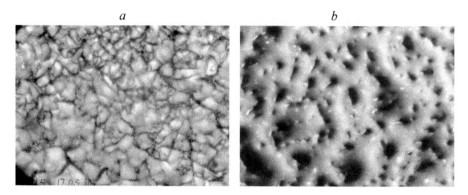

Fig. 6.11. Joint condensation of magnesium and vacuum oil vapours on a copper substrate, ×1000, dark field. a) the mode for Mg: T_0 = 1324 K; T_v = 946–1027 K; for oil T_v = 326 K; for substrate T = 289.5–290.3 K; in the chamber p = $(6–7.6)\cdot10^{-4}$ Pa, b) the mode for Mg: T_0 = 1324 K; T_v = 1009 K; for oil T_v = 347 K; for substrate T = 304 K; in the chamber p = $6.1\cdot10^{-4}$ Pa.

The experiments were carried out with a magnesium vaporizer with a superheater and a sonic nozzle. Figure 6.11 shows images of the condensate for two different regimes.

In Fig. 6.11, the lamellar structure of the magnesium condensate is clearly visible. The cells, formed by plates of magnesium, are filled with the oil condensate. This result proves that the hopes of the possibility of creating a matrix which could serve as a structural support for the metal–organic liquid dispersed system are justified. Figure 6.11 b shows the joint condensate of magnesium vapour and oil vapour at a higher rate of evaporation of oil. It is seen that the cells, formed by the lamellar crystallites of magnesium, are completely filled with the liquid and partially flooded with oil. With further increase in the evaporation rate of oil only oil condenses on the substrate. This mode of joint condensation confirms the concept mentioned above that planning of the regimes joint condensation of different components on a single substrate requires coordination of the modes of evaporators of components.

6.2.5. Condensation of magnesium on solid carbon dioxide

This section discusses the condensation of two components on a single substrate, when a chemical reaction between the components is possible. The study of the interaction of a stream of magnesium vapour with carbon dioxide in the solid state at 77 K was part of [1]. The formulation of these experiments was associated with an attempt to isolate the particles from the magnesium stream, in which homogeneous condensation was supposed to have taken place. Also, as a result of the interaction of the magnesium vapour stream with the layer of a porous carbon dioxide condensate after the experiment and after removal of carbon dioxide by evaporation the layer may contain infiltrated particles of the magnesium condensate.

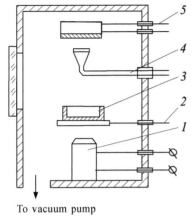

To vacuum pump

Fig. 6.12. Experiment setup. 1) magnesium evaporator; 2) valve; 3) condensate receiver; 4) source of carbon dioxide; 5) cryogenic substrate.

In these experiments, the vacuum chamber was equipped in accordance with the scheme shown in Fig. 6.12. The main elements of the installation were: the metal evaporator 1, the valve 2, the condensate receiver 3, the source of gaseous carbon dioxide 4, the substrate 5 cooled with liquid nitrogen. The valve and the source of carbon dioxide were moved by a rectilinear manipulator.

In the experiments, a target with a diameter of 140 mm was at a distance of 120 mm from the nozzle exit of the evaporator. After reaching a pressure of $1.33 \cdot 10^{-3}$ Pa in the vacuum volume, the diffusion pump gate was closed and carbon dioxide condensed on the target surface cooled with liquid nitrogen. After reaching the desired thickness of the layer of CO_2 on the target the carbon dioxide supply was interrupted, the source of carbon dioxide was shifted to the side and the original pressure was restored in the vacuum chamber. The magnesium vapour source was then activated to the operating mode, the valve was opened and the magnesium vapour flow interacted with the surface of carbon dioxide. During the experiment the chamber pressure slightly increased the evaporation of carbon dioxide in the condenser. In the experiments the interaction time of magnesium vapour with carbon dioxide was varied in the range 1–60 s and the thickness of the carbon dioxide condensate in the range 0.5–3 mm.

It was established that the determining factor in these experiments is the chemical interaction of active magnesium atoms with solid carbon dioxide. The result of their interaction at low temperature is a white powder that covers the whole area of the target and easily peels from the surface. After the interaction with the atmosphere the powder becomes yellow. Consequently, the original idea of trapping of magnesium clusters by the carbon dioxide condensate was not justified.

A special feature of the experiments was the glow of green crystals of carbon dioxide during their interaction with the magnesium vapour and flares in selected areas of the vacuum chamber. The luminous region in the chamber appear immediately after the interruption of the magnesium flow by the valve. No specific plausible explanation of these effects is available at the present

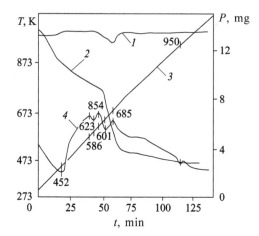

Fig. 6.13. The results of thermal and thermogravimetric analyses. 1) the rate of change of weight on heating; 2) change of the weight of the heated substance with time; 3) heating curve; 4) the differential heating curve.

time. It is difficult to explain the glow by any type of discharge because the pressure in the vacuum chamber was not more than $5 \cdot 10^{-3}$ Pa, the maximum voltage in the studied range was 2.5 V at a frequency of 50 Hz. Under these conditions, the burning of an electric discharge is impossible. Apparently, the observed glow is due to chemiluminescence [2].

The study of the reaction products in the optical and electron microscopes showed that particles of the resulting powder have a developed surface. Some of them are transparent, glassy scales. Electron diffraction analysis shows that the powder particles are amorphous.

Analysis of the resulting powder by X-ray diffraction, spectral emission and thermographic methods revealed that the reaction products are X-ray amorphous, containing 2–3% hydrogen, 13% magnesium, 15–18% carbon. After calcination the ash contains 31.6–33.1% of magnesium oxide. The results are shown in Fig. 6.13. The exact composition of the formed product is difficult to determine. It may be that the powder is a mixture of $MgCO_3$ + MgO.

The method of the joint condensation of vapours of active metals with organic and inorganic reagents on a surface with a cryogenic temperature, refers to cryochemistry which is currently being developed at the interface of chemical physics, thermal physics and materials science. Cryochemical reactions result in the stabilization of metastable products or a chemical reaction may start whose rate is relatively high only at low temperatures. As a result, it is possible to produce new materials which can not be obtained in other conditions [3, 4]. This suggests a promising method for the synthesis of organometallic compounds.

6.3. The formation of the porous structure of the polypropylene condensate suitable for separators of chemical power sources (CPS)

6.3.1. Evaporation of polypropylene

Experiments with the evaporation of organic matter and polymers in vacuum systems have a number of features. First, the polymers require careful degassing before evaporation. Heating the polymers to the evaporation temperature without degassing leads to rapid boiling and emissions from the crucible. Second, due to the high viscosity of the polymer at temperatures of their appreciable evaporation the formation of bubbles in the volume leads to foaming of the polymer which results in the unsteady evaporation process. In this regard, the evaporators of polymers should have a developed evaporation surface. Third, the heating of the polymers can lead to thermal degradation before evaporation. Fourth, the evaporation and condensation of the polymers in the vacuum chamber can lead to a deterioration of the vacuum properties of the system as the evaporation products are difficult to remove from the vacuum tank and their penetration them into the vacuum pumps damages the working body and pump components. In connection with the latter in this paper we confine ourselves to a small number of experiments.

6.3.2. Condensation of polypropylene at normal temperature

The tests on polypropylene condensation on the surface with the temperature close to room temperature were performed using the lock chamber, shown in Fig. 6.1. Granulated polypropylene was evaporated from the evaporator designed for lithium at a temperature of 489±0.2 K. Condensation was carried out on a stainless steel substrate. Evaporation of polypropylene for 10 min resulted in the formation of a continuous polypropylene film with a thickness of about   1 μm on the substrate.

In these experiments conducted at an evaporation temperature of 573±0.1 K no condensate formed on the substrate. Apparently, the condensation of polypropylene in these conditions takes place with low efficiency and increase of the radiant flux on the substrate with increasing evaporation temperature evaporation resulted in the absence of condensation. To improve the efficiency of condensation it is necessary to reduce the substrate temperature.

6.3.3. Condensation of polypropylene at cryogenic temperature

Experiments with evaporation at different temperatures and condensation on the surface having the temperature of liquid nitrogen were performed using a chamber equipped as shown in Fig. 6.3. Polypropylene vapours were deposited on a rotating and a stationary stainless steel condenser. The experiments were conducted at a pressure in the vacuum chamber of about 0.133 Pa.

At an evaporation temperature of 323 K polypropylene did not condense in the form of film. A condensate in the form of individual 'hair' up to 30

a   *b*

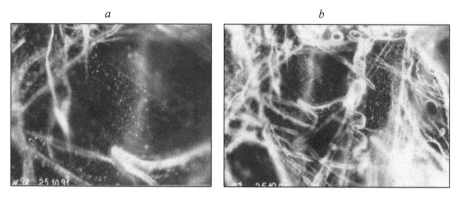

Fig. 6.14. Micrographs of polypropylene films. a) ×500, dark field, polypropylene temperature T_p = 623 K; b) ×200, dark field, polypropylene temperature T_p = 598 K.

a   *b*

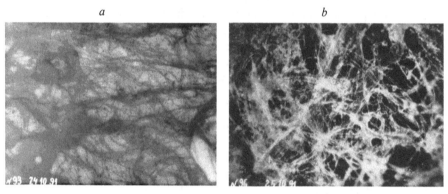

Fig. 6.15. Micrographs of polypropylene 'strands'; ×50, dark field; polypropylene temperature T_p = 623 (a), 598 K (b).

mm long formed on the condenser. When the evaporation temperature was increased to 598–623 K and condensation time to 60 min a polypropylene film, consisting of three layers, formed on the condenser. A thin film of grayish mass, resembling the consistency of clay, settled directly on the condenser surface. On this film there was a second, thin transparent adhesive porous polypropylene film with a thickness of about 30 μm with a pore diameter of 5–60 μm. Fragments of the film are shown in micrographs in Fig. 6.14 a, b. It is seen that on the film there is a layer of polypropylene strands, which consist of interwoven 'hair' with a thickness of 2 to 5 μm. The thickness of the strands reached 50–60 μm (Fig. 6.15 a, b).

At an evaporation temperature of 598 K the strength of the polypropylene film was low and with increasing distance from the condenser fractured to pieces. By increasing the evaporation temperature to 648 K only the plasticine-like mass settled on the condenser formed probably as a result of the condensation of products of thermal degradation of polypropylene.

Thus, experiments have shown the fundamental possibility of the re-evaporation and condensation of polypropylene. The formation of polypropylene products in this way requires the formulation of studies, including the diagnostics of the properties of the resulting material.

6.4. Formation of ultrafine lithium powder in the atmosphere of electrolyte vapours

The work was done at the suggestion of A.F. Naumenko and I.E. Abroskin at the Novosibirsk Chemical Concentrates Plant (NCCP). The general idea is the following. The Laboratory of Thermophysics of Microdisperse Systems has experience in the investigation of processes of the joint condensation of vapours of zinc and magnesium and organic matter with the formation of composite materials, which are based on organic matter, and the filler – a metal in the ultrafine state. The use of these metals determines the range of application of the produced composites. Use of any other metal, of course, extends the range of investigation of the processes of condensation and increases the range of application of condensation products. Lithium has attracted the attention by its 'extreme' physical and chemical properties and the importance of its use in technology and engineering. Therefore, the idea of producing lithium in the ultrafine state and of studying its properties is extremely attractive because it gives qualitatively new options for its effective use, including in the CPS. The interest in this work also results from the fact that the authors of the idea and the authors of this monograph are not aware of anyone getting stable lithium in the ultrafine state.

Since the *a priori* implementation of this concept will meet the difficulties associated with the extreme properties of lithium, the authors felt it necessary to set out the results of a literature review of preface, limited, however, to only a narrow field of application of lithium – as the active electrode material in chemical power sources. A brief review of studies of lithium chemical power sources, prepared by A.F. Naumenko and I.E. Abroskin is essential here for the efficient formulation of the experiments. The slight departure of this part of the monograph from the main content is justified by the authors in the hope that this work is the beginning of an even more interesting continuation in the future.

6.4.1. A brief review of the scientific and applied research of lithium chemical power sources

Lithium is the lightest metal, solid at room temperature. Because of the low density and high standard potential, Li is an ideal electrode material for high-energy galvanic cells. However, lithium is a chemically very active element, which interacts with inert nitrogen already at room temperature. This imposes severe requirements on the electrolytes used in lithium CPSs .Intensive research work related to the development of lithium batteries started only in the 60s. The large-scale manufacture of certain types of similar power sources has now been established. These systems are very flexible to use. By selecting the cathode material the voltage of the CPS cell can be changed from 1.5 to 3.9 V. The specific consumption of lithium is very low: 0.26 g/(A·h), which leads to high values of specific energy, in some cases up to 600 W·h/kg [5]. Current sources with a lithium anode must necessarily be leaktight since the absorption of moisture from the environment leads to severe corrosion of lithium. This complicates their design and manufacture, but is an advantage during operation. Lithium CPSs have a low self-discharge rate and storage time of several years.

Properties of electrolytes
Substances used as electrolytes in lithium chemical power sources must meet the following requirements [6]:
  1. Stability with respect to lithium.
  2. High conductivity.
  3. High ionic mobility.
  4. Compatibility with the cathode material.
  5. Safety
  6. Neutrality with respect to the environment.
  Only the aprotic medium is suitable as the basic material for electrolytes in lithium batteries because lithium reacts violently with water and other protonic solvents. Therefore, the most suitable solvents for the electrolytes in lithium CPSs are, apparently, complex cyclic esters: propylene carbonate, ethylene carbonate, butyrolactone. They have low boiling points, low enough viscosity, high dielectric constant, good chemical and electrochemical stability. Low-viscosity additives commonly used are simple ethers such as diethyl $(C_2H_5–O–C_2H_5)$. The physical constants of the solvents for the electrolytes for lithium CPSs are shown in Table 6.5.
  The characteristics of the CPS are largely dependent on the quality of their components. That is why they are subject to very high requirements. It is important to distinguish the impurities that cause chemical reactions or specific solvation, and impurities, which affect only the physical properties. In particular, the proton matter (water, primary and secondary amines) should be removed because they spontaneously interact with lithium.
  Solvents such as simple lithium salts LiCl, LiF are little soluble in most aprotic solvents. Slightly higher solubility is observed in the case of lithium

Table 6.5. Physical constants of solvents at 25°C

Solvent	Formula	Dielectric constant	Viscosity, cP	Density, g/m³	T_{boil}, °C	T_m, °C
Propylene carbonate	$CH_3–CH–CH_2–C–$ $(O)_2=O$	66.1	2.53	1.198	241	−49.2
Butyrolactone	$(CH_2)_3–O–C–O$	39.1	1.75	1.125	204	−43.5
Tetrahydrofurane	$(CH2)_4=O$	7.4	0.46	0.880	64	−65
Dimethyl sulphooxide	$(CH_3)_2–(O)_2=S=O$	46.7	2.0	1.095	189	18.6

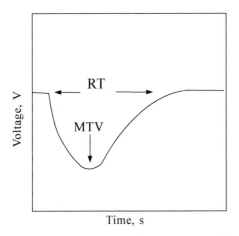

Time, s

Fig. 6.16. Voltage during the discharge of a lithium battery.

perchlorate $LiClO_4$ and lithium salts with complex anions such as $LiAlCl_4$, $LiBF_4$, $LiPF_6$ and others, although for them the solubility is also less than in water – no more than 2 mol/l. The specific electrical conductivity of solutions of these salts is $0.5–2$ ohm/m^2.

The stability of electrolytes with respect to lithium

The most important condition for the electrolytes used in lithium batteries is their stability in contact with metallic lithium. Today it is generally accepted that there are no solvents that are thermodynamically stable with respect to lithium [7]. Stability is kinetic and is based on the formation of a passivating protective film [8]. A protective film is composed of insoluble products of the reaction of lithium with the electrolyte. Its nature depends on the composition and purity of the electrolyte. The film acts as a solid phase of the interface between lithium and the electrolyte. It has the properties of a solid conductor of lithium ions with high internal resistance. Most often, the migration of lithium ions through the passivating film is a stage limiting the discharge rate. During discharge of lithium batteries the voltage drops sharply at first and then rises to a constant value (Fig. 6.16).

The minimum transient voltage (MTV) and recovery time (RT) are functions of variables such as electrolyte composition, temperature, density, discharge current, etc. The rapid voltage drop at the beginning of the discharge is caused by the ohmic voltage drop across the passivating film at the boundary of the lithium with the electrolyte. Constant ion current dissolves mechanically the passivating film which at the time of discharge is immediately restored by the standard lithium cell reaction.

The reaction of lithium with propylene carbonate

The most commonly used electrolytes in the lithium batteries are the electrolytes based on propylene carbonate (PC). This is due to the high solubility of lithium salts in PC, high dielectric constant, low volatility and non-toxicity. In the early 1970s the study of the reaction between PC and lithium amalgam showed that the main reaction products are lithium carbonate and propylene produced by the reaction [9]

(Scheme 1)

$$CH_3 - CH-O \diagdown_{C=O} + 2Li \longrightarrow Li_2CO_3 + CH_3 - CH = CH_2$$
$$CH_2 - O \diagup$$

This has led to a speculation that the major component of the passivating film on the surface of lithium is lithium carbonate. However, more recent studies [10], using more sensitive methods of investigation, in particular infrared Fourier transform spectroscopy, showed that although Li_2CO_3 is indeed present in the passivating film, the main ingredient in it are the alkyl carbonates of lithium with the general formula RCO_3Li. In addition, if the salt for the electrolyte is $LiClO_4$, the film reveals the presence of lithium chloride LiSC, which is obviously formed in the recovery of $LiClO_4$ by lithium. Due to the fact that the passivating film plays an important role in the discharge process of the lithium cell, understanding the mechanism of its formation may be a key to improve the performance of lithium cells.

In [10] the au   rs proposed a scheme of reactions leading to the formation of the passivating film:

(Scheme 2)

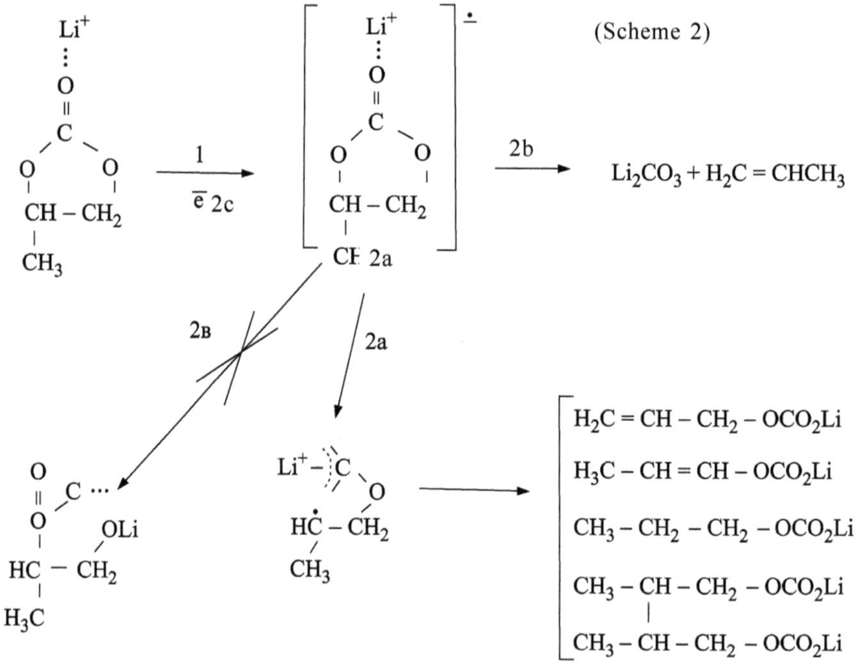

The first stage is a single-electron transfer with the formation of a radical-anion, where the latter can reach the transition state by rupturing the bond C–O, forming a rather stable particle in which the negative charge is compensated by the formation of a common electron pair with the cation from the salt for the electrolyte, a radical is located on the second carbon atom (stage 2a). The subsequent path of the reaction depends on the ease of the second electron transfer in this system. Because of the presence of the passivating film on the surface of lithium the second electron transfer is probably quite slow, in favour of other competing processes, such as the reaction of formation of the radical end (stage 3). Scheme 2 shows four possible compounds that can be formed by the transfer of one electron to the PC molecule with the subsequent closure by the radical. As expected, the competition between the processes of branching and closure leads to formation of a mixture of alkyl carbonates. In the presence of traces of water in the PC, which usually takes place, the passivating film will also contain LiOH, Li_2CO_3, alkoxides of Li, and alcohols.

Water reacts with lithium to form LiON, which in turn reacts with the PC by the following reaction:

(Scheme 3)

$$HO^- + \begin{matrix} CH_2-O \\ | \\ CH-O \\ | \\ CH_3 \end{matrix} {\Large\diagup} C=O \xrightarrow{Li^+} \begin{matrix} CH_2-OH \\ | \\ CH-OCO_2Li \\ | \\ CH_3 \end{matrix} \xrightarrow{Li} \begin{matrix} CH_2-OLi \\ | \\ CH-OCO_2Li \\ | \\ CH_3 \end{matrix} + (1/2)H_2{\uparrow}$$

Thus, the lithium electrodes in the propylene carbonate solution are always covered with a passivating film consisting mainly of lithium alkyl carbonates. This film plays a crucial role in inhibiting the discharge of lithium cells.

The specific conductivity of electrolytes based on PC

Non-aqueous electrolytes are at least 10 times less conductive than aqueous solutions. This is a shortcoming of the current lithium cells, as the increase of the internal resistance of the cell reduces the current and hence the energy density. In this solvent, the conductivity of the electrolyte is determined by the number and mobility of ions and their association in electrically neutral pairs. In the lithium cells the cation of the salt for the electrolyte is Li^+. For high mobility the anion should be small and the solvent should have low viscosity. On the other hand, one can expect a low degree of association for large anions and a high dielectric constant of the solvent.

In practice, the relatively low viscosity and high dielectric constant are achieved only when by mixing the solvents. In propylene carbonate, which has a high dielectric constant, the low viscosity addition is usually represented by dimethoxyethane (DME). For each multi-component mixture there is an optimum where the conductivity reaches a maximum value; for a propylene carbonate–dimethoxyethane mixture this ratio is 42.12:57.87% [11].

The existence of the maximum can be explained by two opposing factors. Adding of the DME to the PC reduces the viscosity of the solvent and thus increases the ionic mobility, i.e. the conductivity of the solution. But at the same time, the dielectric constant of the mixture decreases, which causes an increase in ionic association, and the conductivity decreases due to the smaller number of free charge carriers. Increasing temperature increases the maximum ionic conductivity of the electrolyte.

Other aprotic solvents

Intensive search is being carried out in the world for ways to improve the lithium cells. Different electrolytes and additives to electrolytes and salt–electrolyte–solvent combinations are studied. In addition, it is planned to replace the now widely used manganese dioxide MnO_2 by the cathode material. In particular, the latest developments in this area are the cathodes made of a polymer material.

The most promising solvent for lithium cells is 1,3-dioxolane [12]. This solvent is used as a basis for creating a rechargeable lithium battery [13]. Although the 1,3-dioxolane also reacts with lithium and a passivating film, consisting mainly of lithium alkoxides, forms on the lithium electrode surface, the film in the dioxolane has the best electrochemical properties than in other aprotonic media and is more homogeneous. In addition, LiOH (formed on the surface of lithium by reaction with unavoidable traces of water) reacts with dioxolanes to form soluble products.

The cyclability of lithium cells in dioxolane–$LiClO_4$ systems is much higher than in cells with other solvents. The disadvantage of the 1,3-dioxolane – $LiClO_4$ system is its explosiveness caused by a sharp increase in the resistance of the solution due to an unexpected polymerization of the electrolyte [14].

In addition to dioxolane another promising electrolyte for lithium batteries is 2-methyl tetrahydrofurane (cycling efficiency 96%) [6].

Thus, at present studies of the lithium batteries are carried in the following interrelated areas:

1) study of the chemical composition and morphology of the passivating film formed on the lithium electrode in aprotic solvents, and the mechanism of its formation;

2) search for new electrolyte compositions having a low resistivity;

3) search for a better cathode material;

4) attempts to improve the lithium battery.

Based on the above review, it was decided that the experiments to obtain the ultrafine composite of lithium should be carried out using propylene carbonate.

6.4.2. The experiments

The experiments required additional training of the pilot equipment.

Features of the experimental setup

To organize the flow of metal vapour and organic electrolyte, joint condensation and collection of the products an experimental setup based on the Metagus-2 metal gas-dynamic installation was constructed. The apparatus is shown in Fig. 6.17.

The installation is assemble in the airlock chamber 7 equipped with a vacuum pumping system, consisting of a high-vacuum oil vapour pump 3 with a capacity of 2.15 m³/s, the vacuum pumps 1 and 4, with a capacity of 0.04 m³/s and a rotary vane pump designed for degassing of the solvent

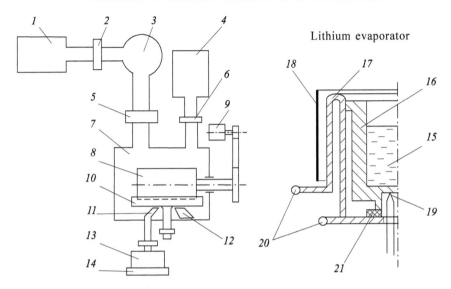

Fig. 6.17. Scheme of a vacuum unit and lithium evaporator. 1) backing pump, 2) solenoid valve, 3) diffusion pump, 4) booster backing pump, 5) valve, 6) valve, 7) airlock, 8) condenser, 9) gear motor, 10) collection of product, 11) nozzle of the electrolyte, 12) lithium evaporator, 13) evaporator of the electrolyte, 14) electric heater, 15) melt, 16) the crucible, 17) heater, 18) screen, 19) thermocouple, 20) current leads, 21) insulator.

and the pumping the Wilson seal of the condenser. The pumps are connected to the chamber through the gate 5 and the valve block 6. In the right cover of the chamber there is the Wilson seal in which the hollow condenser 8 rotates at 4 rpm. The condenser is made of a machined stainless steel tube 165 mm in diameter with thermally insulated (with Teflon) plugged ends. The condenser is 110 mm long and is suspended on a cantilever hollow rod through which liquid nitrogen is delivered into the condenser and the boiling point of nitrogen is stabilized at 78 K. The condenser is driven through a belt drive by the gear motor 9. The chamber has high-current inputs on which the metal evaporator 12 is mounted. It is possible to feed DV voltage on the evaporator. The chamber also includes a pipeline to feed the vapour of the organic electrolyte 11, connecting two nozzles with a critical diameter of 4 mm with the evaporator 13, located on the electric heater 14.

The resulting composite is collected using the collector 10, made of copper sheet 0.5 mm thick. The upper part of the collector touches the condenser and scrapes the resulting product, thus ensuring constant conditions in the condenser. To reduce wear and tear, the upper part of the collector is joined along its full length by soldering to a thin stainless sheet (thickness 0.2 mm, width 9 mm) with a sharpened sawtooth cutting edge. The lower part of the collector is soldered to a copper pipe to drain the collected product in the flask. The design of the collector provides a continuous collection of the composite and minimizes the evaporation of the more volatile component from it.

Table 6.6. Characteristics of the vapour source

Required power, kW	to 1.5
Maximum temperature, K	1300
Maximum loading of the metal melt, ml	7
Time to establishment of the stationary mode at T = 1000 K, min	18
Stability of Li temperature, K	±0.4
Temperature oscillations period, s	0.3
Area of the metal melt, cm²	11.6

The pressure in the chamber during experiments was maintained at 0.1–0.5 Pa. The temperature of the thick walls of the chamber made of stainless steel matched room temperature, 17–18°C.

Characterization of lithium vapour source
The lithium vapour flow was created using the open stainless steel crucible 16 installed in the evaporator of the metal (see Fig. 6.17). The evaporator is a cylindrical coaxial heating element 17 made of stainless steel strip 0.3 mm thick. The desired temperature of the crucible was maintained by radiation of the heating element and controlled by a chromel–alumel thermocouple 19, recessed in the bottom of the crucible. To minimize heat losses by radiation the metal shield 18 was placed coaxially in relation to the heater. Automatic maintenance of temperature of the melt at a constant level was carried out by furnace thermostat R-133 defining the feed voltage of the heating element with the signal amplifier U13 through a step-down transformer. The main characteristics of the lithium vapour source are shown in Table 6.6.

The crucible was loaded with lithium from a sealed container in an argon environment. The dependence of the rate of evaporation of metal on the

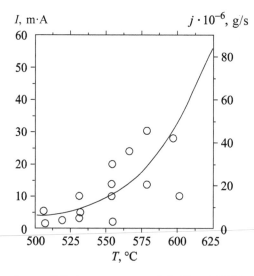

Fig. 6.18. The dependence of discharge current and the flow of lithium on the temperature of the melt. Line – the flow of vaporized lithium, points – glow discharge current.

crucible temperature is determined as follows. The Knudsen formula is used to estimate the vapour flow:

$$j = \left(p / (2\pi mkT)^{1/2}\right) K_T K_m S. \tag{6.9}$$

The vapour pressure of lithium approximated by the formula

$$p = K_1 T^{-K_2} \exp(K_3 / kT).$$

In a specific case $K_1 = 1.47505 \cdot 10^{13}$, $K_2 = 1.4393406$, $K_3 = 19\,251$, $K_T = 0.194$, $K_m = 1.53 \cdot 10^{-21}$, $S = 22.5 \cdot 10^{-6}$ m². The result of the calculation of the evaporated flux by (6.9) is depicted by the solid line in Fig. 6.18.

In the experiment in the evaporation of lithium a negative potential of −17 V was applied to the crucible and the resultant glow discharge current was measured. The positive discharge column glowed bright carmine-red light (670.8 nm), which indicates the presence of excited atoms and lithium ions. The currents of the glow discharge for different temperatures of lithium are shown by dots in Fig. 6.18. The graph shows the proportional cross-correlation function of the discharge current and the evaporation rate with respect to the evaporation temperature of the lithium melt. This correlation arises for the following reason. The glow discharge current at a constant voltage is proportional to the concentration of neutral, ionized (by discharge) lithium atoms in the vapours and is completely determined by the cathode discharge area, spreading approximately 1–2 mm above the surface of the molten lithium. This zone is located inside the crucible, and in the surface area it can be assumed that the lithium particle concentration is proportional to the vapourized flow [15, 16]. It follows that in a first approximation, the discharge current should be proportional to the flow of evaporated lithium. This suggests the possibility of determining the evaporation rate and ability to manage the vapour flow guided by the value of the glow discharge current.

Characterization of the organic electrolyte evaporator
The evaporator of propylene carbonate consists of a water bath with a flask with the electrolyte 13 (see Fig. 6.17). The desired bath temperature is maintained by the electric heater 14, which is powered by the power amplifier U13. Automatic maintenance of the desired temperature is ensure using the furnace thermostat R-133 and readings of a chromel–alumel thermocouple, sending a control signal to the amplifier. The flask is connected by vacuum-tight joints to copper tube 11, which is introduced into the chamber through a vacuum seal. The large area of heat exchange between water and the flask allows to hold the desired temperature for more than 100 ml of propylene carbonate.

Before the experiment the propylene carbonate is carefully degassed through the vacuum line of the booster pump, and the volatile dimethoxyethane is almost completely evacuated, so only the propylene carbonate is evaporated in the operating conditions of the experiment. The vapour source (PC) has the following main characteristics:

Power consumption, kW	to 1.5
The maximum temperature, K	360
The maximum loading of the organic electrolyte, l	1
Time to steady state at 373 K, min	27
The amplitude of the fluctuations in temperature, K	±1,1
The oscillation period, min	6

In all experiments, the temperature of the water bath was maintained at 100°C. The temperature of the electrolyte in the flask was 98°C.

6.4.3. Joint condensation of lithium vapours and propylene carbonate on the cryogenic surface

All experiments were performed in the following sequence. After downloading and preparing to work the camera evaporators evacuated to the residual atmosphere of less than 0.1 Pa. Then the capacitor is rotated at 4 rpm and frozen in liquid nitrogen. After freezing the evaporator condenser metal output at a given temperature. When a trace of condensation of lithium vapour propylene feed was opened. At the same time consumption of liquid nitrogen through the condenser should be increased to maintain the temperature of the condenser. The chamber pressure is not greater than 0.5 Pa. The operating mode is kept constant during the consumption of lithium from the evaporator. Upon completion of the operation the heating of the lithium evaporator is switched off, the supply of the propylene carbonate vapours is stopped, the condenser thawed, and the product collected in the receiving flask.

When submitting propylene carbonate vapours the area of the lithium vapour flow shows easily observed chemiluminescence, and the brightness and region of emission increase with increasing lithium flow. In the visible range the main emission band is situated in the range 550–580 nm. The condenser and the chamber walls show precipitation of a green–yellow reaction product of the interaction of the electrolyte and lithium. The resulting product is almost completely free from pure lithium. In accordance with the information presented in the survey in section 6.4.1 it is assumed that lithium alkyl carbonates form at a high rate in the vapour phase and then settle on the condenser. In most experiments, propylene carbonate was found in the vapour phase in large excess. Therefore, the composite consisting of propylene carbonate and ultrafine insoluble products of the reaction of lithium with PC formed in the condenser.

At a slight excess of lithium in the vapour phase the product contains traces of pure lithium passivated by the electrolyte. The case where lithium vapours are present in large excess is discussed below.

The condensation of lithium in the atmosphere of low-pressure pressure electrolyte vapour

Experiments were performed at a chamber pressure of 0.133–0.665 Pa. Because it is not possible to completely from the vacuum chamber the propylene carbonate which remains on the walls from previous experiments, the residual atmosphere of the vacuum chamber in these experiments is almost entirely composed of propylene carbonate vapour, which condenses on the condenser, is collected by the collector, evaporates from it and from the flask and condenses again on the condenser. Thus, the walls of the vacuum chamber were the source of propylene carbonate vapours which circulated in the wall–condenser-collector–walls–condenser circuit. The main difference from previous experiments is that there was no additional supply of the propylene carbonate vapour from the evaporator in this case. The pressure in the chamber was regulated by moving the slide valve of the vacuum gate at the pumps operating in a stable mode. Under these conditions no glow was observed in the chamber and the unoxidized lithium dispersion was produced.

With increasing Li evaporation temperature the colour of the condensing product varies from black, which indicates the formation of a nanodispersed composite, to opaque, indicating the coarsening of lithium particles. With a very small stream of lithium vapour the condenser has the colour of the rainbow from orange to green, indicating the deposition of a thin transparent film of only propylene carbonate.

The black and matte colours of the condensate in the experiment indicates the formation on the condenser of disperse structures with characteristic dimensions less than 1 μm. The condensate on the knife of the collector shatters into black powder and as a solid falls into the receiving flask. In an atmosphere of propylene carbonate vapour and in the packing of solid propylene carbonate the lithium powder is passivated by reaction products of lithium and the electrolyte with generation of a small amount of gas pumped away by the pump. Gas flows (presumably hydrogen) and evaporating propylene carbonate create eddies rising in the receiving flask.

Unfortunately, no reliable method for the determination of the particle size and the concentration of lithium was available absent. Therefore, based on the colour of the condensate it was *apriori* assumed that the deterioration of vacuum and reduction of the lithium flow leads to an increase in the particle size of lithium. The same trend was observed at the joint condensation of zinc and butanol.

To prevent the ignition of lithium powder and the possibility of storing the powder without coagulation at the end of the experiment, a small amount of propylene carbonate from the electrolyte evaporator was re-condensed to the receiving flask to form a composite containing 1–4 wt.% of lithium. Greenish-yellow and carmine-red flashes were observed in the receiver of the product and on the knife. Apparently, there is intense passivation of the surface of Li clusters which has a relative low rate at cryogenic temperature.

It is established that the state of the vacuum chamber walls plays an important role in the condensation processes. The porous condensate with the

Table 6.7. Characteristics of specimens of produced composite

No.	Type of condensation	Pressure in chamber	Lithium melt temperature, °C	Mass concentration of Li in composite, %
1	Joint	5	557	
2	In PC vapours	7	580	1
3	As above	5	600	1
4	As above»	1	582	4

columnar structure of lithium, which is formed on the chamber walls during the experiments, can absorb the propylene carbonate and affect its flow to the condenser. The destruction of this condensate may cause the generation of Li clusters to form an additional source of flow of lithium on the condenser. Complete stability and repeatability of experiments and modes are possible, if the condensate is not damaged and removed from the chamber walls. Complete removal from the vacuum chamber walls of the Li condensate and products of its reaction with the air atmosphere is difficult, especially since the condensate on the walls of the chamber is renewed in the course of each experiment.

The experiments produced four samples of the ultrafine Li–propylene carbonate composite for testing in experimental chemical power sources (Table 6.7).

6.4.4. Prediction of the size of lithium clusters in a composite

Condensation of Li vapours in the atmosphere of the electrolyte vapour at low pressures (see section 2.5.2) may be described by a model of heterogeneous nucleation and growth in the presence of impurities, taking into account the surface diffusion of atoms and molecules of the components adsorbed on the substrate [17]. The effect of diffusion on the nucleation and growth of particles and the inclusion of diffusion migration in the model at cryogenic temperatures is justified in several experiments [18, 19] and reviews [20]. The model of nucleation of two-component vapours on a solid substrate, taking into account surface diffusion, is described in Section 3.3 of this monograph.

The kinetics of the concentration of adsorbed (on a flat grid) vacancies of particles of the two-component vapour mixture (1st component – Li, 2nd components – PC) in one-dimensional formulation can be represented by two equations (3.17) (see section 3.3.1 in Chapter 3). Analysis of the solutions of these equations in environments where heterogeneous nucleation of clusters of lithium and the formation of capillary nuclei of propylene carbonate around them are possible gives an insight into the mechanism of formation of the composite at the initial stage. The composite forms as a result of the chain mechanism of formation of nuclei of lithium and propylene carbonate, as shown schematically in Fig. 3.17 in chapter 3. This is the basis for the growth of the condensate of the composite in the form of columnar structures of lithium filled in the gaps with propylene carbonate. Cutting the condensate from the condenser and defrosting results in the formation of liquid propylene carbonate containing lithium clusters.

The lower limit of the typical mean mass size of the lithium cluster, which is formed on the condenser can be determined as follows. The flow of the condensed particles of lithium on one vacancy per unit time is expressed by the equation

$$\Pi_1 = c_1 \left[\Theta_1(0) - \Theta_{h1} \right] / \tau_0. \tag{6.10}$$

Here c_1 is the probability of a local exchange of the adsorbed lithium atoms on the condensation surface; Θ_1 is the surface concentration of adsorbed lithium atoms close to the capillary nucleus of propylene carbonate made dimensionless on the basis of the density of vacancies on the condensation surface n_a; Θ_{h1} is the same in the equilibrium conditions at saturation of the surface traps in the state where the body is long enough in a saturated atmosphere of lithium vapour; τ_0 is the characteristic time of processes.

At the same time the frequency of nucleation of lithium for one vacancy is given by (3.19)

$$J = \zeta v \Theta_1 n \, (N^*) \, \exp \, (-u_{D1}/kT), \tag{6.11}$$

where $\varsigma = (1 - \Theta_2)/2$ is the proportion of the vacancies at the border of the nucleus, which can serve as a way to transport particles of lithium; v is the frequency of thermal vibrations of the adsorbate along the surface $n(N^*) = n_a \exp(-\delta\Phi(N^*)/kT)$ is the density of aggregates containing particles on the surface, which form and break due to fluctuations $\delta\Phi(N^*)$ is the energy barrier of aggregate formation; u_{D1} is the activation energy of diffusion. With increasing Θ_1 when the J value of about 1 is reached a nucleus of a cluster of the first component forms on the unit surface. The concentration of lithium Θ_1 is taken at the border of the PC nucleus. From this the estimate of the size is obtained by dividing the condensing flow (6.10) by the nucleation rate (6.11).

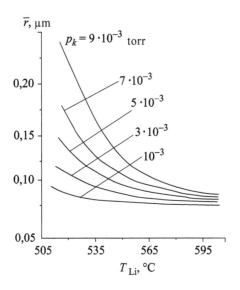

Fig. 6.19. The dependence of mean mass cluster size of lithium on the temperature of molten lithium and the chamber pressure.

Fig. 6.20. A suspension of lithium in propylene carbonate, stored for 5 days in air. The suspension was prepared as follows: a drop of the composite was placed between two slides; ×500.

The typical mean mass size is the radius of a circular cluster, having the same mass as the aggregate formed on the condenser.

Numerical estimates of the condensation of lithium vapours in the atmosphere of PC were obtained using the following input data: $n_a = 2.24$ m^{-2}, $\tau_0 = 9.1 \cdot 10^{27}$ s, $D_1 = 1.38 \cdot 10^{-18}$ m^2/s, $D_2 = 8.14 \cdot 10^{-13}$ m^2/s, $a_1 = 0.5$, $a_2 = 9.56 \cdot 10^{18}$, $c_1 = 1.08 \cdot 10^{10}$, $c_2 = 6.37 \cdot 10^5$, $b_1 = 3.97 \cdot 10^{43} \cdot T^{-1,4393}$ exp $(-19251/T)$, $b_2 = p_b \cdot 3.70 \cdot 10^{30}$, where p_b is the pressure in the chamber, the other notations are the same as in chapter 3.

The results of assessment calculations on cited models are presented in Fig. 6.19 in the coordinates 'typical mean mass size–the melt temperature of lithium' for different pressures in the chamber. It is evident that with the deterioration of vacuum and a decrease in melt temperature (decrease of the incident vapour flow of lithium) the lithium cluster size increases, as observed in experiments. Our calculations show that in this process it is not possible to obtain clusters with a characteristic radius of less than 837 Å:

$$\langle \overline{r} \rangle = 8{,}37 \cdot 10^{-8} \sqrt[3]{1/(1 - \Theta_2)}.$$

Studies of the various experimentally produced polyethylene–magnesium and butanol–zinc composites with an electron microscope showed that the particles have an elongated shape with a length to diameter ratio of (3–5):1, which are aggregated into larger particles through a layer of an organic component. Analysis of the lithium–propylene carbonate composite with an electron microscope is greatly hampered by its strong reactivity in air. Therefore, Fig. 6.20 shows a photograph of only products of the interaction of the lithium–propylene carbonate composite with the air atmosphere. The picture shows the presence of particles of micron and submicron size. Of course, this is the result of oxidation and aggregation of the primary particles.

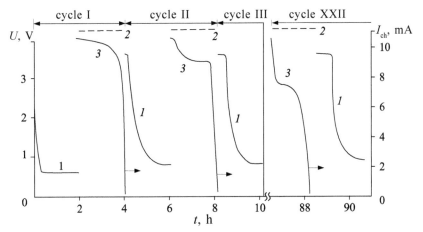

Fig. 6.21. The voltage across the resistor of 201 ohms and the charging current during cycling cell No. 1. 1) discharge; 2) charge; 3) charging current.

6.4.5. Tests of the manganese-ion cell of the chemical power source with a 'liquid electrodes'

Two types of experimental CPS assembled in a laboratory were tested. In the preparation of the first type of CPS an Orion manganese–zinc cell was opened and the zinc powder with the suspended electrolyte was removed. The cell was washed and dried with ethyl acetate. Approximately 1 ml of the composite, which contained about 10 mg of lithium, was placed in the cell in air. This was followed by adding about 7 ml of the electrolyte with propylene carbonate. The cell was closed with a tight but not hermetic cap with a built-in nickel electrode. The theoretical capacity of the cell was about 40 mA·h

The results of cycling through the load impedance of 201 ohms are shown in Fig. 6.21. It is seen that during the discharge of the element after assembling the discharge current falls rapidly to a value of 3.6 mA and remains the same for more than two hours. The capacity of the cell after assembly is less than 15 mA·h but after charging the cell for two hours the discharge current in the first 8 minutes is 13 mA, and then decreases exponentially to 4 mA and stabilizes. The graph shows that when cycling the initial value of the discharge current, the discharge time at this maximum current and the final value of the current increase from cycle to cycle. The type of the time dependence of charging current also changes. A total of 22 cycles were performed. At the last cycle the measured battery capacity was 36 mA·h. Improved performance with increasing number of operating cycles can be related to the structuring of the composite, which improves the electrical contact between the particles themselves and between the electrode and particles. When the cell was dismantled deposition of the dense layer of lithium on the electrode was not observed. The work of this cell, despite the fact that it was not hermetic, was found to be satisfactory.

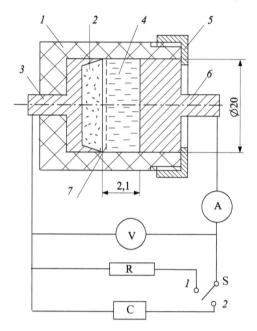

Fig. 6.22. The test setup of the lithium power source. 1) housing; 2) manganese dioxide powder; 3) cathode (stainless steel); 4) electrolyte with the lithium particles; 5) nut; 6) anode (stainless steel); 7) separator (tissue paper); A) ammeter; V) voltmeter; R) load resistor; S) mode switch; C) DC source.

The second type of the modelling CPS cell was based on the DML-120 cell by replacing the lithium plate by a 'liquid' electrode. At the same time the following points were taken into account:

1) the ability to seal the cell;

2) compatibility of the characteristics with industrial CPS;

3) recording of the dynamic characteristics of the CPS with a 'liquid' electrode at the discharge to the load, simulating the MK-51 calculator.

Figure 6.22 shows a diagram of the model element (cell) and the measuring circuit of the test bench. The hermetically sealed PTFE casing 1 contains stainless steel cathode 1, with manganese dioxide powder 2 pressed into the recess in the cathode. At the cathode there is the separator 7 consisting of one layer of tissue paper. On the other side the battery is sealed with a stainless steel anode 6 pressed with the nut 5. The gap between the anode and cathode was filled with 0.66 ml of the composite, mixed with the standard electrolyte in a 1:1 ratio. The theoretical capacity of the cell was 3.16 mA·h. The cell was discharged through the load resistance of 46 kiloohms, the voltage drop under load was measured with a voltmeter V-G1212 and recorded by a recorder, the discharge and charging current were measured by an ammeter A-B7-35.When th cell was charged the charging current was stabilized (stabilization error 16%) using a DC source. The cells were assembled in an argon atmosphere from components previously dried under vacuum for 20 min.

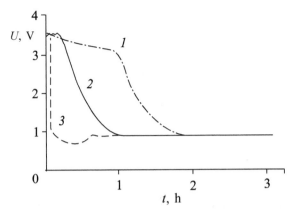

Fig. 6.23. Comparison of the dynamics of the voltage across the resistor R with power supplied from: 1) cell number 1 with a resistor $R = 201$ ohms; 2) cell number 2 after charging with a resistor $R = 46$ kiloohms; 3) cell number 2 after three days of storage.

The dependence of the characteristics of the CPS in cycling is similar to the previous cell. Figure 6.23 shows a comparison of the characteristic dynamics of the voltage across the load in time for the first and second cell. Curve 2 corresponds to the fourth discharge of the second cell. It is seen that in the first 15 min the discharge current value is around 74 mA, which is 19% higher than the operating current of the calculator. Then within 40 min the current drops to 20 µA, which is two times less than the permissible value. Further, the current stabilizes at this level and remains unchanged for more than 14 hours of continuous operation so that the real capacity of the modelling cell could not be determined. The third curve shows the discharge characteristic of the same cell after three days of storage with no load. It clearly shows a dip in voltage caused by the surface passivation of lithium.

Small steady-state values of the discharge current and voltage of the second cell may be due to the following reasons:

a) insufficient salt concentration in the electrolyte – half the factory value;

b) insufficient ether concentration in the electrolyte – half the factory value;

c) the possible presence of water in the electrolyte due to its long-term storage;

d) inefficient sealing of the cell, which led to the formation of poorly conducting passivating films on the anode surface, visible to the naked eye.

The main result of the tests was the proof that the rechargeable CPSs with a liquid active anode in the form of a composite of ultrafine lithium, packed in propylene carbonate, are functional. Research should be continued in this promising area.

6.4.6. Summing-up

The composite of dispersed lithium with a particle size of 0.2 µm in propylene carbonate was produced as a result. The process of its production was justified by the earlier models of two-component vapour adsorption. It was concluded that it is possible to control the size of clusters in the range greater than

0.1 μm. The technology for producing lithium powder, apparently, which is a valuable commercial product, requires only a vacuum system, loading and evaporation of lithium, collection of the powder and its packaging, so it can be quite simple. However, the processes involved in this technology are far from simple.

At sufficiently strong flows of the evaporated PC almost all lithium is oxidized by the vapours in the gas phase with the formation of lithium alkyl carbonates. The oxidation reaction is accompanied by chemiluminescence. In the visible range the emission spectrum covers the yellow–green region. Apparently, chemiluminescence accelerates the chemical oxidation of lithium, exciting the molecules of propylene carbonate, its radicals, and possibly lithium monomers and dimers.

When applying a potential of −117 V to the crucible in the presence of the lithium flow we ignite the glow discharge glowing carmine-red. The current of the glow discharge can be used to control the lithium vapour flow and organize the feedback between the discharge current and the vapour flow in regulating the process.

Preliminary tests of the 'PC-based electrolyte–dispersed Li' composite were conducted for its use as a 'liquid active anode'. Tests have shown that this composite material is workable in rechargeable CPSs, which opens up prospects for the development of renewable 'electric petrol' for electric vehicles.

Using the composite of domestic and industrial secondary elements requires more detailed working out the concentration of salt and other ingredients, their careful dehydration and sealing of the cell, and also the possibility of changing the basis of the electrolyte. Using 1,3-dioxolane as the electrolyte will improve cycling and thinning of the passivating film that can reduce the internal resistance of CPS.

Of special scientific interest is the generation of clusters by the cryogenic drum and the chamber walls and the role of the walls in the condensation process. This question has been poorly studied and, at the same time, this process is ubiquitous – from the formation of fog to a variety of processes in chemical reactors and crystallizers.

6.5. Study of photodesorption processes with periodic exposure to light

Studies of magnesium vapour condensation on the cold surface of magnesium with a resonant quartz balance in relatively shallow vacuum ($\sim 10^{-4}$ Pa) showed that at a low constant vapour density at the initial time after opening the evaporator the quartz balance showed a decrease of coating thickness, and the rate of condensation was a negative value. After closing the evaporator with a shutter the growth of the thickness of the coating did not stop immediately, but after a while. This effect was repeated consistently in multiple opening and closing the evaporator, indicating that this phenomenon is not related to the condensation process in the usual sense.

To explain this phenomenon, it was assumed that after the opening of the evaporator the atoms adsorbed on the surface of the quartz probe are desorbed mainly under the influence of the luminous flux emitted by the heated evaporator and falling on the sensor surface, which is recorded as a reduction in the weights of quartz thickness. In contrast, after closing the evaporator shutter residual gas atoms of the vacuum chamber are adsorbed on the sensor surface and the sensor shows a further increase in coating thickness. This fact has stimulated research in mass transfer processes of photodesorption and subsequent adsorption in pure form, i.e. in the absence of condensed matter vapour. Particular attention was paid to the investigation under periodic illumination of the surface.

6.5.1. Photodesorption experiments

Problems of the photodesorption of adsorbed atoms on the surface of various materials have been studied in a relatively large number of studies [21–23]. However, most work on this issue has been carried out either at constant surface illumination or with pulsed effects using single pulses. The question of mass transfer of the surface with the surrounding gas space with periodic exposure, when gas adsorption can take place the intervals between flashes of light, has been neglected. In this work attention was given to the photodesorption with periodic impulse excitation of light on the surface of the quartz sensor, which also served as the object of study and the measuring device. The inertia of the mass measurement system using the MSV-1841 quartz balance was determined by the type of recording apparatus. In our case, the readings of the quartz balance were recorded using recorder Endim-620.02 with a lag of about 0.2 s, as the light source – a 400 W heating lamp of a film projector, installed outside the vacuum chamber. The light beam was interrupted using a metal slotted disk mounted on the axle of the motor. The minimum period of interruption of the light is chosen an order of magnitude longer than the time lag of the recording

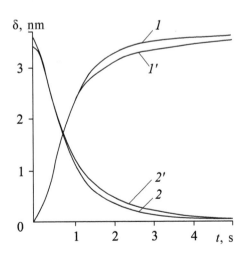

Fig. 6.24. The dependence of the effective thickness of the adsorbed layer on the time of exposure. 1, 1' – adsorption, 2, 2' – desorption. The values of p, Pa: $8 \cdot 10^{-3}$ (1, 2), $1 \cdot 10^5$ (1', 2'), $J = 615$ W/m².

equipment. The temperature of the sensor and the radiant heat flux on its surface were measured with a calorimeter [24], installed near the quartz sensor. In order to eliminate differences in the reflection coefficients the surfaces of the sensor and the calorimeter were coated with a layer of magnesium 100 nm thick under identical conditions. The experiments were performed in a wide range of gas pressures: from atmospheric to $1 \cdot 10^{-4}$ Pa when the repetition rate and duty cycle of the light pulses were varied. To preserve the spectral composition of radiation the intensity of the light flux was reduced by increasing the distance between the lamp and the sensor.

At a low repetition rate of light pulses, when the effective thickness of the adsorbed layer reached the steady state, the readings of the quartz balance, the surface temperature of the calorimeter and the heat flux were recorded by the method described in [24]. This technique is independent of the heating mechanism of the calorimeter and can be used for measuring the density of the luminous flux in the absence and presence of the vapour of condensed matter. Figure 6.24 shows the evolution of the effective thickness of the coating quartz sensor illuminated with light pulses of long duration. As can be seen from this figure, the process of photodesorption from the surface of the sensor and the subsequent adsorption take a sufficiently long period of time equalling few seconds, and the effective thickness varies by a few nanometers. After turning off the light beam the quartz balance shows an increase in the thickness of the adsorbed layer to the original value during the same period of time, so that the time dependences of the photodesorption and subsequent adsorption are like mirror images of each other and do not depend on the gas pressure in the vacuum chamber up to atmospheric pressure.

As follows from the readings of the calorimeter, the temperature increase of the quartz sensor did not exceed 4 K for the maximum heat flux density, which indicates practically the absence of thermal desorption. In addition, it

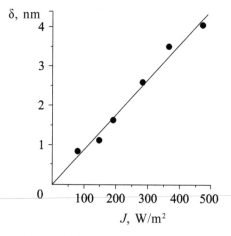

Fig. 6.25. The dependence of the effective thickness of the adsorbed layer on the intensity of exposure.

is known that heating of the sensor usually leads to deviations of the readings of the quartz balance in the direction of the apparent increase in the effective thickness. Therefore, a decrease of the measured thickness under illumination can not be explained by thermal desorption. This fact is an indication that the readings of the quartz balance correspond to a change in the effective thickness of the adsorbed layer under the action of radiation.

By increasing the density of the luminous flux the effective thickness of the desorbed layer increased. Figure 6.25 shows the dependence of the effective thickness of the desorbed layer on the density of the luminous flux. The linearity of this relationship indicates that in the conditions of this experiment, the optical emission does not change the binding energy of the adsorbed atoms with the surface [23].

6.5.2. Model of Langmuir photodesorption

For a theoretical analysis of mass transfer in photodesorption – adsorption with periodic exposure of the light flux on the surface is taken as a model of monolayer Langmuir adsorption. According to this model [25]

$$\frac{d\xi}{dt} = \frac{\alpha p}{(2\pi mkT)^{1/2}}\left(1 - \frac{\xi}{n_a}\right) - \frac{\xi}{\tau_a}, \tag{6.12}$$

where ξ is the number of particles adsorbed on the surface of 1 m²; $p/(2\pi mkT)^{1/2}$ is the number of collisions of gas molecules with mass m with the unit surface area at a gas pressure p and temperature T; n_a is the adsorption vacancy concentration per unit surface area; τ_a is the lifetime of adsorbed molecules on the surface; α is the adhesion coefficient.

Making (6.12) dimensionless by dividing by τ_a and n_a, and setting $\alpha = 1$, we obtain

$$\frac{d\Theta}{d\tau} = b - (1+b)\Theta, \tag{6.13}$$

where $\Theta = \xi/n_a$ is the degree of surface coverage by the adsorbate; τ is dimensionless time; $b = p\tau_a/[n_a(2\pi mkT)^{1/2}]$ is the dimensionless flux of molecules to the surface from the gas phase.

Assuming that the luminous flux incident on the surface does not change the binding energy of adsorbed particles to the surface and only changes their lifetime on the surface of (6.13), we obtain

$$\frac{d\Theta^E}{d\tau} = b - (1+b+\beta E)\Theta^E, \tag{6.14}$$

where E is the density of the luminous flux; β is the coefficient of proportionality.

Let the surface, in a state of equilibrium, to receive at the time $\tau_0 = 0$ a constant flux of density E.

Then the general solution of (6.14) has the form

$$\Theta^E = \frac{b}{1+b+\beta E} + C_0 \exp[-(1+b+\beta E)\tau], \tag{6.15}$$

where C_0 is an arbitrary constant.

Due to the fact that the layer adsorbed on the surface layer at a time $\tau_0 = 0$ in a state of equilibrium is described by equation (6.13) in which $d\Theta/d\tau = 0$, then from (6.13) we have $\Theta_0 = b/(a + b)$ at $\tau = 0$. Defining C_0 from this condition, we obtain

$$\Theta^E = \frac{b}{1+b+\beta E}\left\{1 + \frac{\beta E}{1+b}\exp[-(1+b+\beta E)\tau]\right\}. \tag{6.16}$$

Let the luminous flux fall on the surface indefinitely, so the adsorbed layer comes to equilibrium, after which the light beam is turned off. In the absence of light exposure on the surface the state of the adsorbed layer is described by equation (6.13) with the initial condition obtained from (6.16) for $\tau \to \infty$. Then from the solution of equation (6.13) we have

$$\Theta^0 = \frac{b}{1+b}\left\{1 - \frac{\beta E}{1+b+\beta E}\exp[-(1+b)\tau]\right\}. \tag{6.17}$$

It follows from a comparison of the expressions (6.16) and (6.17) that a ta low density of the luminous flux when $\beta E \ll (1 + b)$, the time dependences of Q^E and Q^0 are mirror-symmetrical, which is consistent with the experimental results (see Fig. 6.24). The difference between the values of (6.17) and (6.16) when Q^0 and Q^E reached the equilibrium values, yields the value the layer desorbed under the effect of light, i.e. mass transfer in a single on-off cycle:

$$\Delta\Theta = \Theta^0 - \Theta^E = \frac{b\beta E}{(1+b)(1+b+\beta E)}. \tag{6.18}$$

From the expression (6.18) it follows that under the condition of low-intensity light exposure ($\beta E \ll (1 + b)$) the maximum mass transfer of the surface with the surrounding space in the process of desorption and subsequent adsorption depends linearly on the density of the luminous flux, which also agrees with the experimental results (see Fig. 6.25).

Lack of dependence of the experimentally measured layer thicknesses of photodesorption and adsorption on the gas pressure and the presence of such a dependence in the results of theoretical analysis is explained, apparently, by the fact that the experiments revealed sorption processes of vacuum oil vapours that are always present in the vacuum chamber. Indeed, at room temperature, the partial vapour pressure of the vacuum oil is about $1\cdot10^{-5}$ Pa, and it does not depend on the pressure of the gases simultaneously present in the chamber. The estimates based on the Langmuir adsorption model show the characteristic time of establishment of adsorption equilibrium of the vacuum oil vapours as 2–3 s which is almost completely independent of the ambient gas pressure. The time of establishment of sorption equilibrium of atmospheric gases is several orders less, approximately 10^{-7} s, which can not be measured by the equipment used in the experiments [25].

6.5.3. Sorption processes with periodic illumination

To analyze the effect of periodic exposure of the adsorbed layer to light, we assume that the luminous flux incident on the surface is periodically switched on and off [26]. Let the light pulses have a duration τ_p and the time interval between τ_c. The intensity of the light in the pulse is constant and the rates of rise and fall are infinite. Further, let us assume that an infinite number of pulses have been applied at the beginning of counting t_0, so that the amplitude of fluctuations of the function Θ has reached a steady-state value. In this case the general solution of (6.14) for the section $t_0 + t_p$ is described by (6.15), in we can arbitrarily set $t_0 = 0$. Turning to real time, we have

$$\Theta^E = \frac{b}{1+b+\beta E} + C_1 \exp\left(-\frac{1+b+\beta E}{\tau_a}t\right). \tag{6.19}$$

Similarly, for the section τ_c from the solution of (6.13) we write

$$\Theta^0 = \frac{b}{1+b} + C_2 \exp\left(-\frac{1+\beta}{\tau_a}t\right). \tag{6.20}$$

The integration constants C_1 and C_2 are determined from the initial conditions, which in this case have the form

$$\begin{aligned}\Theta^E &= \Theta^0_S \text{ at } t = 0,\\ \Theta^0 &= \Theta^E_S \text{ at } t = t_p,\end{aligned} \tag{6.21}$$

where Θ^0_S and Θ^E_S are the steady-state values of the amplitudes of pulsations of the functions Θ^0 and Θ^E. The values Θ^0_S and Θ^E_S are determined using the conjugation conditions: $\Theta^E = \Theta^0 = \Theta^E_S$ at $t = \tau_p$, $\Theta^0 = \Theta^E = \Theta^0_S$ at $t = \tau_p + \tau_c$ due to the periodicity of the process with the unattenuated amplitude. Using these and the initial conditions (6.21), to determine Θ^0_S and Θ^E_S we obtain a system of linear algebraic equations:

$$\Theta^E_S = \frac{b}{1+b+\beta E} + \left(\Theta^0_S - \frac{b}{1+b+\beta E}\right)\exp\left(-\frac{1+b+\beta E}{\tau_a}\tau_è\right),$$

$$\Theta^E_S = \frac{b}{1+b} + \left(\Theta_S - \frac{b}{1+b}\right)\exp\left(-\frac{1+b}{\tau_a}\tau_c\right), \tag{6.22}$$

the solution of which allows to determine the type of functions Θ^0 and Θ^E the periodic effects of light on the surface. However, to determine the mass transfer in photodesorption and subsequent adsorption in the intervals between the flashes of light explicit form of the functions (6.19) and (6.20) does not matter. It suffices to define the difference between the values of Θ_0^S and Θ_0^E. Solving (6.22), we obtain

$$\Delta\Theta_S = \frac{b\beta E}{(1+b)(1+b+\beta E)} \times$$

$$\times \left\{ \frac{\left(1-\exp\left(-\frac{1+b}{\tau_a}T_0C\right)\right) - \exp\left[-\frac{1+b+\beta E}{\tau_a}T_0(1-C)\right] + \exp\left[-\frac{T_0}{\tau_a}(1+b)-\frac{\beta E}{\tau_a}T_0(1-C)\right]}{1+\exp\left[-\frac{T_0}{\tau_a}(1+b)-\frac{\beta E}{\tau_a}T_0(1-C)\right]} \right\},$$

$$(6.23)$$

where $C = \tau_i/T_0$ is the filling factor; $T_0 = \tau_p + \tau_0$ is the pulse repetition period.

The expression (6.23) is a function of one variable C, in which experimentally measurable quantities βE and T_0 are the parameters, and it can be easily seen that $\Delta\Theta_S = 0$ and when $C = 0$ and $C = 1$. Therefore, the $\Delta\Theta_S(C)$ function should have a maximum at some value C. In determining the extreme value of the function $\Delta\Theta_S(C)$, we find that this quantity is maximum at a value of C, satisfying the equation

$$(1+b+\beta E)\,\text{ch}\left(\frac{1+b}{\tau_a}T_0C\right) - (1+b)\,\text{ch}\left[\frac{1+b+\beta E}{\tau_a}T_0(1-C)\right] = \beta E. \quad (6.24)$$

From the expression (6.24) it follows that the mass transfer in a periodic pulse excitation has a maximum whose position depends on both the pulse repetition frequency and the efficiency of action.

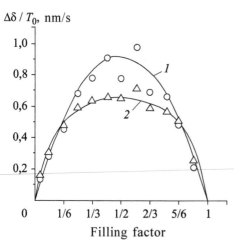

$\Delta\delta / T_0$, nm/s

Fig. 6.26. The dependence of the mean rate of mass transfer on the filling factor. Solid line – calculation, points – experiment. The values of T_0, s: 1.7 (1), 5 (2). $E = 615$ W/m².

Figure 6.26 shows the calculated values of the average rate of mass transfer for the period for the parameters b, βE, τ_a and T_0, close to the values realized in the experiment. This figure also shows the points of the experimental values of the rate of mass transfer. At the same time due to the fact that in the experiment we directly measured the change in the thickness of the adsorbed layer, and calculations by the formula (6.23) give the change in surface coverage by the adsorbate, a direct comparison of calculated values with the experiment is impossible. In addition, the calculation of changes in the thickness of the coating is rather arbitrary due to lack of knowledge about the values of the adsorbed layer capacity and lifetime of adsorbed atoms on the surface. Therefore, the calculated curve shows the experimental value at the point $C = 5/12$ determined by the usual procedure.

As seen in Fig. 6.26, there is good agreement between the calculated and experimental mass transfer rates in the whole range of the filling factor. Moreover, this agreement is observed only for the quantities $\tau_a \approx 1$ s, which indicates a significant lifetime of the particles adsorbed on the surface.

Thus, the experimental studies and theoretical analysis of mass transfer processes in periodic photodesorption and adsorption show a qualitative agreement with the Langmuir monolayer model. At the same time due to the fact that the theoretical analysis was carried out in general terms, the results will be valid for any periodically changing physical effects on the surface. The mass transfer maximum observed in experiments and theoretical analysis at a certain pulse ratio can be important when analyzing the processes of heterogeneous crystal growth and deposition on the surface of various coatings.

6.5.4. Accelerating the growth of plants under periodic illumination

The material presented in this section is beyond the scope defined in the planning of this monograph. However, these developments are a direct continuation of the previous section and contain an interesting idea.

It is well known that the rate of plant growth is defined as the rate of mass transfer with the soil through the root system and by the mass transfer rate from the atmosphere through the leaves. If we assume that mass transfer occurs through the leaves through a phase of adsorption of air molecules on the surface of the leaves, the findings obtained in the preceding section shall be valid in this case. It follows that the periodic illumination of plants should affect the rate of growth. In view of the absence of any data on the molecular properties of the surface of plant leaves we cannot assess the optimal frequency pulse of light to accelerate growth. However, we know that the leaves exchange with the atmosphere mainly gas components, for which the characteristic times for the establishment of sorption equilibrium are small. This gives hope that the optimum frequency, accelerating growth, is in any case higher than the frequency of changes of day and night. Therefore, for the formulation of principal experiment to test the outlines ideas it is sufficient to ensure a reasonable frequency of pulsed light which can be obtained using the available technology. The frequency of 1 Hz was selected from these considerations.

Tomato seeds were planted in two boxes. The soil for both boxes was prepared in one container and then divided between them. Watering the soil in both boxes was carried out in doses and at the same time. The boxes were placed in a darkened room, divided into two halves. The experimental and references boxes were selected at random. Identical lights – incandescent bulbs – were placed above the boxes. The lamp above the experimental box was powered from the network through the interruption circuit, the lamp above the reference box operated continuously. Thermocouples were placed in the soil of the boxes for temperature control. The lamp above the reference box was removed from the box to a distance greater than the distance of the lamp above the experimental box so that the soil temperatures in both boxes were equal. This meant that both boxes received the same average energy flux from the lamps.

The experiment lasted continuously without stopping for the night. The first shoots of tomatoes were found in the reference box. Shoots in the experimental box appeared two days later. This result is not understood and has not been explained. Systematic measurements of the plant height over a period of several days showed that the rate of growth of plants in the experimental box is about twice the rate of growth of reference plants. Within three days the growth of the experimental plants surpassed that of the reference plants. The rate of growth of the experimental plants progressed over time. The experiment ended when the experimental plants left to grow throughout the night grew so close to the lamp that their tops were destroyed by high temperature and their growth in height.

Since the experimenters were not experts in the field of agricultural sciences, the staging and results of the described experiment can not be considered professional. Therefore, experiments were terminated, and the results have remained unpublished. Nevertheless, the authors believe that this was yet another confirmation of the validity of the previously described theory of photodesorption with occasional illumination and believe that these promising results will find understanding among agricultural scientists.

Conclusions

1. Lock chambers of the metal–gas dynamic installation, suitable for obtaining samples of anodes of chemical power sources (CPS) by evaporation and condensation of the initial metals – lithium and aluminum – were designed and constructed.

2. The method of the joint condensation of two metal for CPS electrodes from metal alloys was developed.

3. An example of the joint condensation of metal vapour and organic vapour on magnesium and vacuum oil vapours was described. The tendency of the metal vapour to condense with the formation of a lamellar structures

was shown. Organic matter vapours filled cells of the lamellar structure with the liquid condensate.

4. The example of magnesium vapour condensation on the surface of the carbon dioxide condensate has been described. This condensation is accompanied by difficult to explain chemiluminescence.

5. The possibility of evaporation and condensation processes polypropylene for use in the manufacture of separators of the CPS by the condensation of propylene to produce a porous structure was discussed.

6. The possibility of production of propylene carbonate–ultrafine lithium composite was verified. Samples with a liquid renewable anode were produced and tested at the NCCP. The tests showed that the characteristics are lower that those of the already available CPS, but in principle the possibility of this trend of producing the CPS has been proved.

7. The effect of photodesorption at periodic illumination on the rate of mass transfer in the adsorbed phase has been studied. The maximum mass transfer observed in the experiments and theoretical analysis at a certain pulse ratio can be important when analyzing the processes of heterogeneous crystal growth and deposition on the surface of various coatings. It is also shown that the artificial periodic illumination accelerates the growth of plants.

List of Symbols

A_1, A_2 - scaling factors that reflect the real properties of the evaporators;
B_1, B_2 - geometric form factors of the evaporators;
b - dimensionless flux of atoms on the surface, equal $p\tau_a / [n_a (2\pi_{mkT}) 1 / 2]$;
C - the fill factor, $C = t_i/T_0$;
C_0 - constant;
C_1, C_2 - the constants of integration;
E - the density of luminous flux, lumen;
f_2 - the function of the type shown in Fig. 6.5 and 6.6;
f_k - the area of the site being deposited on the substrate, m^2;
$\dot{H}$ - Linear velocity of condensation or evaporation, m / s;
J_1, J_2 - flux couple between two metals on the substrate, m$^{-2} \cdot$s^{-1};
J_v - Atomic flux evaporation in a vacuum, m$^{-2} \cdot$s^{-1};
J_k - Rate of rise of the mass of condensate on the substrate, kg / s;
K_1, K_2, K_3 - the coefficients in the approximation of the saturation pressure;
k - Boltzmann constant, J / K;
m - mass of the atom, kg;
n_a - the vacancy concentration of adsorption per unit area surface, m^{-2};
P_h - vapour pressure, Pa;
p - pressure of gas;
Q_1, Q_2 - radiative fluxes from the evaporator to the substrate, J/m^2;
r_1, r_2 - the distance from the substrate by evaporation, m;
T_v - evaporation temperature, K;

T_0 - the pulse repetition period, $T_0 = t_p + t_0$;

t_0 - the time of the origin, c;

α - coefficient of adhesion;

β - coefficient of proportionality;

$\Delta\delta/T_0$ - average rate of mass transfer in the adsorbed layer, nm / s;

$\Delta\Theta_S$ - Difference between steady-state values of the amplitudes of pulsations with light and without light;

δ - effective thickness of the adsorbed layer, nm;

$\Theta = \xi / n_a$ - degree of surface coverage by the adsorbate;

$\Theta = \xi / n_a$ - degree of surface coverage by the adsorbate;

Θ^0 - degree of surface coverage by the adsorbate in the absence of light;

Θ^E - degree of surface coverage by the adsorbate in the presence of light;

Θ_S^0 and Θ_S^E - steady-state values of the amplitudes of pulsations functions;

μ_{Mg}, μ_{Li} - atomic mass of magnesium and lithium, as well. e.;

ξ - the number of particles adsorbed on the surface of 1 m^2 m^{-2};

ρ_k - condensate density, kg/m^3;

τ - dimensionless time;

τ_a - the lifetime of adsorbed molecules on the surface, c;

t_p - the pulse of light, c;

T_s - time interval between pulses, with.

Literature

1.   Bochkarev A.A., et al., in: Phase transitions in pure metals and binary alloys: Ed. S.S. Kutateladze, V.E. Nakoryakova, Institute of Thermal Physics, Academy of Sciences, Novosibirsk, 1980, 133–145.

2.   Sokolov V.A., Cando-luminescence, Tomsk, Tomsk State University, 1967, 42–66, (Sokolov V.A., Gorban' A.N., Luminescence and adsorption. Moscow, Nauka, 1969, 51).

3.   Moskowitz M., Ozin G. Cryochemistry, Springer-Verlag, 1979.

4.   Sergeev G.B., Batyuk B.A., Cryochemistry, Moscow, Khimiya, 1978.

5.   Bagotsky V.S., Skundin A.M., Chemical current sources. Moscow, Energoatomizdat, 1981.

6.   Herr R., ElectroChim Acta, 1990, V. 35, NO. 8, 1257–1265.

7.   Brummet S.B., ET AL., Energy storage with ambient temperature rechargeable lithium batteries, Final Report, 1 Apr. 1975, 31 Dec. 1977, EIC, Inc., Newton, MA, 01/1978, 123.

8.   Dey A.N., Thin Solid Films, 1977, V. 43, No. 2, 131–171.

9.   Dousek F.P., Jansta J., Rihaz J., J. Electroanal. Chem., 1973, V. 46, 281.

10.   Aurbach D., et al., J. Electrochem. Soc., 1987, V. 134, 1611–1620.

11.   Barthel J., et al., in: Modern Aspects of Electrochemistry, No. 13, Ed. B.E. Conway, N.Y., Plenum Press, 1979, 1–79.

12.   Aurbach D., et al., J. Electrochem. Soc., 1989, V. 136, 3198–3205.

13.   Aurbach D., et al., Electrochim. Acta, 1990, V. 35, No. 3, 625–638.

14.   Aurbach D., J. Electrochem. Soc., 1988, V. 135, No. 8, 1863–1871.

15.   Luikov A.V., Int. J. Heat Mass Transfer, 1971, V. 14, No. 2, 177–184.

16.   Landau L.D., Lifshitz E.M., Statistical Physics, Part 1, Moscow, Nauka, 1976.

17.   Bochkarev A.A., Polyakova V.I., Proc. Reports. 2nd Conf. Modeling of crystal

growth, 1987, V. 1, 84–86.

18. King D.A., CRC Critical Reviews, Solid State Mater. Sci., October 1978, 167–208.

19. Robins J.L., Appl. Surf. Sci., 1988, V. 33/34, 379–394.

20. Geguzin Ya.E, Kaganovsky Yu.S., Diffusion processes on the surface of the crystal, Moscow, Nauka.

21. Shapira Y., Cox S.M., Lichtman D., Surf. Sci., 1975, V. 50.

22. Ekwelundu E., Ignatiev A., Surf. Sci., 1987, V. 179.

23. Tagirov R.B., Dokl. AN SSSR, 1985, V. 285, No. 2.

24. Bochkarev A.A., et al., Izv. SO AN SSSR, Ser. tekh. nauk, 1987, V. 18, No. 5.

25. Frolov Yu.G., The course of colloidal chemistry. Surface phenomena and disperse systems, Moscow, Khimiya, 1982.

26. Bochkarev A.A., et al., Izv. SO AN SSSR, Ser. tekh. nauk, 1989, No. 3. S. 25–30.

7

Nanodisperse zinc–butanol composites

At present there is great practical interest in materials containing metal clusters in organic matter, due to their unique physical and chemical properties. In energy terms the best method of creating such materials with dispersed particles smaller than 0.1 μm is the evaporation–condensation method [1]. The number of articles devoted to this method steadily increases but there is only a small number of studied that describe the relationship between the compositions of the final product of condensation and the initial vapour mixture. This is probably due to the assumption that the composition of the condensate must be the same as that of the vapour mixture fed to the condensation surface.

However, the efficiency of condensation for a large number of metals varies considerably not only in high-temperature zone of 'collapse of condensation', where thermal desorption acts as a principal process, but also in the intermediate low-temperature zone [2–6]. This fact is the result of the transition from the vapour–liquid–solid condensation mechanism to the vapour-solid mechanism, or is caused by adsorption of residual atmospheric products (oxygen, nitrogen, vacuum oil and other organic matter) on the condensation surface. In [7–9] it is shown both theoretically and experimentally, using examples of condensation of magnesium in the residual atmosphere at a pressure of $9.3 \cdot 10^{-2}$–$2.7 \cdot 10^{-1}$ Pa and condensation of heavy water vapour in a helium atmosphere that a small change in the concentration of impurity vapour or gas can lead to significant changes in the efficiency of condensation for the other component. This is the reason why the study of condensation of binary mixtures of steam relating to the composition and morphology of the condensate are so complex that some of the results are not yet understood, are not explained and therefore are of great practical and scientific interest.

Next, we present some features of the vacuum condensation of a binary mixture of zinc vapour with the vapour of butanol-1 on the cryogenic surface.

7.1. Joint condensation of zinc and butanol vapours

In this section we study the effect of condensation conditions on the formation of a metastable ultrafine colloid. The characteristic dependences, as measured during the process, turned out to be non-monotonic, with a variable coefficient of condensation of zinc vapour varying from 0.2 to 1. Therefore, analysis of the results was not simple. The analysis showed processes significant for colloid formation:

1) formation of amorphous or polycrystalline columnar structures of condensed butanol;

2) capture of zinc atoms by condensing butanol and the formation of zinc condensation centres;

3) heating the condensation surface, often local, accompanied by a change in the morphology of the condensate;

4) the likely breaking of bonds between zinc clusters and the condensation surface during butanol evaporation followed by withdrawal of zinc clusters in the vapour phase. It is established that under certain difficult to control conditions the structure, condensing on the chamber walls, spontaneously decomposes and generates zinc clusters.

7.1.1. Experiment

A simplified diagram of the experimental setup is shown in Fig. 7.1. The installation is similar to that used for joint condensation of lithium and propylene carbonate vapour in chapter 6 (see Fig. 6.17). Vapours of components are supplied to the lateral surface of a rotating hollow condenser 8. The end of the condenser is closed and insulated with Teflon. The condenser is cooled with liquid nitrogen. Condensate collector 10 is installed in order to gather the dispersed condensate, which is formed on the condenser. The collector is made of copper with a soldered blade made of a wear-resistant material. The upper part of the collector with the blade touches the condenser and collects the cut condensate. Thus, the renewable conditions in the condenser are maintained. The cut condensate softens on the collector and drains into the flask 15 (see Fig. 6.17).

During the experiments, the chamber pressure was maintained at 4–7 Pa, the temperature of the thick walls of the chamber made of stainless steel remained approximately equal to room temperature, 17–18°C.

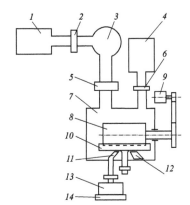

Fig. 7.1. Scheme of a vacuum unit. 1) the main vacuum pump; 2) solenoid valve; 3) diffusion pump; 4) auxiliary vacuum pump; 5) slide; 6) valve; 7) chamber; 8) condenser; 9) drive; 10) collector; 11) butanol evaporator nozzle; 12) zinc evaporator; 13) zinc evaporator; 14) heater.

Zinc and butanol evaporators were controlled by a programmable thermostat. In preliminary experiments the temperature dependence of the evaporation rates was calibrated. Errors in determining the evaporation rate with respect to temperature did not exceed ±3% for zinc and ±8% for butanol. The calculated evaporation rates were used for capacity planning of loading the crucibles of the evaporators, the expected duration of the experiment and the composition of the resulting composite. The actual mass concentration of zinc in the produced real composite colloid was measured after the experiment by the 'hydrogen' method [10] described in section 7.2. This method is based on measuring the volume of hydrogen released in the reaction of zinc and hydrochloric acid. The method was calibrate by a set of comparative tests in which the composite sample with ultrafine zinc was weighed. The composite was then heated to evaporate butanol and weighed a second time. The results of the two weighting experiments were used to calculated the mass concentration of zinc in which the statistical error did not exceed ±10%.

7.1.2. Experimental results

The dependence of the condensation rate of butanol K_{bu} on its evaporation rate J_{bu} is shown in Fig. 7.2. The dashed line corresponds to the complete condensation of evaporated butanol. We can see that the flow of condensing butanol (solid line in Fig. 7.2) weakly depends on the zinc evaporator temperature and is proportional to the butanol vapour flow. Approximation of experimental data gives $K_{bu} = J_{bu} - 8.338 \cdot 10^{-3}$.

Physically, this formula means that regardless of the strength of the evaporated flow during condensation of butanol losses occur with an approximately constant speed of $8.338 \cdot 10^{-3}$ g/s. The atmosphere in the vacuum chamber is largely made up of butanol vapours and the pressure is controlled by the valve of the diffusion pump, which explains apparently the partial evacuation of butanol from the chamber.

Figure 7.3 shows the experimental dependence of the rate of zinc vapour condensation on the size of the evaporated flow. The dashed line corresponds

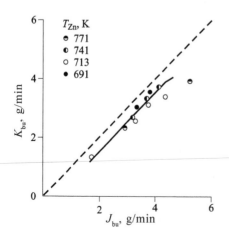

Fig. 7.2. The condensation rate of butanol, depending on its evaporation rate.

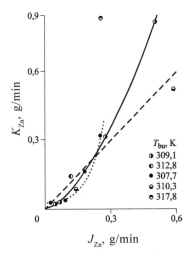

Fig. 7.3. The condensation rate of zinc depending on the its evaporation rate.

to equality of the evaporated and condensed flows. The solid line connects the points where the butanol vapour temperature was 309.1 K, and the dashed line – 307.7 K. Here we can see that the condensing flow of zinc also depends on the flow of the condensing butanol. The points lying above the line of complete condensation of the graph indicate that the condensing flow exceeds the evaporated flow. Further condensation of zinc can be explained by the following considerations.

The regimes, which were investigated when setting up the experiment, correspond to points located below the line of complete condensation. In these experiments, the vacuum chamber walls were covered with a thin porous layer of condensate which under mechanical action turned into a powder with a zinc particle size smaller than 1 μm. Often during the experiments the walls of the chamber showed the formation of a film with good adhesion. This powder and these films had a clear x-ray amorphous structure. It was almost impossible to remove the entire film from the chamber walls, but in the experiments the condensate covering the walls of the chamber spontaneously generated free zinc clusters. Their subsequent migration to the condenser is an additional source of zinc. Obviously, the mechanism by which the condensate is destroyed and the flow rate of zinc from the walls of the chamber are defined by the following factors: the oxidation of the condensate between the experiments, the evaporation of butanol through the micropores of the condensate, relaxation of the condensate and microdeformation of the condensate arising due to capillary forces, which in turn are determined by the release of latent heat during condensation of butanol and zinc vapours.

The initially pure condenser cooled in high vacuum and with unheated zinc and butanol evaporators was sometimes covered with a gray deposit, indicating the condensation on it of some volatile products coming from the chamber walls. A very fine powder was produced after cutting off the deposit from the condenser. This is the proof of cluster generation by the chamber

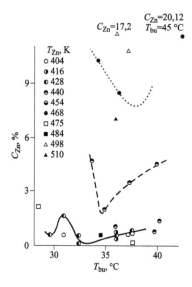

Fig. 7.4. Mass concentration of zinc in the composite depending on the evaporator temperature of butanol. The solid line connects the points corresponding to the temperature of the evaporator of zinc – T_{Zn} = 428.4°C, dashed – T_{Zn} = 468°C, dashed-dotted – T_{Zn} = 440.2°C.

walls. The generation of clusters by the solid surface in vacuum is described in [11–15] and in section 4.4.2.

Figure 7.4 illustrates the dependence of the concentration of zinc in the resulting experimental composite on the evaporation temperature of butanol at a fixed temperature of evaporation of zinc. These dependences are clearly non-monotonic, although the opposite was expected. Points of each isotherm were obtained during a single series of experiments: the chamber is completely cleared before the experiments, zinc and butanol were taken from the same batches and external experimental conditions were constant (room temperature, humidity, and pump parameters, etc.). To help to determine the scale two points with a concentration greater than 12% are also shown at the top of the graph. The isotherms of the mass concentration of zinc at a constant temperature of butanol are shown in Fig. 7.5.

When the condensate of the composite is cut from the condenser during the experiment, the surface of the condenser is covered with the remains of the condensate. Since the condensate was hard and porous, it may be expected that the remaining condensate will create roughness on the surface of the condenser. This roughness is changed, when the butanol condensate is again applied to the condenser after cutting, but before zinc vapour condensation. The structure of the newly formed butanol condensate will essentially depend on the condensation mode. Convective heat transfer between the butanol flow and the condenser is essential for the butanol vapour flow with high density. This can lead to heating of the roughness peaks of the condensate, a decrease in the efficiency of condensation, smoothing out irregularities on the condensation surface and even asperity melting of the condensate. These phenomena can be observed on cooling the condenser with liquid nitrogen. In this case an ice-like condensate forms the condenser. When the density of the butanol vapour flow correspond to the experimental points shown in Figs. 7.4 and 7.5 white 'snow' appears on the condenser. This means that the

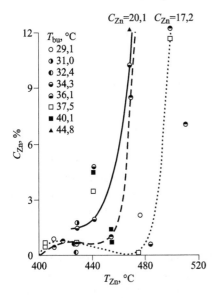

$C_{Zn}=20,1$ $C_{Zn}=17,2$

T_{bu}, °C
o 29,1
ⱺ 31,0
ⓞ 32,4
ⱺ 34,3
ⱺ 36,1
□ 37,5
■ 40,1
▲ 44,8

Fig. 7.5. The dependence of the mass concentration of zinc in the composite on the temperature of the evaporator. The solid line connects the points corresponding to the temperature of the butanol evaporator $T = 34.3$°C, dotted line – $T = 468$°C, dash-dotted – $T = 440.2$°C.

operating modes, the process of co-deposition of zinc and butanol forms a polycrystalline porous condensate with a clearly oriented columnar structure. Such a matrix structure of butanol is a substrate for subsequent condensation of zinc, and it defines the structure of the condensation of zinc.

The characteristic supersaturation of butanol with respect to the surface of the condenser is determined by the ratio of the vapour pressure of butanol in the chamber to the saturation pressure of butanol at the boiling point of liquid nitrogen. In the cited experiments, the value of the supersaturation is greater than 10. In such circumstances, the calculated critical radius of the nucleus of the butanol condensate is smaller than the size of the butanol molecules. Therefore it was necessary to expect an amorphous structure for the butanol condensate. The discrepancy of the experimental facts expectation suggests that the actual structure of the condensate is determined not by nucleation of butanol vapour on the condenser but by the initial roughness on the condenser after removing the condensate and the development of inhomogeneities in the course of subsequent condensation.

During rotation of the condenser the butanol condensate with the columnar structure is included in the visibility range of the zinc source and the stage of the joint condensation of vapours of zinc and butanol starts. When the evaporator of zinc can be seen from the condensation surface under oblique angles, the concentration of zinc atoms at the condensation front is small. In these conditions, the capture of single zinc atoms by the butanol condensate takes place. This process promotes the formation of the solid solution of zinc in butanol. During further rotation of the condenser the angle of visibility of the zinc evaporator increases. This leads to an increase in the density of the zinc vapour flow and zinc concentration in the adsorbate. Although the speed distribution of zinc atoms and butanol molecules is different, we should expect the inhomogeneous distribution of zinc concentration in the adsorbate on the surface of the condenser. Some parts of the condensation surface have an increased concentration of zinc adatoms, so nucleation of zinc can take place there.

When the local concentration of the zinc atoms in the adsorbate exceeds the saturation value, the probability of formation of a nucleus of metallic zinc differs from nonzero. The value of this probability depends on both the supersaturation and the surface area with higher concentration. As soon as the local probability of nucleation of metallic zinc 1, this part of the surface shows the formation of the centre of condensation of zinc. As a consequence, the zinc atoms in the adsorbate are beginning to flow to the established centres of condensation and the supersaturation in the adsorbate decreases. Metallic zinc particles grow in the process of the joint condensation of vapours of zinc and butanol, in which a solid solution of zinc in butanol forms, and in the parallel process of nucleation.

Due to surface diffusion the condensation centres receive Zn atoms from adjacent areas. The area of regions is bounded roughly by the mean free diffusion path of zinc atoms on the condensation surface. In other areas of the condensation surface the behaviour of condensation centres of metallic zinc is independent, and the nucleation of zinc is possible here under the same scheme. The distribution of condensation centres of zinc, resulting from nucleation, and their growth rate are determined by the geometry of the condensation surface, the parameters of the vapour flow, temperature and thickness of the condensate.

The thickness of the condensate is a significant factor as there is a need to remove the heat that occurs during condensation and the heat received by the surface of the condensate from the evaporators. These two types of heat are fed to the condenser. If the thickness of the condensate is increased, its thermal resistance leads to an increase in the surface temperature of the condensate, which leads to a decrease in supersaturation in the adsorbate. This, in turn, affects the structure of condensed matter, which is transformed from amorphous to polycrystalline.

When the temperature varies, the kinetics of sorption processes is also changing. Butanol as the more energy-dependent substance is more sensitive to temperature changes. When the surface temperature of condensation is higher than 5103 K, re-evaporation of butanol from the condensate from becoming a significant factor. Butanol evaporation absorbs heat and this is an additional process which cools the columnar structure of condensate. This evaporation process is more efficient to cool the condensate than the conductive heat transfer through the layer of condensate in the condenser surface. The evaporation rate is determined by the butanol vapour pressure near the condensation surface. The evaporation of butanol should affect the process of condensate structure formation and the final condensation product.

Critical cases of the influence of heat from the evaporator of zinc and the latent heat of condensation are the complete evaporation of butanol or the possible melting of the butanol condensate. Obviously, the zinc particles, which have lost contact with butanol due to its evaporation, may lose contact with the condenser, then go into the chamber and condense on the unheated parts of the chamber. This phenomenon is analogous to the mechanism of dust formation from the chamber walls, which is always observed in experiments. The individual processes, described above, can also take place on the chamber walls, for example, the zinc particles may lose contact with the walls of the chamber their temperature or local temperature of any parts inside the chamber increased during the experiment.

Melting the butanol condensate is another critical case. This process is more likely when relatively high vapour pressure of butanol forms inside the chamber. The liquid phase, formed on the surface of the condensate, also radically alters the structure of the condensate. Due to the low surface tension of butanol the zinc particles are wetted and the likelihood of their leaving the chamber in a gaseous medium is minimal. The particles show an increased mobility in the liquid phase compared to the solid condensate and this opens the way for their coagulation. Because the surface area of the condensate decreases after the melting of the solid columnar structure, the effective rate of condensation can be changed. We can expect a decrease in the concentration of the zinc particles in the final product, especially fine particles. During rotation of the condenser its area moves away from the zone of the zinc evaporator. The concentration of the zinc atoms in the adsorbate decreases as the zinc vapour flow on the condensate surface. Once the zinc adatom concentration is below the saturation level, the condensation of zinc on the metal particles stops and there is condensation of butanol. During this process, the zinc atoms are trapped by butanol and a zinc solution in butanol again forms. The last condensate layer formed, as well as some old coating covered by the butanol polycrystalline condensate, are cut off.

As noted above, the process of formation of the dispersions by the joint condensation of vapours of zinc and butanol shows that at different stages, we have different processes and different modes, which are responsible for a complex mosaic of the experimental data. We can definitely say that the dispersion produced during the experiment, must contain some organometallic components consisting of a solid solution of zinc in butanol. Due to the fact that the proportion of the organometallic substance contained in the dispersion is proportional to the flow densities of zinc vapour and butanol, it is expected that the concentration of zinc increases with increase in these flows, together or separately.

The behaviour of zinc concentration in the condensate at a stage of the nucleation of condensation of metallic zinc is also qualitatively clear. Because the concentration of the zinc adatoms on the surface of the condensate depends on the zinc vapour flow, the zinc concentration in the dispersion also depends on the density of the zinc vapour flow and this behaviour should be regarded as normal. When the zinc vapour flow density increases, the increase in the zinc concentration in the composite is mainly due to the increase of the metallic zinc content.

Drastic reduction in the zinc concentration in the composite can be expected in the condensation modes which are associated with partial evaporation of butanol from the condensate. This reduction will be more powerful if the butanol melts. In this situation, the flow of zinc atoms from the evaporator is directed against the flow of evaporating butanol, which should lead to a drastic decrease in the zinc concentration. In contrast, in the butanol condensation mode it is expected that the concentration of zinc in the final product will be determined by the capture of the zinc adatoms with the formation of the solid solution of zinc in butanol in areas free of metallic particles.

The foregoing description of the processes allows us to qualitatively explain the experimental data. These processes may be described quantitatively using a generalization of systematic experiments or simulation of separate stages. As the result of the process under study, as expected, is a unique ultrafine system, which has a number of new properties, efforts to describe these phenomena are clearly warranted. At the present time, the processes of sorption, nucleation and structure growth dynamics can be studied by the method of mathematical modelling. It is more difficult to model the processes occurring in the condensate after its removal from the cryogenic condenser.

7.1.3. Some properties of the zinc–butanol composite

In the process of obtaining the ultrafine zinc–butanol composite we developed a specific procedure for safe shutdown of the experiment.

1. A flask with the composite used in a specific is cut off and disconnected from the vacuum chamber and heated naturally to room temperature.

2. The heaters of the zinc and butanol evaporators are turned off. Deposition of the composite on the condenser slows down and stops, the residues of the composite are cut off from the condenser and fall into the receiver. This condensate is not obtained in the regular mode, so it is not placed into the flask with the composite.

3. Cooling the condenser with liquid nitrogen is stopped.

4. The vacuum chamber is cut off from the evacuation system, filled with argon to atmospheric pressure and heated to room temperature. Argon is needed for passivation of ultrafine zinc condensates and splashes of the composite on the chamber walls.

5. After complete thermal accommodation of the vacuum chamber in a room its cover can be opened. Once opened, there is a natural slow replacement of the atmosphere within the chamber – argon is replaced by air. If the change of argon with air occurs rapidly, then flashes are observed on the chamber walls – burning of the nanodispersed part of the condensate, accompanied by the formation of white smoke.

6. The work in the vacuum chamber can be performed only after all the relaxation processes are over. When cleaning the walls of the chamber it is necessary to use a grounded tool made of soft metals.

Failure to comply with this order of completion of the experiment may result in possible spontaneous ignition of the evaporating composite accompanied by a pulsed increase in pressure that can cause injury to personnel. Spontaneous combustion of the composite at evaporation of butanol is due to denudation of the nanosized zinc particles with a high physical and chemical activity.

The composite collected in the flask and heated to room temperature is a viscous black liquid. In the closed flask the composite in real time is almost stable. Within six months of storage pieces of metallic zinc form at the bottom of the flask as a result of coagulation of nanodispersed zinc. After pouring the composite in an open container its behaviour becomes 'strange'. Every five minutes the liquid the open container 'boils up' with intense convection of the entire volume. Boiling lasts 10–20 seconds, then the liquid settles down. With time the intervals between the 'boiling points' increase and terminate after 2–3 h. A grayish tint appears in the liquid. With prolonged storage of the composite in the open container the composite becomes stratified. A layer of transparent butanol forms in the upper part of the container and a white precipitate forms in the lower part.

'Boiling' of the composite in the open container is explained as follows. The surface layer of the liquid composite absorbs oxygen and water from the atmosphere. When a certain concentration of these oxidants is reached the accelerated oxidation reactions of nanodispersed zinc reach the speed at which a 'non-isothermality' forms locally in the composite which in turn accelerates the oxidation reaction due to the increase of local temperature and is sufficient for the initiation of convection, despite the high viscosity of the liquid. Convection, and hence mixing of the composite accelerate the oxidation reaction even further, and all the oxidant absorbed by the composite is used. The liquid settles down and the absorption of oxidants from the atmosphere starts again. The mechanism of these periodic reactions has been described by A.M. Zhabotinsky [16].

Study of condensation using a transmission electron microscope after evaporation of butanol showed that the final dispersed zinc particles are aggregates of single crystals. The size of the latter lies in the range between 80 and 500 Å. The ratio of the characteristic dimensions of single crystals of is 4:1. Reversible coagulation is likely to occur during the move of the condensate cut off from the condenser and into the receiving container. This is accompanied by heating and melting of the condensate. This creates favorable conditions for the aggregation of clusters of zinc and for the nucleation of additional clusters from zinc solution in butanol. These processes are obviously critical for the formation of zinc clusters in the final product.

Noted that if two dissimilar electrodes are lowered into a freshly prepared composite placed in an open vessel, then an electromotive force (EMF) forms on them. With connection of the load resistance to the electrodes the composite acts as a chemical current source. In long-term discharge of this chemical power source (CPS), the black colour of the composite becomes lighter. If a reversed charging current is connected to the CPS the composite color returns back to black. This means that the ultrafine zinc–butanol composite is able to operate as a renewable electrolyte for CPS. Measurement of the electrical capacitance of the CPS, the calculation of the specific capacitance of zinc contained in the composite and a comparison with data for CPS showed that the capacitance of the composite corresponds to the oxidation reaction of zinc by oxygen.

7.1.4. The 'shagreen skin' phenomenon

If the evaporators of zinc and butanol and switched during the experiment and cooling the condenser with liquid nitrogen is interrupted, then an unusual phenomenon will be observed on the condenser. In the slow heating of the condenser melting and evaporation of the butanol from the condensate on the condenser is accompanied by the 'shagreen skin' phenomenon. During the evaporation of butanol cracks appear on the melting black uniform condensate layer. Cracks in the condensate remain wet from the presence of butanol. Black islands, formed during this process, continue to shrink, decreasing from a fifth to a tenth part of their former size. This phenomenon is characteristic for the variance-weighted concentration of zinc more than 1%.

This process is more intense for a colloidal dispersion of zinc and benzene. Typical structures, formed during the evaporation of benzene, are shown in Fig. 7.6.

The following explanation of the 'shagreen skin' phenomenon was proposed. In the course of melting and evaporation of butanol from the composite and the associated increase in zinc concentration the interval between the zinc particles is reduced and the probability of their contact and interaction during Brownian motion increases. Because of the high activity the zinc particles are 'welded' together by the transport diffusion mechanism. Atomic zinc, dissolved in butanol, also plays an important role. The appearance of atomic zinc in the dispersion is the result of two mechanisms: 1) atomic zinc trapped by butanol molecules during butanol condensation is present in the composite at the initial stage; 2) in the next stages the atomic zinc appears as a result from the decay of smaller clusters due to the 'isothermal distillation' [8]. The concentration of the zinc atoms trapped by butanol can exceed the equilibrium value, which contributes to the nucleation of additional particles, growth and consolidation

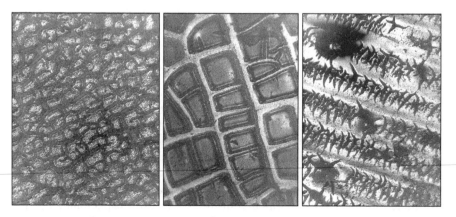

Fig. 7.6. The 'shagreen skin' phenomenon.

of existing ones, which had previously formed due to coalescence. Thus, a zinc frame, filled with liquid butanol, forms The evaporation of butanol from this frame is accompanied by the formation of compressive stresses caused by capillary forces of butanol.

7.1.5. Discussion of the results

Experimental studies of the joint condensation of vapours of zinc and butanol to form the composite showed a non-monotonic dependence of the concentration of zinc in the final dispersion on the components of the vapour flow. This clearly indicates the abundance of the complex processes involved in the monitoring process. It was determined the following processes are important for the production of dispersions: the formation of an amorphous condensate or polycrystalline columnar structures; capture of zinc atoms by the condensing butanol and the formation of zinc condensation centres, local heating of the condensation surface and, consequently, changes in the morphology of the condensate; the emergence of local surface cooling by evaporation of butanol; possible loss of bonds between zinc clusters and the condensation surface during butanol evaporation and departure of clusters into the vapour environment.

It is established that there is some difficulty in controlling the conditions under which the condensed structures on the chamber walls are destroyed and spontaneously generate zinc clusters.

The unusual properties of the composite of the butanol–nanodisperse zinc composite were discovered. The freshly prepared composite, placed in an open vessel, periodically 'boils', which indicates the occurrence of the periodic oxidation of nanodispersed zinc by the oxidants, absorbed from the atmosphere. It was also found that the composite is loaded with an electrolyte and is capable of generating an emf. The discharge of the electrolyte corresponds to the oxidation of zinc with oxygen. Finally, the 'shagreen skin' phenomenon was observed – shrinkage of the organic–ultrafine zinc condensate due to the aggregation of metal particles in the frame and the compression of the frame by capillary forces generated by the evaporation of organic matter.

7.2. The hydrogen method of measuring the concentration of the dispersed phase of zinc in colloids

This section presents a gas volume method for measuring the mass metal content in their sols, based on the ability of metals to interact with the acids with the evolution of hydrogen. The possibility of applying this method to disperse systems is shown. The circuit of a volume meter is presented. The method allows to measure the concentration of a metal in the range of 0.1–15 wt.% with an accuracy of <10%. With this method we determined only the concentration of free metal, so it can be used, for example, to study the kinetics of oxidation of metals in their sols.

The dispersed systems are characterized mainly by two indicators: the average size of the dispersed particles and the function of their size distribution. The mass concentration of the dispersed phase is usually known in advance from the concentration of soluble chemicals if the dispersion is obtained by means of their interaction [17] or determined by weighing, if the sol is obtained by dispersing a known amount of a substance in a certain quantity of solvent [18]. For sols of metals produced by joint condensation of metal vapours in a vacuum and solvent [15], along with the particle size the metal concentration in the formed product is an important characteristic by which to judge the process of obtaining a sol. The mass concentration of metal in this case is not known in advance. It can be determined by evaporation of the solvent and weighing the residue of the metal. This method is used when measuring the concentration, but its disadvantages are the considerable duration and the error due to the difficulty of complete removal of the solvent from the dispersed phase. In addition, the size of dispersed particles and their concentration in dilute dispersions are widely determined using optical methods [19]. The disadvantages of the optical methods include the fact that when measuring the concentration in them it is either assumed of that the particles are spherical and monodispersed, which in reality does not hold, or the polydispersed shape and shape factor are taken into account by a complex mathematical processing with the introduction of correction factors. All this affects the reliability of analytical results.

We propose the gas volume rapid method for measuring the mass content of metals in their sols. Volumetry is widely used in analytical chemistry [20], but measurements in dispersed sols have their own characteristics related to, firstly, the presence of a liquid dispersed medium and, secondly, with high chemical activity of ultrafine metals. The first feature influenced the proposed scheme of the device, the second one affected the limits of applicability of the method, since an increase in the dispersion of the metal may shift the balance in a dispersed system toward the formation of reaction products of the metal with the solvent. In addition, the results of the analysis should not be affected by the dispersion of the medium.

The proposed method is based on the ability of metals react with acids with the evolution of hydrogen. By measuring the volume of evolved gas, we can determine the amount of unreacted metal, and knowing the volume or mass of the sample taken for analysis also the mass concentration of the metal. In the analysis method described below the dispersion medium is n-butanol, the dispersed metal – zinc, the reagent served is hydrochloric acid. The dispersion medium does not affect the results of the analysis, since in this system the only possible reaction taking place with the release of gas is the reaction of the metal and the acid ($Zn + 2HCl = ZnCl_2 + H_2$). The diagram of the device for 'hydrogen' analysis is shown in Fig. 7.7. One bend of the instrument is filled with the sample (0.18 ml), the other with 0.5 ml of hydrochloric acid (1:2). Once both bends of the device are filled with the appropriate reagents, the device is sealed, tilted, and the sample is mixed with HCl. The volume of hydrogen released by the reaction of hydrogen is determined by changes in the level of water in a calibrated cylinder. The use of water as the sealing liquid in this case is justified, since the solubility of hydrogen in water is very low (at 18°C 0.0185 of the volume of hydrogen in 1 l water).

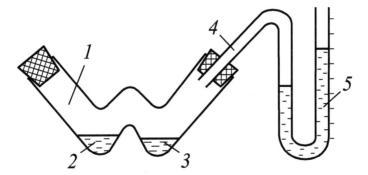

Fig. 7.7. The scheme of the device for 'hydrogen' analysis. 1) a reaction vessel, 2) hydrochloric acid, 3) sample, 4) rubber hose, 5) measuring cylinder.

The mass of unreacted zinc was determined by the formula

$$m_1 = 65{,}38V_0/22.416,$$

where 65.38 is the molecular weight of zinc; V_0 is the volume of hydrogen, reduced to standard conditions, 1; 22.416 is the volume of one gram-mole of gas under normal conditions.

The mass content of zinc in the sample was determined, knowing the volume of the sample, according to the formula

$$f_{Zn} \; (wt\%) = m_1 \cdot 100/[V_1\rho_2 + m_1 \; (1 - \rho_1/\rho_2)],$$

where V_1 is the volume of the sample; ρ_1, ρ_2 are the densities of zinc and butanol, respectively. The duration of one measurement is about 10 min.

To test the 'hydrogen' method, the sample in the instrument was replaced a known amount of zinc shavings, a fixed amount of butanol equal to the volume of the sample typically used for analysis (0.18 ml) was added, and the amount of generated hydrogen was measured. As can be seen from Fig. 7.8, the experimental points lie slightly below the theoretical curve, which is quite natural, since the zinc shavings in the air are always covered with an oxide film. Therefore, the proposed method can be used to determine the content of free zinc in the dispersion.

The method tested for zinc sols in butanol with a mass content of zinc from 0.2 to 30%. Table 7.1 presents data on the magnitude of scatter of the mass content of zinc in butanol for different concentrations. It should be noted that at zinc concentrations greater than 10% there are difficulties associated with sampling, since at higher concentrations the zinc particles coagulate and settle. In the selection of samples for analysis using a pipette at high zinc concentrations of zinc the pipette tip was clogged with large zinc aggregates. In addition, at high concentrations during the time required for sampling there is a partial deposition of zinc, which affects the representativeness of the sample.

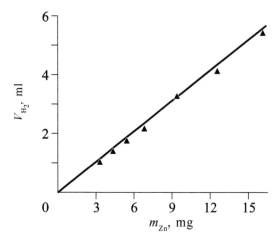

Fig. 7.8. The dependence of the volume of released hydrogen on the mass of initial zinc (the amount of hydrogen is given for the normal conditions).

Table 7.1. Standard deviation of the measurement of the mass content of Zn in butanol for different concentrations

f_{Zn}, wt.%	Number of measuremements	Δf, %	$\Delta f \, 100/f_{Zn}$, %
0.21	7	0.02	9.5
0.99	8	0.04	4.2
2.09	5	0.08	4.0
2.28	5	0.10	4.4
7.21	7	0.52	7.2
14.2	4	1.2	8.7

When the zinc concentration is less than 0.3% the zinc sol in butanol is resistant to sedimentation and therefore there are no problems in taking samples, and the increase of the standard deviation is associated with the given method coming closer to the sensitivity limit. Thus, the table shows that in the investigated concentration range the standard deviation of the method is less than 10%.

The reaction of zinc with hydrochloric acid is exothermic but at the Zn concentrations and the volume of the sample used in this study we can neglect the influence of temperature changes during the reaction on the evolved gas volume. Thus, at a concentration of 30% the heat effect is 33.2 cal (with a sample volume of 0.18 ml and the of HCl of 0.5 ml), which causes a temperature increase of about 5°C. The related systematic error in the determination of released hydrogen is only 2%.

Thus, the proposed rapid method can be used to measure the concentration of dispersed free metals, which react with acids to release hydrogen. Application of this method is limited to metals, which are in the electrochemical series of voltage to the left of hydrogen and react with acids, releasing hydrogen. The method allows to measure the metal content in a wide range of concentrations (0.1–15%) with an accuracy of <10%. The method is simple to apply and not time-consuming and expensive. It can be used, for example, to study the kinetics of oxidation of metals in their sols, because with this method we measure only the concentration of metal in the free state.

7.3. The kinetics of deposition of colloidal particles on a hard surface

This section presents the results of studying the process of forming a metal film on a steel substrate of a colloidal zinc–butanol composite. Some experimentally obtained relations for the kinetics of film growth are presented. Fairly thick coatings were produced by two methods: dipping a substrate into the colloid and dripping the colloid dripping onto the substrate surface. The dependence of the thickness of the deposited film on the initial weight fraction of zinc particles, which greatly differ for these two methods, were determined. A theoretical model for film growth, based on the Smoluchowski coagulation theory, is proposed

In recent years, interest in the synthesis of stable non-aqueous metal colloids has been growing; several methods for producing metal organosols have been developed [21–24]. These nanosystems are of practical importance and of scientific interest, due mainly to their strong catalytic action [25], the unique optical and electromagnetic properties of thin layers of such particles [26] and film-forming properties. Here we demonstrate the ability to create fairly thick coatings on metal colloidal dispersions, present some of the experimentally obtained relationship for a 'thick' coating growth kinetics and describe a growth mechanism on the phenomenological level. The used preparation conditions clearly affect the properties of colloidal particles, determining the composition of the surface layer of the particle, which, in turn, controls the stability and structure formation in colloidal dispersions. The most promising method, in our opinion, is the production of metal colloidal dispersions by co-precipitation of metal vapours and solvent, because this method makes it possible to control the dispersion and the composition of the surface layer of particles directly during the production process of the colloid [15, 27–29]. The method was used to prepare metastable zinc sols in toluene, benzene, and butanol.

When a metastable metallic organosol comes into contact with the solid surface a thin layer of metal particles is deposited, which leads to the formation of a film. Deposition of metal films from non-aqueous sols of palladium, silver, gold and indium on various substrates is described in [29-32]. The films were prepared by dripping a colloidal solution on a substrate and evaporating the solvent. Here we describe the obtaining of metallic films of zinc sol in butanol on a steel surface by two methods. In the first method the substrate was immersed in the metal dispersion and left there for a preset time. In the second method a certain amount of colloid was dripped on the substrate and butanol was evaporated. A mathematical model and physical treatment for colloid particle deposition by immersion, taking into account the block coagulation and deposition of particles on the surface, are also proposed This theoretical approach is based on the kinetics of Smoluchowski coagulation. The need to develop a theoretical model arose from the need to categorize and summarize the experimental data that have not yet given a physical picture of the growth kinetics of the film.

7.3.1. A mathematical model and physical regime of the process

When applying a film of a colloidal dispersion there are two important competitive process, namely: volumetric particle coagulation and adhesion of particles to the solid boundary, which is in contact with this colloidal dispersion. On this basis, the rate of change of the concentration of particles I is equal to the sum of the rate of coagulation I_c and the rate of settling of particles on the surface I_d:

$$I = -dv_\Sigma/dt = I_c + I_d. \tag{7.1}$$

According to Smoluchowski' theory [33], the instantaneous rate of coagulation is determined by the equation

$$I_c = dv_c / dt = 8\pi DRP^c \exp\left[-\Delta E^c / (k_B T)\right]v_\Sigma^2, \tag{7.2}$$

where v_c is the number of coagulated particles per unit volume of the colloid; D is the diffusion coefficient; R is the average particle size; P^c is the steric factor, taking into account the favourable spatial positions of the colliding particles, their shapes and sizes; ΔE^c is the activation energy of coagulation; k_B is the Boltzmann constant; T is the temperature of the environment; v_Σ is the number of particles per unit volume.

We assume that the particles deposited on a flat hard surface do not affect each other, and deposition occurs at 'coagulation vacancies', whereas in Smoluchowski' theory the flux of particles from half space on a single vacancy is determined by the formula $J_d = 2\pi DRv_\Sigma$. The number of surface vacancies is assumed to be directly proportional to the solid surface area S and inversely proportional to the cross-section of the particle $n = S/(\pi R_2)$ Then the rate of particle deposition per unit volume on a flat surface is defined as

$$I_d = dv_d / dt = 2DS(RV)^{-1}P^d \exp\left[-\Delta E^d / (k_B T)\right]v_\Sigma, \tag{7.3}$$

where v_d is the number of particles deposited on the surface from unit volume of the colloid; V is the total volume of colloid; P^d and ΔE^d are the steric factor and activation energy for adsorption of particles, respectively.

To make the equation (7.1) dimensionless, we introduce the dimensionless concentration

$$v = v_\Sigma / v_0.$$

Here v_0 is the initial concentration of particles:

$$v_0 = \frac{3}{4\pi} \frac{c}{(1-c)} \frac{\rho_1}{\rho_p} \frac{1}{R_3},$$

where c is the initial mass concentration of particles in the colloid; ρ_1 and ρ_p are the density of the liquid phase and the substance of the colloidal particles, respectively.

We represent the dimensionless time τ as a

$$\tau = t / t_0,$$

where t is the current time and

$$t_0 = \left\{ 8\pi DRP^c \exp\left[-\Delta E^c / (k_B T) \right] \right\}$$

is the characteristic time of coagulation of the first particles. According to [33], equation (7.1) can be rewritten in dimensionless form:

$$\overline{I} = -dv / d\tau = v^2 + A_k V_m v. \tag{7.4}$$

Here $A_k = \left\{ P^d \exp\left[-\Delta E^d / (k_B T) \right] \right\} / \left\{ P^c \exp\left[-\Delta E^c / (k_B T) \right] \right\}$ is the dimensionless criterion equal to the ratio of the probability of attachment of particles to a solid surface to the probability of one act of coagulation, $V_m = \left[V_1 / (6 V_p) \right] (2 RS / V)$, where V_1 and V_p are the volumes of the liquid phase and particles, respectively. The second term in V_m is the volume ratio of the monomolecular layer of particles on the surface to the whole of the colloidal mixture. Scale parameter V_m itself is a volumetric ratio of the various components. The solution of equation (7.4) with initial conditions $\tau = 0$, $v = 1$ gives the formula

$$v = \frac{A_k V_m}{(1 + A_k V_m) \exp(A_k V_m \tau) - 1}. \tag{7.5}$$

This expression describes the dimensionless concentration of particles in the colloid in time τ, which in Smoluchowski' theory gives $v = 1 / (1 + \tau)$. Formally, the Smoluchowski coagulation theory is a special case of the studied problems in which the influence of the free solid surface is negligible, i.e., $A_k V_m = 0$, and can be obtained by a limiting transition $A_k V_m \rightarrow 0$.

Equation (7.5) can be used to determine the number of particles coagulated in a volume colloid, and the number of particles strongly bonded to the solid surface.

After making (7.2) dimensionless and substituting (7.5) into it we obtain a differential equation of volume coagulation kinetics:

$$\overline{I}_c = d\overline{v}_c / d\tau = \left[\frac{A_k V_m}{(1 + A_k V_m) \exp(A_k V_m \tau) - 1} \right]^2.$$

Solving this equation with the initial conditions: $\tau = 0$, $v_c = 0$, we obtain the expression

$$\overline{v}_c = (1 + A_k V_m) \left\{ 1 - \left[1 + \frac{1 - \exp(-A_k V_m \tau)}{A_k V_m} \right]^{-1} \right\} - A_k V_m \ln\left| 1 + \frac{1 - \exp(-A_k V_m \tau)}{A_k V_m} \right|,$$

while in Smoluchowski' theory $\overline{v}_c = \tau / (1 + \tau)$.

Similarly, the equation for the deposition rate of particles can be derived:

$$\bar{I}_d = d\bar{v}_d / d\tau = \frac{(A_k V_m)^2}{(1 + A_k V_m) \exp(A_k V_m \tau) - 1}.$$

Given the initial conditions $\tau = 0$, $v_d = 0$, this equation takes the form

$$\bar{v}_d = A_k V_m \ln \left| 1 + \frac{1 - \exp(-A_k V_m \tau)}{A_k V_m} \right|.$$

Of interest are several special cases of the solution.

1) When $(A_k V_m \tau) \ll 1$, which corresponds to either the initial stage of the process of coagulation or to a sufficiently low proportion of particles deposited on the surface. Then

$$\left[1 + \frac{1 - \exp(-A_k V_m \tau)}{A_k V_m} \right] \cong (1 + \tau)$$

and

$$v \cong \frac{1}{1+\tau} - \frac{A_k V_m \tau}{1+\tau},$$

$$\bar{v}_c \cong (1 + A_k V_m) \frac{\tau}{1+\tau} - A_k V_m \ln|1 + \tau|,$$

$$\bar{v}_d \cong A_k V_m \ln|1 + \tau|.$$

One can see that the concentration of particles at time τ is smaller than the corresponding concentration in the Smoluchowski equation by the value of $A_k V_m \tau/(1 + \tau)$, which refers to the influence of the solid surface on the dynamics of the process.

2) At $(A_k V_m \tau) \ll 1$ and $\tau \gg 1$, which corresponds to a sufficiently small effect of the deposition process on the process of establishing dispersion equilibrium $(A_k V_m) \ll 1/\tau$, where $\tau \gg 1$, we have

$$\left[1 + \frac{1 - \exp(-A_k V_m \tau)}{A_k V_m} \right] \cong \tau,$$

and

$$v \cong \frac{1}{1+\tau} - A_k V_m,$$

(7.6)

$$\bar{v}_c \cong \frac{\tau}{1+\tau} + A_k V_m - A_k V_m \ln|\tau|,$$

$$\bar{v}_d \cong A_k V_m \ln|\tau|.$$

(7.7)

Because $A_k V_m \ll 1/\tau \cong 1/(1+\tau)$, from (7.6) we see that $(A_k V_m)$ is sufficiently small compared with v. From this it follows that the term (7.1) and (7.4) which takes into account the rate of deposition of particles on the surface, is small in comparison with the coagulation rate. In this case, the correction in comparison with the concentration of Smoluchowski particles must be of the same order as $A_k V_m$. The number of the coagulated aggregates of particles, deposited on the surface per unit volume, is a function of the logarithm of time.

It is interesting to express the total number of particles deposited per unit surface area for the last case in dimensional coordinates, i.e.

$$v_d \cong \frac{1}{4\pi} \frac{1}{R^2} A_k \ln \left| \frac{c}{1-c} \frac{\rho_1}{\rho_p} \frac{6}{R_2} Dp^c t \right|,$$

where $p^c = P^c \exp(-\Delta E^c / k_B T)$, and the proportion of particles deposited from the colloid volume:

$$\Delta m \cong \frac{1}{3}\rho_p R A_k \ln \left| \frac{c}{1-c} \frac{\rho_1}{\rho_p} \frac{6}{R_2} Dp^c t \right|.$$

It is seen that the proportion of deposited particles is proportional to $\ln \left| \frac{c}{1-c} \frac{\rho_1}{\rho_p} \frac{6}{R_2} Dp^c t \right|$ the diffusion coefficient is inversely proportional to the size of the particle, $\approx R \ln \left| R^{-3}, ... \right|$. In this case, the reduction in the viscosity of the colloid or the average particle size or increase of the initial weight fraction leads uniquely to an increase in the proportion of deposited particles relative to the bulk particle coagulation, v_d / v_c (7.7) The increase in A_k leads to the same result. One can show that the effect of the average particle size on the viscosity of the colloid at $c < 0.24$ gives a correction of the second order for these results.

When $(A_k V_m \tau) \ll 1$ and $\tau \gg 1$ it is convenient to study this process experimentally by measuring the weight of the deposited dense coating as a function of the initial concentration of particles, their average size and process time. Experimental data for one pair of substances should be presented in the generalized coordinates

$$\ln \left| \frac{c}{1-c} R^{-3} t \right|, \quad \Delta m / R. \tag{7.8}$$

7.3.2. Experiment

As mentioned above, the experiments investigated the weight of the particles adhered to the substrate, i.e. the weight of the coating. Preparation of metal films having good adhesion to the substrate in a vacuum is a well-known process. Our goal is to prepare the zinc coating on the surface of mild steel in the open air by 'cold wet galvanizing'. It was assumed that using a composite of zinc in butanol for this purpose and disrupting its stability on the substrate, we can produce a dense zinc coating, firmly adhering to the steel surface to a degree commensurate with the cohesion of zinc.

The ultrafine composite, used in these experiments, was prepared by the joint condensation of vapors of zinc and butanol on the cryogenic surfaces in vacuum [15]. After heating and melting of the solid metal–organic condensate composite removed from the cryogenic condenser, one obtains a liquid dispersion of nanosized zinc in butanol. The average size of colloidal particles, measured in a transmission electron microscope, was 10 nm.

Nanosized zinc in butanol is metastable with respect to aggregation due to solvation of the particles (steric effect) and electrostatic effects (charge of the particles). When the weight percentage of zinc is less than 0.3% this nanosystem is stable with respect to sedimentation. Slow segregation takes place at higher proportions of zinc: a spongy zinc precipitate is formed and the composite with the weight fraction of zinc, approximately equal to 0.3%, is produced As the layers of organic matter on the particle surface prevent their irreversible coagulation, the zinc residue can be converted back to a suspended state by simple dilution, dispersion or peptization. In nanosystems with high concentrations of zinc particle aggregation produces thin metal films.

The surface of the substrate is rough polished, washed with benzene and acetone, etched for 15 min with hydrochloric acid HCl (0.5 mol/liter). The substrate is then washed with distilled water, dried and weighed on an analytical balance with an accuracy of 0.01 mg. The weight error should not exceed 0.1%.

The experiments were performed in two different ways. First: the substrate was placed in two premixed zinc–butanol composites with a known weight fraction of zinc particles (from 0.01 to 14%) and kept there for a specified time from 10 s to 4 h, which is a rather short time compared to the sedimentation time of over 120 h. The substrate was then left in the air for 24 hours, washed with distilled water, dried, and the zinc deposit with insufficient adhesion to the substrate was removed. As a result, the substrate was coated with a zinc film which has a strong adhesion, 0.4–0.6 of the volume cohesion of zinc. The colour of the coating was metallic, gray, dark gray or almost black depending on the thickness and density of the films. The coating weight was determined by second weighing. The thickness of the coating, calculated according to the weight of zinc and zinc bulk density, varied in the range $3 \cdot 10^{-8} – 4 \cdot 10^{-6}$ m.

In the second method, the same amount of the colloid, 0.2 ml, was on sections of the substrate surface, equal in size $100 \, \Delta S/S = 3\%$, and then dried at 20°C under normal atmospheric conditions for 24 h. The conditions were kept identical for all series of experiments. The mass concentration of the particles in the colloid ranged from 0.01 to 14% with two different dispersion formulations. Subsequent measurements of the weight of the coating were carried out by the above scheme. The coating thickness was varied in the range $3 \cdot 10^{-7}$ to $8 \cdot 10^{-6}$ m.

7.3.3. The experimental results

Figure 7.9 presents data for the coatings produced by immersion of the substrate into the colloid. Here, the specific weight of the zinc coating depends on the logarithm of the product of holding time of the substrate in the colloid by the concentration term. This graph shows the summary data for zinc plating in zinc–butanol colloids of approximately equal dispersion compositions and properties. They are presented for two cases: curve 1 corresponds to the case where the substrate has been treated previously, as described above, and curve 2 to the case where the substrate was washed with a detergent before the final wash. The graphs show that the experimental data can be generalized in the coordinates

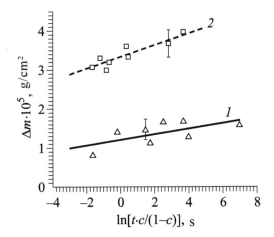

Fig. 7.9. The dependence of the specific weight of the zinc coating on the product exposure time by the concentration term. Immersion, for notations see the text.

(7.8) recommended for the case of a small influence of the deposition process on the process of establishing the colloidal equilibrium. This fact qualitatively confirms the physical representation of the considered processes. Increase of the specific weight of the deposited coatings for the second curve is correlated with the removal of greasy dirt, insoluble in butanol, from the substrate surface after its treatment with a surfactant and, therefore, with the improvement of the surface activity of particles.

The dependence of the specific weight of zinc in the coating fabricated by evaporation of butanol on the square root of the concentration term is illustrated in Fig. 7.10. This relationship includes the summary data for zinc plating in the zinc colloid in butanol with approximately the same properties at the same rate of evaporation of butanol and the same process time for two cases: curve 1 corresponds to the dispersion with small and curve 2 with large colloidal particles.

7.3.4. Discussion

The essence of the process of deposition of zinc on the solid surface of the colloid consists of the joint action of two processes, namely, the volume coagulation and adsorption of particles, as demonstrated in accordance with the qualitative behaviour of the experimental data and calculations in accordance with the existing model. The appearance in Fig. 7.9 of the logarithmic dependence on the number of deposited particles, which is absent in the Smoluchowski theory, is due to differences in the kinetics of volume coagulation and surface processes. The main difference is as follows. In the volume process the coagulation vacancies are located on similar particles; as a result, their concentration is proportional to the concentration of the particles. In the surface process the number of vacancies is limited in accordance with the distribution of the particles on the surface and is constant, and the concentration of vacancies

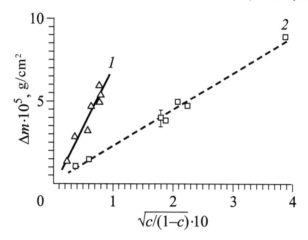

Fig. 7.10. The dependence of the specific weight of zinc in the coating on the square root of the concentration term. Dripping of a colloid on the substrate. For other notations see the text.

with respect to the volume depends on the volume and the average particle size. The inclusion of these features is enough to describe the experimental data.

However, we must note several limitations of the presented model of deposition of colloidal particles on the surface. First, the model suggests that the mechanism of transport of particles to the surface is the diffusion mechanism and it is not a factor limiting the rate of deposition. Obviously, this assumption is valid only for colloids with a high diffusion coefficient of the particles or when the colloid is mixed with a moderate speed. Secondly, the constant steric factor and activation energy for the coagulation in (7.3) limit the potential possibilities of the model to describe two real cases: 1) the deposition of a monomolecular layer and 2) the deposition of particles on the surface, created by the particles, i.e. when the time of the initial coating of the substrate by the particles is negligible. Third, this model does not take into account possible collective interactions of particles during the deposition process.

The conclusion which can be deduced from a comparison of (7.2) and (7.3) is of independent interest. The comparison shows that the volume coagulation and precipitation on the surface depend on the concentration of colloidal particles in different ways. The volume coagulation is quadratic with respect to concentration, while precipitation is linear. This means that the deposition on the surface would be significant only at low concentrations of particles This conclusion is useful in developing the technology for manufacturing thin-film coatings from the colloid, but this suggests possible ways to speed up the deposition process.

In the case when the particles were deposited with the simultaneous evaporation of butanol the results were used to determined the quadratic dependence of the deposited particles on the concentration term. Such dependence is probably related to the fact that evaporation not only affects the concentration but also supports some other processes, helping them to influence the kinetics of deposition. Among these related processes are: the emergence of the profile of the particle concentrations, arising due to the evaporation of the surface of the colloid and sedimentation of the particles; variation of the diffusion coefficient due to strong changes in the volume concentration of particles in time and enrichment of the colloid with soluble gaseous and foreign contaminants during evaporation and thinning of the colloidal film. All this affects the likelihood of coagulation and adsorption.

The latter case considerably enhances the growth of the coating of the colloid, but the ratio of the number of particles coagulated in the volume to the deposited ones appears to increase with increasing evaporation rate.

Thus, the possibility of applying a rather thick coating of a colloidal dispersion has been confirmed. The exponential dependence of the film weight on the process time and the initial concentration of the dispersed particles were found, depending on the method of producing the film. It was also suggested that the phenomenological concept of the relative competition between the processes of block coagulation of metastable colloidal particles and their deposition on the solid surfaces. It is assumed that this competition controls the kinetics of formation of the coating.

7.4. Information about the technology of 'cold wet galvanizing' of steel

This section describes the results of the experimental study of the process of obtaining dense hard zinc coatings on steel by deposition of ultrafine clusters from butanol. The rate of deposition in dependence on the presence of impurities is studied. The end of the section gives preliminary information about the process of forming the coating.

7.4.1. Characteristics of zinc sol in butanol

The method of cool 'wet' galvanizing of steel products is based on the principle of stabilizing the suspension of ultrafine zinc in butanol followed by violation of the stabilization on the surface of the product.

A suspension of zinc in butanol is a lyophobic disperse system, i.e. having an excess of surface energy. The lyophobic systems are characterized by spontaneously processes of consolidation of the particles, i.e. there is a reduction of surface energy as a result of the reduction of the surface. The enlargement of the particles takes place in two ways:

1) isothermal distillation – the mass transfer from small to large particle, since the chemical potential of the latter is lower. As a result the smaller particles dissolve and the large one grow;

2) coagulation, which consists in the coalescence and bonding of the particles.

There are several factors of stability of disperse systems:

1. The electrostatic factor reduces the interfacial tension as a result of the formation of the electrical double layer on the surface of the particles in accordance with the Lippmann equation. The appearance of the electric

potential at the interface is determined by the surface electrolytic dissociation or adsorption of electrolytes.

2. The adsorption–solvation factor reduces the interfacial tension in the interaction of the dispersed phase particles with the environment due to adsorption and solvation.

3. The entropic factor describes the tendency of the dispersed phase to a uniform distribution over the volume of the system.

4. The hydrodynamic factor reduces the rate of coagulation due to changes in the viscosity of the medium and the density of the dispersed phase and dispersed medium.

Typically, aggregate stability is ensured by several factors simultaneously. Each factor corresponds to the stability of a specific method to neutralize. For example, the effect of the electrostatic factor can be disrupted by the introduction of electrolytes into the system which cause compression of the electrical double layer.

The dispersed system of zinc in butanol, obtained in our experiments, belongs to sols (liquid dispersion medium, the size of the dispersed phase is less than 100 μm). Here zinc in butanol is an aggregate stable system as a result of the entropy and adsorption–solvation factors. The role of the entropy factor is the involvement of both the particles and their surface layers in thermal motion, which ensures uniform distribution of particles by volume. The adsorption–solvation factor is determined by the adsorption and solvation of the butanol molecules on the surface of zinc particles. At low concentrations of zinc (less than 0.1%), the system is also stable to sedimentation of zinc particles. When the zinc concentration increases slow stratification take place: a dense precipitate of zinc forms and a sol with a zinc concentration of 0.1% forms above it. Since the surface layers of the solvent on the surface of the zinc particles prevent their aggregation, the zinc precipitate can be transferred back to the suspended state by simple agitation or addition of a flocculent.

Thus, galvanizing with a sol with low zinc concentration or additionally stabilized and disrupting the stability of the system at the surface of the product, we can obtain dense zinc layers with good adhesion.

7.4.2. Precipitation of the zinc layer with ionic surface-active substances

The method of deposition of the zinc coating with aqueous solutions of ionic surface active agents (surfactants) is based on the destruction in the surface layer of the sample of adsorption–solvate layers on the surface of the zinc particles. The role of ionic surfactants is twofold: 1) clean the surface of a sample to remove adsorbed impurities, and 2) actually the destabilizing role.

The galvanizing process occurs in two stages: 1) deposition of an aqueous solution of the surfactant on the surface of the galvanized sample; 2) exposure of the sample in the zinc suspension.

Thus, the zinc coatings with a thickness of 0.7 μm was produced. As in the experiments it is not yet possible to cause coagulation of the particles only in the surface layer, and coagulation takes place partly in volume, an excess of

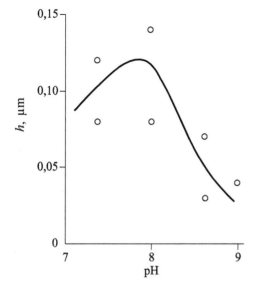

Fig. 7.11. Dependence of coating thickness (in microns) on the pH of the solution of surfactants. Points – experimental data, line – the calculations.

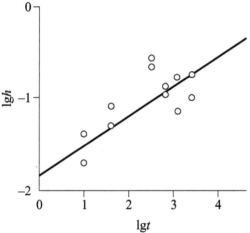

Fig. 7.12. The thickness of zinc coating on the exposure time of the sample in zinc suspension. The concentration of Zn = 1.76%, pH = 7.5 solution of surfactant. Straight line – calculation, points - experiment.

zinc forms on the surface after galvanizing, i.e. a layer with poor adhesion, which is easily removed by washing with water or by a simple wipe.

The thickness of the zinc coating also depends to a large extent on the acidity of the surfactant solution. Figure 7.11 shows the dependence of the thickness of the zinc coating on the pH of the solution of surfactants. It is evident that there is an optimum pH at which the maximum thickness of the coating is obtained with other conditions being equal.

Figure 7.12 shows the dependence of the thickness of the zinc coating on the time of exposure of the samples in the zinc suspension. The experiment was conducted as follows. All samples were pre-incubated for 15 s in an aqueous solution of surfactant (pH 7.5) and then at different times in a zinc suspension with a zinc concentration of 1.76%. The figure shows that the exposure time of 5 min 20 s results in an abnormally rapid growth of coatings and further exposure in a decrease of thickness. If we do not take into account the value of the 'anomalous' point, then all the other experimental data can be described by the dependence is given $h \cong t^{1/3}$. A preliminary explanation of these dependences is presented in section 7.4.6..

7.4.3. Precipitation with the addition of a micelle-forming component

The addition of a micelle-forming component additionally stabilizes the system due to the formation of micelles and, consequently, increase the strength of the surface layer on the surface of zinc particles. In addition, this supplement, like the surfactant addition, helps to clean the surface from contamination.

The experiment was conducted as follows. On the surface of the samples deposited zinc suspension with different concentrations of additives, the amount of zinc for all the samples remained the same. After drying, the samples are washed off with excess zinc, and the samples were weighed. The the zinc suspension was deposited with the next layer. Similarly, the experiment was repeated several times. The experimental results are presented in Table 7.2.

It can now be assumed that the micelle-forming additions have a complicated effect on the growth rate of the coating. Further experiments are required for more specific conclusions.

7.4.4. Precipitation with the addition of lyophobic component

The deposition of zinc on the surface was carried out also with the addition of a lyophobic component. This additive activates the hydrodynamic stability factor. In the evaporation of the volatile additive its stabilizing effect is reduced and coagulation of particles of zinc in the bulk and on the surface of the zinc-plated product takes place. Experiments were carried out by the scheme described in the previous section – the concentration of the additive was varied but the amount of zinc remained constant.

Figure 7.13 shows the dependence of film thickness on time for various concentrations of zinc supplements. The figure shows that the curves are divided into three sections: exponential growth $h \sim (t - t_0)^{1/3}$, a proportionate

Table 7.2. Dependence of the thickness of the zinc coating on the number of treatments of one sample

No.	Volume addition, %	h_1, μm	h_2, μm	h_3, μm
1	0	0.06	0.09	0.16
2	9.1	0.04	0.07	0.13
3	16.7	0.05	0.09	0.14
4	23.1	0.06	0.08	0.13
5	28.6	0.06	0.11	0.15
6	33.3	0.08	0.11	0.15
7	37.5	0.05	0.07	0.13

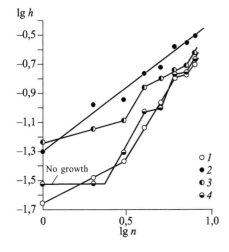

Fig. 7.13. The dependence of the thickness of the zinc coating (μm) on the multiplicity of treatment in a suspension with a lyophobic addition. The concentration of zinc 4.4%. The concentration of the lyophobic additive, vol.%: 1) 16.7; 2) 23.1; 3) 33.3; 4) 37.5; points – experiment.

increase in $h \sim (t - t_0)$ and intermediate with increasing $h \sim (t - t_0)^{0.76}$. In addition, on the curve 4 there is an area where the coating does not grow.

7.4.5. The laws of growth and the factors determining their

The experimental results revealed four characteristic coating growth laws.

1. Proportional growth: $h = P (t - t_0)$, where h is the thickness of the coating, P is the growth rate of the coating, t is the time of growth, t_0 is the time to reach the given mode.

2. 'Inhibited' growth: $h = C (t - t_0)^{1/3}$, where C is a constant determined by the nature of materials, t_0 is the time of establishment of this growth law.

3. Growth according to the 'competitive' law: $h = C_1 (t - t_0)^{0.76}$.

4. 'Dissolution' of the coating: $\partial h / \partial t < 0$, i.e., decrease in the thickness of the coating.

The experiment showed the change of the laws of growth with time. Before discussing the dynamics of change in the law of growth, we elucidate the physical processes that determine a law of growth.

Proportional growth. The proportional growth law is determined by the kinetics of deposition of ultrafine zinc from the adsorbed layer. The adsorbed layer is understood to be a thin liquid film separating the solid phase and the liquid volume of the solution. The state of the adsorbed layer may be very different from the state of the bulk phases. The deposition process involves four components, which can be adsorbed on the steel: butanol, ultrafine zinc, surfactant or stabilizer additions, impurities which initially contaminate the substrate and have a high binding energy with the substrate (oxygen, salt acid or alkali impurities).

We write the equations for the dynamics of the adsorbed layer without considering chemical reactions but taking into account the condensation of one component:

$$\dot{\Theta}_1 = b_1 (1 - \Theta_1 - \Theta_2 - \Theta_3 - \Theta_4) - a_1 \Theta_1 - c_1 (\Theta_1 - \Theta_{h1}),$$

$$\dot{\Theta}_2 = b_2 (1 - \Theta_1 - \Theta_2 - \Theta_3 - \Theta_4) - a_2 \Theta_2,$$
$$\dot{\Theta}_3 = b_3 (1 - \Theta_1 - \Theta_2 - \Theta_3 - \Theta_4) - a_3 \Theta_3, \qquad (7.9)$$
$$\dot{\Theta}_4 = b_4 (1 - \Theta_1 - \Theta_2 - \Theta_3 - \Theta_4) - a_4 \Theta_4.$$

Here the first equation describes the rate of change in the concentration of ultrafine zinc in the adsorbate, the second equation – the rate of change in the butanol concentration, the third – the effect of surfactant or stabilizer additions, the fourth – the effect of impurities. At the initial time ($t = 0$) the adsorbed layer contains only impurities chemisorbed on the sample. Then the initial conditions are formulated as follows:

$$t = 0, \ \Theta_{10} = 0, \ \Theta_{20} = 0, \ \Theta_{40} = \Theta_4^0.$$

The first term on the right side of equation (7.9) contains the factor b_i, which has the meaning of the particle flow of this component from the bulk phase of the solution in the adsorbate. The components $b_i \Theta_j$ in the first term of each equation describe the occupation of adsorption vacancies by the already adsorbed particles of components and describe the reflected beam of particles, inverse to the flow of particles coming from the bulk phase of the solution in the adsorbate. The second term in each equation determines the thermal desorption flow of the particles of the components; factor a_i is inversely proportional to the lifetime of these particles on the surface. The third term in the first equation determines the deposition flow of ultrafine zinc from the to the coating. In (7.9) c_1 is the dimensionless condensation rate, Θ_{h1} is the equilibrium concentration of zinc particles in the adsorbed layer. Θ_{h1} is determined by the thermodynamic stability of the interface and the concept of 'of ' for the interfaces.

The flow b_1 is determined by the diffusion flow of the dispersed particles to the interface, and depends on the sedimentation stability of the suspension of zinc and micelle-forming additions in butanol and the Brownian motion of the particles. In addition, the particles that fall to the interface must overcome the energy barrier preadsorption defined by the existence of 'Gibbs excess'.

The desorption term is determined by the energy barrier (the binding energy of the adsorbate to the sample ε_{si}) which the particle must overcome during desorption:

$$a_i \approx v_\perp \exp(-\varepsilon_i / k_B T),$$

where $v_\perp$ is the frequency of thermal vibrations in the plane perpendicular to the sample surface; k_B is the Boltzmann constant, T is the temperature of the sample.

The condensate flow in the coating at the current time is given by

$$\Pi' = c_1 (\Theta_1 - \Theta_{h1}) m_1 m_0,$$

where m_1 is the mass of a cluster of zinc; m_0 is a scale factor that transforms the dimensionless flow of the kinetic equations to a real system. In the case of steady-state mode $\Pi'(n) = $ const, the weight gain is $\Delta M = \Pi'\Delta t = (t-t_0)$. However, the coating thickness is associated with the weight gain by the equation $h = \Delta M/\rho S$, where ρ is the density of zinc, S is the area of the sample. Then, $h = \Pi = (t-t_0)$, where $\Pi = \Pi'/\rho S$.

Thus, the heterogeneous adsorption in the presence of impurities gives a proportional growth law of the coating. The growth rate Π depends on the condensation mode and conditions on the substrate by a factor in the case $c_1 = (\Theta_1 - \Theta_{h1})$. If $c_1 = $ const, the rate of condensation is determined by the saturation in the adsorbed phase.

Saturation depends strongly on the incident flow b_1 and the occupation of the vacancies, determined by the flow of all components and the system parameters.

It can be concluded that with all other parameters constant:

a) aggregation stability reduces the flow b_1 and thus reduces Π;

b) the introduction of immiscible and insoluble additives reduces Θ_1 and thus reduces Π;

c) the introduction of additives that reduce the viscosity increases the Brownian flows, i.e. increases Π;

d) presence chemisorbed contaminants on the substrate, poorly soluble in butanol, leads to a decrease of Θ_1 due to the occupation of the vacancies.

'Inhibited' growth. In [34] it was shown that the law of inhibited growth in classical condensation is written in the form $h = C(t - t_0)^{1/3}$ and is associated with the presence of repulsive forces of neighbouring fractures of the same name in the stage of growth, theoretically predicted by Landau [35]. The impurity chemisorbed on the face of the single crystal inhibits the growth of the first of m moving fracture of the same name which prevents the Landau forces and the growth of the remaining $(m - 1)$ fracture. This is because the neighbouring fracture, having come together to a distance L^*, are arrested, since joining of the particle to the fracture becomes unsuitable because of the presence of the repulsive energy of the Landau forces [34, 35].

We assume that the growth of the coating is determined by this law when $\Theta_4 > \Theta_4^*$ (the concentration of the chemisorbed contaminant). Here, the critical concentration Θ_4^* is assumed to be the concentration at which the inhibitory effect of the Landau force prevails over the saturation of the adsorbate with zinc clusters, which indirectly increases the effective concentration Θ_{h1}. It is concluded at $\Theta_1 > \Theta_4^*$ the growth rate of the coating is determined by the law of the inhibited growth, and at $\Theta_4 < \Theta_4^*$ by the law of proportional growth.

In case, if $\Theta_4 \cong \Theta_4^*$ both growth mechanisms will be competing, we will try to use this case to explain the growth of the coating by the 'competitive' law.

It is possible to overcome the inhibitory effect of the stoppers by the following mechanism: the growth stage can penetrate through the chain of the stoppers spaced at a distance of L_c. At the same time protrusion with radius r_b and the arc length $l_b = \pi r_b$ grow between the adjacent stoppers. If the value of L^* is equal to several periods of the crystal lattice a, and the value $m \cong 10 - 20$ [34], for the growth of the projection it is necessary to satisfy the condition $L_c \geq 2r_k$ condition which is sufficient [34]. Here r_k is the radius of the deposited clusters.

The distance between two adjacent stoppers can be determined if we know the distance between two neighbouring adsorption vacancies (a) and the concentration of the components, causing inhibition of growth $L_c = a/\Theta_4$. Hence we obtain the condition under which the inhibition does not occur: $\Theta_4^* < a/2r_k$.

It follows that 1) the mechanism of inhibition is determining when the impurity concentration reaches the value greater than Θ_4^*; 2) while keeping other conditions constant ($\Theta_4^* = $ const) the decrease of the cluster radius below the critical cluster size $r_k = a/2\Theta_4$ provides growth of the coating by the proportional law.

Growth by the 'competitive' law. The growth law for some of our experiments can be written as

$$h = C_1(t - t_0)^\nu,\tag{7.10}$$

where $\nu \cong 0.76$. It is assumed that the concentration of stoppers in the adsorbate is approximately equal to the critical concentration $\Theta_4 \cong \Theta_4^*$. Then the condensation process is determined by two processes: the condensation by the proportional growth low and the condensation by the inhibited growth law. This process can be described by a power law, whose parameters can be determined from the strength of the effect of each process on total condensation.

Let λ be the strength of the effect of the inhibition process on the final result ($0 \leq \lambda \leq 1$). If we assume two mutually exclusive laws under consideration in elementary acts of condensation, the thickness of the layer can be written as

$$h = \left[C(t - t_0)^{1/3}\right]^\lambda \left[\Pi(t - t_0)\right]^{1-\lambda}.\tag{7.11}$$

Comparing (7.10) and (7.11) and separating the time-dependent variables, we obtain

$$C_1 = C^\lambda \Pi^{(1-\lambda)}, \quad \nu = 1 - 2\lambda/3.$$

In our experiments $\nu = 0.76 \cong 3/4$, then the strength of the effect of the inhibition factor is $\lambda = 3/8$ and $C_1 = C^{3/8}\Pi^{5/8}$.

Thus, it was shown that at $\Theta_4 \cong a/2r_k$ the competitive growth law can be justified by the joint action of the proportional and inhibition laws and can be determined by the strength of the effect of each of them.

7.4.6. Mechanisms of destruction of the coating during their deposition

In the experiments we observed the modes in which there was a decrease of the coating thickness or when the growth of the coating was stopped (see Fig. 7.12 and the section without growth 1 of curve 4 in Fig. 7.13). The invariance of the coating thickness can be explained by the fact that under the given conditions saturation in the concentration of zinc is not reached in the adsorbed layer. But the withdrawal of ultrafine zinc from the surface can be explained only by the formation conditions when the coating becomes unstable and collapses. One of the mechanisms of destruction can be chemical etching of the coating. If the adsorbed layer accumulates a sufficient amount of oxygen or acid, these substances may not only retard the growth of coatings, but also destroy the already formed coatings. In the case where etching dominates over the condensation process, the coating thickness decreases.

Another mechanism for the destruction of the coating may operate in the case when a porous coating forms and a liquid, such as butanol, is trapped in the pores of the coating. The relaxation and recrystallization of the coating may be accompanied by the formation of disjoining pressure sufficient for its destruction.

7.4.7. Recommendations for the development of 'cold wet galvanizing' technology

The most efficient method is the growth of the coating by the proportional law. Its implementation can be achieved by the following actions:

1) decrease in the concentration of impurities using a surfactant;

2) decreasing the size of the deposited zinc clusters.

An example of a coating obtained after cold 'wet' galvanizing is shown in Fig. 7.14.

7.5. Pilot plant for the production of nanocomposites

Figure 7.15 shows the appearance of a Disma-1 pilot plant, created for the production of a colloidal mixture of butanol with metastable zinc nanoparticles. This composite is designed for cold 'wet' galvanizing the surfaces of steel products. The characteristics of the installation are shown in Table 7.3.

Conclusions

1. Experimental and industrial units for the realization of joint of condensation processes of vapours of organic solvents and metals have been constructed.

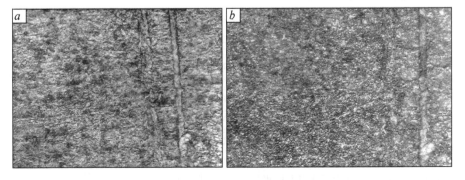

Fig. 7.14. Appearance of the coating produced by cold 'wet' galvanizing with a micelle-forming addition; ×1000. a) sample before galvanizing; b) after galvanizing.

Fig. 7.15. Pilot plant Disma-1.

Table 7.3. Technical characteristics of equipment

Parameter	Value
Dimensions of working chamber, mm	1700×1500×1450
Weight of working chamber, kg	460
Area occupied by equipment, m²	36
Total weight of equipment, kg	1500
Maximum required power, kW	63
Working vacuum, Pa	0.106–13.3
Zinc charge in a single production cycle, kg	18
Consumption of cooling water, kg/h	3,2
Consumption of liquid nitrogen, kg/h	450
Productivity of equipment at a zinc concentration in composite of 21%, t/year	100
Range of regulation of the mean zinc particle size in butanol, m	10^{-9}–$2\cdot10^{-6}$

2. The effect of pressure in the vacuum chamber, the condenser temperature and the flow rate of the components on the concentration and particle size of the ultrafine zinc suspension obtained in butanol was investigated.

3. A significant effect of the composition of the atmosphere of the chamber, the cleanness of the walls and the presence of the flow of zinc clusters, generated by the condenser and the chamber walls, on the composition of the resultant ultrafine composite was shown.

4. A high physico-chemical activity of the liquid and jelly-like produced composite was observed.

5. The 'shagreen' skin phenomenon in the cluster-containing composite films, formed during the phase transition in the matrix of the composite was clarified.

6. The appearance of hysteresis phenomena in adsorption processes at small periodic oscillations of component vapour flows was noted. These phenomena are manifested in the variations of the condensation rate and productivity of the process, with periods of oscillations of the flows greater than the characteristic time of the system.

7. A method for measuring the zinc content in the ultrafine zinc–butanol composite was developed.

8. Systematic studies of the process of joint condensation of vapours of zinc and butanol have been carried out. The condensation regimes for the production of the ultrafine zinc–butanol composite have been determined.

9. The properties of the zinc–butanol composite and its physico-chemical activity were investigated and fundamentals of the technology of cold 'wet' galvanizing steel products for the purpose of active corrosion protection have been developed.

List of Symbols

A_k - dimensionless criterion is the ratio of the probability attachment of particles to a solid surface to the probability one act of consolidation;

a_i - describes the employment vacancies adsorption is adsorbed particles and the components it makes sense the reflected flux of the adsorbed particles;

b_p - dimensionless flux of particles of the component from the bulk phase solution of the adsorbate;

C_{Zn} - mass concentration of zinc,%;

c - the initial mass concentration of particles in the colloid,%;

c_1 - the dimensionless rate of condensation;

D - diffusion coefficient, m / s;

f_{Zn} - the mass of zinc in the sample, kg/m^3;

h - thickness of the coating, m;

$I = -dv_\Sigma/dt$ - rate of change of particle concentration, c^{-1}·m^{-3};

$\overline{I}$ - The dimensionless rate of change of particle concentration;

$\overline{I}_c$ - The dimensionless rate of coagulation;

I_c - the rate of coagulation;

I_d - the speed of deposition of particles on the surface;

$\overline{I}_d$ - Dimensionless deposition rate on the surface;

J_{bu} - the rate of evaporation of butanol, g / min;

J_{Zn} - the rate of evaporation of zinc, g / min;

J_d - The flow of particles from half to one vacancy, s^{-1};

K_{bu} - the rate of condensation of butanol, g / min;

K_{Zn} - the rate of condensation of zinc, g / min;

k_B - Boltzmann constant, J / K;

L_c - distance between the surfactant-stoppers in the adsorbate;

L^* - the same criticism, m;

m_1 - mass of reacted zinc, mg;

m_{Zn} - the mass of the source of zinc, mg;

m_1 - mass of a cluster of zinc, kg;

m_0 - a scale factor that takes the dimensionless flow kinetic equations in a real system, s^{-1};

n - the number of surface vacancies;

P^c - the steric factor;

P^d - steric factor;

p^c - The probability of coagulation;

R - the average size of particles, m;

r_b and l_b - the radius and arc length of the projections between the condensate two stoppers, m;

r_k - the radius of the deposited clusters, m;

S - area of the solid surface, m^2;

T - ambient temperature, K;

T_{bu} - butanol evaporation temperature, K;

T_{Zn} - zinc evaporation temperature, K;

t - current time, s;

t_0 - The characteristic time for coagulation of the first particle, c;

V - total volume of colloid, m^3;

V_1 and V_p - volume of the liquid phase and particles, respectively, m^3;

V_m - Volume ratio of a monomolecular layer of particles on the surface of the colloidal mixture to the whole;

V_{H_2} - The volume of hydrogen liberated in the reaction with hydrochloric acid, Jr.;

V_0 - volume of hydrogen, reduced to normal conditions, l;

V_1 - volume of sample, mL;

v_0 - initial concentration of particles, m^{-3};

$\overline{v}_c$ - A dimensionless number of coagulated particles of unit volume of colloid;

$\overline{v}_d$ - A dimensionless number of particles deposited per unit volume count Lois;

v_Σ - total number of particles per unit volume of a colloid, m^{-3};

V_c - the number of coagulated particles per unit volume of the colloid, m^{-3};

v_d - the number of particles deposited on the surface of a unit volume colloid, m^{-3};

$v = v_\Sigma / v_0$ - Dimensionless concentration;

ΔE^c - coagulation activation energy, J;

ΔE^d - the activation energy for adsorption of particles, J;

Δm - Specific fraction of particles deposited from the bulk of the colloid;

ΔM - The weight gain of coating, kg;

Θ_1 - the degree of filling of vacancies of adsorption particles of zinc;

Θ_2 - the degree of filling of vacancies adsorption of molecules of butanol;

Θ_3 - the degree of filling of vacancies adsorption surfactant molecules or stabilizer;

Θ_4 - the degree of filling of vacancies adsorption of impurities;

Θ_{h1} - the degree of filling of vacancies of adsorption of zinc particles in equilibrium;

Θ_4^* - the critical concentration of polluting hemoadsorbirovannoy impurities;

λ - the degree of influence on the deposition process of inhibition;

$v_\perp$ - Frequency of thermal vibrations in the plane perpendicular to the sample surface, s^{-1};

P - the growth rate of coverage, m / s;

Π' - The flow of condensate into the coating at the moment, kg / s;

ρ_1, ρ_2 - the density of zinc and butanol, respectively, kg/m^3;

ρ_1 and ρ_p - the density of the liquid phase and the colloidal particles of matter respectively, kg/m^3;

$\tau = t / t_0$ - Dimensionless time.

Literature

1.   Bochkarev A.A., Poroshk. Metall., 1982, No. 4, 40–47.
2.   Palatnik L.S., Komnik Yu.F., Fiz. Met. Metalloved., 1960, V. 10, No. 4, 632-633.
3.   Gladkikh N.T., Palatnik L.S., Dokl. AN SSSR, 1961, V. 140, No. 3, 567–575.
4.   Palatnik L.S., et al., Fiz. Tverd. Tela, 1962, V. 4, No. 1, 202–206.
5.   Palatnik L.S., Gladkikh N.T., Fiz. Tverd. Tela, 1962, V. 4, No. 2, 424-428.
6.   Palatnik L.S., Gladkikh N.T., Fiz. Tverd. Tela, 1962, V. 4, No. 8, 2227–2332.
7.   Bochkarev A.A., et al., in: Phase transitions in pure metals and binary alloys, Novosibirsk Institute of Thermal Physics, USSR Academy of Sciences, 1980, 133–145.
8.   Bochkarev A.A.,in: Proc. Int. Workshop, 6-8 Sept. 1988, Novosibirsk, USSR, Netherlands, VSP, 1989, 499–525.
9.   Bochkarev A.A., Polyakova V.I., Teplofiz. Vysokikh Temperatur, 1989, V. 27, No. 3, 472–474.
10.   Bochkarev A.A., et al., Zavod. Lab., 1994, No. 4, 34.
11.   Bochkarev A.A., et al., in: Boiling and Condensation (Hydrodynamics and heat exchange). Novosibirsk, Institute of Thermal Physics, USSR Academy of Sciences, 1986, 102–110.
12.   Knauer W., J. Appl. Phys., 1987, V. 63, No. 3, P. 841.
13.   Hawley J.H., Ficalora P.J., J. Appl. Phys., 1988, V. 63, No. 8, 2884–2885.
14.   Bochkarev A.A., PolyakovA V.I., in: Proc. Reports. on X Conf. on the Dynamics of rarefied gases, Moscow, 1989.
15.   Bochkarev A.A., Pukhovoi M.V., Sib. Fiz.-Tekh. Zh., 1993, No. 5, 1–11.
16.   Zhabotinsky A.M., Concentration oscillations, Moscow, Nauka, 1974.
17.   Barnickel P., Wokaun A., Mol. Phys., 1990, V. 69, No. 1, 1–9.
18.   Tano T., J. Coll. Interface Sci., 1989, V. 133, No. 2, 530–533.
19.   Frolov Yu.G., The course of colloidal chemistry, Moscow, Khimiya, 1988.
20.   Babko A.K., Pyatnitsky I.V., Quantitative analysis, Moscow, Vysshaya shkola, 1962.
21.   Kimura K., Bandow S., Bull. Chem. Soc. Jpn., 1983, V. 56, 3578.
22.   Andrews M.P., Ozin G.A., J. Phys. Chem., 1986, V. 90, 2929–2935.
23.   Zeiri L., Efrima S., J. Phys. Chem., 1992, V. 96, No. 14, 5908–5917.
24.   Lin S.T., et al., Langmuir, 1986, V. 2, No. 2, 259–260.
25.   Henglein A., Ber. Bunsenges. Phys. Chem., 1980, V. 84, 253–259.
26.   Abe H., et al., Chem. Phys., 1980, V. 47, 95–104.
27.   Kernizan C.F., et al., Chem. Mater., 1990, No. 2, 70–74.
28.   Kilner M., et al., J. Chem. Soc. Chem. Commun., 1987, V. 5, 356–357.
29.   Gardenas-Trivino G., et al., Langmuir, 1987, V. 3, No. 6, 986–992.
31.   Tan B.J., et al., Langmuir, 1990, V. 6, No. 1, 105–113.
33.   Smoluchowski M.V., Phys. Z., 1916, B. 17, 593.
34.   Sorokin V.K., Palatnik L.S., Izv. AN SSSR, Neorg. Mater., 1990, V. 26, No. 3.
35.   Landau L.D., Equilibrium form of crystals, V. 26, Moscowm Nauka, 1969.

8

Numerical simulation of molecular processes on the substrate

In the 1990's the authors developed an experimental technology to produce various nanosized metal–organic composites [1] with unique properties [2] by joint vacuum deposition of two components of the vapours on a cryogenic condenser. During the experiments, the authors were surprised by two circumstances. First, the inability to provide long-term stable operation of the process without manual adjustment. Second, even when applying a mixture of vapours with a low content of metal vapour in comparison with the vapour of organic matter the final product turned out to have a high concentration of the metal [3]. These two circumstances have not found a sufficient understanding and radical solutions. Therefore, in order to revive work on nanodispersed composites the authors carried out mathematical modelling of the condensation of two-component vapour on the cryogenic surface. The condensation of water and silver vapours on the surface with a temperature of 223 K was modelled. The calculations showed a number of new and unexpected phenomena that help explain the strange behaviour of processes in the experiments and develop a technology of production of composites on a new level.

8.1. Stimulated adsorption and capillary condensation

The condensation process begins with adsorption. The physical polymolecular adsorption on a non-porous homogeneous surface is described in most cases using the BET model [4], based on the model of monolayer Langmuir adsorption [5]. Despite some limitations, the BET model is used to create methods to measure the specific surface area of disperse systems. These methods are based on the interpretation of adsorption isotherms measured under equilibrium conditions, when the system under investigation – absorbent – is held long enough in the atmosphere of vapour or gas. These methods are not applicable for cases where there is no equilibrium. Such situations arise, for example, in the process of degassing vacuum systems and in thin film deposition processes in the gas or vapour environment. In the technology of production of dispersed composites by co-condensation of two components [1] it is also not possible to apply the BET model to estimate the current situation at the condensation front. Therefore in this study it is attempted to carry out direct numerical simulation of sorption processes in the conditions when the situation on the surface of the adsorbent is non-stationary. Modelling physical adsorption under these conditions has revealed a number of previously unstudied phenomena. In particular, this study shows the molecular aspects, which are responsible for capillary condensation.

8.1.1. The model of molecular processes on the surface an ideal substrate

Langmuir [6] suggested that his single-layer adsorption model can be applied to the description of the condensation processes. Later, this idea was consistently developed by many researchers and a simple phenomenological model of the molecular processes at the ideal surface was described in [7]. This model is adopted as the basis for modelling in this paper. Previously this approach was successfully applied to describe the process of forming ultrafine composites [8], to search for the conditions for growth of whisker crystals [9], and modelling the features of the molecular processes at the liquid surface [10]. The essence of the model is as follows. The surface of the substrate is a homogeneous lattice of adsorption vacancies with a characteristic distance between the adsorption vacancies adequate to the parameters of the crystal lattice of the vapour condensate. The vapour molecule incident on the surface of the crystal–substrate is fixed in one of the adsorption vacancies with the adsorption energy ε_S and is instantly accommodated with the surface of the crystal as regards temperature. As a result of thermal fluctuations and fluctuations of the lattice vibrations of vacancies, this molecule can either move to the next adsorption vacancy, which indicates surface diffusion, or desorb back into the vapour environment. The lifetime of the molecule to desorption can be written as

$$\tau_S \approx v^{-1} \exp\left(\varepsilon_S/kT_S\right),$$

where v is the frequency of thermal vibrations of the molecule in the adsorbed state, $v \approx 10^{13}$–10^{14}; k is the Boltzmann constant; T_S is the temperature of the substrate. The movement of the adsorbate molecule to a neighbouring vacancy requires less intense thermal fluctuations than for desorption. Therefore, the lifetime of the molecule to movement to the next position can be written as

$$\tau_{SD} \approx v^{-1} \exp\left(\beta\varepsilon_S/kT_S\right),$$

where β is the fraction of the activation energy of surface diffusion U_D in the adsorption energy ε_S, $U_D = \beta\varepsilon_S$. The value of β is determined by the geometry of the lattice of the vacancies. In [7] the value of $\beta = 0.3$–0.4 is recommended. For specific calculations, the value of β is specified by the calculations of stationary situations on the substrate. Adsorption energy ε_S is calculated from the substrate surface energy or Young's hypothesis [11, 12].

8.1.2. Application of the model for the two-component vapour

The model for the two-component vapour is complicated by the fact that the adsorbed molecule may have multiple heterogeneous neighbours in adjacent adsorption vacancies. Therefore, when calculating the lifetime of each molecule should take into account its interaction with its nearest neighbors with the corresponding binding energies. It is assumed that the crystal lattices of the substrate and the condensate are the same.

If the adsorbed molecule has several neighbours in the neighbouring adsorption vacancy, the lifetimes of the molecules of the first and second vapour components are evaluated as follows

$$\tau_{SD,1} \approx v^{-1} \exp \left(\beta(n_1\varepsilon_1 + n_{12}\varepsilon_{12} + \varepsilon_{S,1})/kT_s \right); \ \tau_{S,1} \approx v^{-1} \exp \left((n_1\varepsilon_1 + n_{12}\varepsilon_{12} + \varepsilon_{S,1})/kT_s \right),$$

$$\tau_{SD,2} \approx v^{-1} \exp \left(\beta(n_2\varepsilon_2 + n_{21}\varepsilon_{21} + \varepsilon_{S,2})/kT_s \right);$$
$$\tau_{S,2} \approx v^{-1} \exp \left((n_2\varepsilon_2 + n_{21}\varepsilon_{12} + \varepsilon_{S,2})/kT_s \right),$$

where n_1, n_2 are the numbers of the nearest homogeneous and n_{12}, n_{21} – inhomogeneous neighbours. For the cubic lattice the number of nearest neighbours is in the range 0–4. The probabilities of the events for the adsorbate molecule – surface diffusion or desorption – are computed as the reciprocals of the corresponding lifetimes. The share of the activation energy of surface diffusion β in the binding energy is taken equal for both components and for the diffusion in the condensate. This is natural, since it is assumed that the crystal lattices of the substrate and the condensate are the same.

8.1.3. Sorption equilibrium

As a result of desorption of previously adsorbed molecules, the specific flow of the molecules leaving the surface of the substrate or condensate is

$$J_{S,i} = n_{M,i} \ \Sigma \ (1/\tau_{S,i}). \tag{8.1}$$

Here the summation of the probabilities of desorption is performed for all $n_{M,i}$ molecules located on the unit surface area.

From the molecular theory of gases is known that the flow of the molecules passing through the unit area of any section of the gas or vapour space is

$$J_{K,i} = P_{K,i}/(2\pi \ M_i \ kT_K)^{1/2}. \tag{8.2}$$

Here $P_{K,i}$ is the partial pressure of the vapour; M_i is molecular weight; T_K is temperature of the vapour mixture.

The equilibrium state, namely, the equilibrium concentration of the adsorbate of the vapour components on the surface of the substrate or a crystal is defined by $J_{S,i} = J_{K,i}$. The dynamics of establishment of the equilibrium state of the adsorbate on the surface is formally defined by the dynamic equation

$$dn_{M,i}/dt = J_{K,i} - J_{S,i}, \tag{8.3}$$

The physical meaning of this equation is the instantaneous rate of sorption.

In view of (8.1) and (8.2) we see that the equations (8.3) have the form of relaxation equations. Their solutions for sufficiently large values of time should give the concentrations of the adsorbate of the vapour components on the substrate or the condensate.

Model calculations in terms of the equilibrium state $J_{S,i} = J_{K,i}$ were used to refine the values of adsorption energies, bonding and the proportion of the activation energy of surface diffusion. In the non-equilibrium state the fraction of molecules remaining on the surface of all the molecules present on the surface, $\alpha_{C,i} = 1 - J_{V,i}/J_{K,i}$, is a measure of the rate of sorption, the relative rate of adsorption or condensation.

Vacancies on the substrate may be partially or completely filled with adsorbate or condensate. The mechanism of sorption of molecules, again coming to the surface of the condensate, is similar to what we saw on the substrate surface. The only difference is that the molecules are arranged in a layer of homogeneous molecules interacting not with the substrate but with the condensate molecules.

Figure 8.1 shows a portion of the condensate on the surface of the crystal when the surface molecules $A4$ and $B2$ have the maximum number of neighbours.

The binding energy of the molecule $B2$ is calculated as the sum

$$E_{B2,i} = \sum_{i=1}^{2} E_{BC,i} n_{BC,i} + \sum_{i=1}^{2} E_{BB,i} n_{BB,i} + \sum_{i=1}^{2} E_{BS,i} n_{BA,i} + E_{B2A2,i},$$

where, depending on the configuration of surrounding atoms $n_{BC,i} = 0–4$, $n_{BB,i} = 0–4$, $n_{BA,i} = 0–4$.

The summation is carried out separately for homogeneous and heterogeneous surrounding molecules. For the case when the molecules $A1$, $A2$, $A3$, $B1$, $B2$, $B3$, $C1$, $C3$ are homogeneous

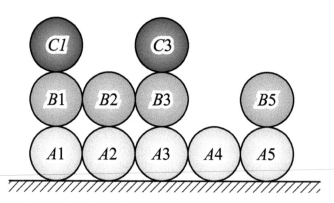

Fig. 8.1. The scheme of interaction of the molecules $B2$ and $A4$ with the nearest surrounding molecules.

$$E_{BB} = E_{B2A2} = \varepsilon_i,$$

then

$$E_{B2,i} = \varepsilon_i(n_{BB,i} + L\,(n_{BC,i} + n_{BA,i}) + 1),$$

where $L = \exp\,(6 \cdot \ln\,(\mathrm{sqrt}\,(2) - 1))$ is a coefficient taking into account the interaction of molecules arranged diagonally in the principal planes of the crystal lattice. The interaction of molecules in the diagonal planes is not considered.

The activation energy for surface diffusion is

$$U_{DB2,i} = \beta \cdot E_{B2,i}.$$

The characteristic time of desorption and the desorption probability per time δt is

$$\tau_{B2,i} = \nu^{-1} \exp\,(E_{B2,i}\,/kT), \quad P_{B2,i} = \delta t/\tau_{B2,i}.$$

The characteristic time to the start of the diffusive motion of the atom and its probability for time δt

$$\tau_{B2D,i} = \nu^{-1} \exp\,(U_{DB2,i}\,/kT), \quad P_{B2D,i} = \delta t/\tau_{B2D,i}.$$

The binding energy of the molecule $A4$ is calculated as the sum

$$E_{A4,i} = \sum_{i=1}^{2} E_{AB,i} n_{AB,i} + \sum_{i=1}^{2} E_{AA,i} n_{AA,i} + E_{A4S,i},$$

where, depending on the configuration of surrounding atoms $n_{AB,i} = 0\text{--}4$, $n_{AA,i} = 0\text{--}4$.

For the case when the molecules $A3$, $A4$, $A5$, $B3$, B are homogeneous,

$$E_{AA,i} = \varepsilon_i,$$

then

$$E_{A4,i} = \varepsilon_i(n_{AA,i} + Ln_{AB,i} + \varepsilon_{S,i}/\varepsilon_i).$$

The activation energy for surface diffusion of the molecule $A4$

$$U_{DA4,i} = \beta \cdot E_{A4,i}.$$

The characteristic time of desorption and the desorption probability per time δt

$$\tau_{A4,i} = \nu^{-1} \exp\,(E_{A4,i}\,/kT), \quad P_{A4,i} = \delta t/\tau_{A4,i}.$$

The characteristic time prior to the start of the diffusive motion of the atom and its probability for time δt is

$$\tau_{A4D,i} = \nu^{-1} \exp\,(U_{DA4,i}\,/kT), \quad P_{A4D,i} = \delta t/\tau_{A4D,i}.$$

It is clear that the described model of molecular sorption processes can only be used for numerical processes playing out on the substrate surface.

8.1.4. The numerical calculation algorithm

The authors have developed a program that for the processes described above. The field for the calculations is a 70×70 grid of adsorption vacancies with a characteristic mesh size equal to the lattice constant. The time step for drawing a single act of the events on the grid is chosen to be the minimum mean time of two components between arrivals of molecules from the vapour environment on the grid. The cell which receives the molecule which again arrived is chosen randomly. At one time step all of the cells are analyzed successively. The calculation algorithm is described in detail in [8]. In the presence of the molecule in this cell, the following calculations are carried out.

1. To determine which molecular layer contains the surface molecule in this cell.

2. Analyze the nearest grid cells. Depending on the configuration of filling the adjacent cells we calculate the binding energy of the surface molecule in this cell.

3. Calculation of the activation energy of surface diffusion.

4. Determination of the probability of surface diffusion events and desorption of the surface molecule in this cell.

5. If the total probability of these events is much less than unity, then the sum of probabilities is normalized to unity.

6. The event is played using a random number generator, specifically programmed to reduce the error of the computer generator.

7. If there is a probability of occurrence of desorption, the surface molecule is removed from the examined call..

8. If there is a probability of occurrence of molecule migration, the random migration direction is modelled and the molecule is removed from the examined cell and added to a neighbouring cell.

9. The lifetime of the molecule before the above-mentioned event is calculated and then deducted from the time step. The remaining time is used for re-drawing in this cell starting from step 1, taking into account the changed configuration of its filling and filling of adjacent cells.

10. When the analysis time of the considered cell exhausts the time step we transfer to the analysis of the next cell.

After completing the analysis of all cells the fall of the next molecule in a random cell is modelled. All current results are stored in computer memory and then in the output files. The current situation on the crystal surface is shown on the monitor. When the number of arrivals in this class of molecules and the total time become equal to the time between arrivals of molecules of another type, the arrival of a foreign molecule at a random location is modelled.

When the program starts flows of dissimilar molecules are applied to the initially clean surface of the crystal–substrate. The first vapour component is supplied from the heat source. The flow of molecules of this vapour onto the substrate is calculated from the geometry of the source and the substrate under the assumption of spherical dispersal of molecules from the nozzle of the source. The second component is fed from the half-space filled with the vapour at given values of pressure and temperature. We compute the dynamics of the numbers of the molecules of the components in each layer on the condensation surface, as well as the total number of the molecules of the components on the substrate.

8.1.5. The results of the calculations. Filling of the substrate by the adsorbate

Figure 8.2 shows the images of successive stages of filling of the substrate with the water adsorbate and silver condensate. The calculations were performed with the parameters listed in Table 8.1. The table shows that water vapour is in an unsaturated state with respect to the substrate. In the absence of silver vapour, water vapour in these conditions creates a dynamic equilibrium of the adsorbate on the substrate (15.6 ± 3.4 molecules) with no signs of condensation. The silver vapours are strongly supersaturated with respect to the temperature of the substrate, the silver can condense. From Fig. 8.2 a it is clear that in the presence of silver vapours in the supersaturated state each silver is the centre of the nucleation of an island of the adsorbate of the water molecule. That is, the presence of a silver cluster promotes the adsorption of water vapour. The silver cluster, surrounded by the water adsorbate, loses the ability to grow by the mechanism of joining these atoms. However, on the periphery of some islets of adsorbed water rise to new clusters of silver. This means that the adsorbate of the water molecules stimulates in turn the nucleation of silver. Emerging silver clusters are also surrounded by the water adsorbate. A similar mechanism of chain nucleation has been shown previously by the authors in the phenomenological consideration of sorption processes [13]. This mechanism leads to a complete filling of the substrate with the silver clusters, the intervals between which are filled with the water adsorbate (see Fig. 8.2, a–d). Thus, the adsorption of water vapour, stimulated by the presence of silver clusters, leads to the formation of a dispersed silver condensate flooded by the capillary water condensate.

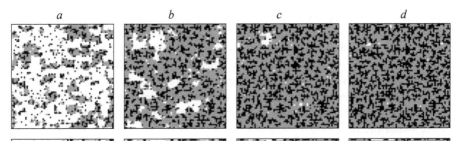

a   b   c   d

Fig. 8.2. Stages of filling of the 70 ×70 substrate with vacancies. Black pixels indicate the atoms of silver, gray – water molecules. The top image – the substrate in plan, the bottom – cross-sections on the 10th line from the top. The parameters adopted for calculations are shown in Table 8.1, the distribution of particles in layers is shown in Table 8.2.

Table 8.1. Parameters for calculating the results shown in Fig. 8.2

Silver		Water	
Atom flux on substrate, $m^{-2} \cdot s^{-1}$	$9.65 \cdot 10^{24}$	Atom flux on substrate, $m^{-2} \cdot s^{-1}$	$1.30 \cdot 10^{27}$
Saturation	$8.07 \cdot 10^{30}$	Saturation	0.392
Adsorption energy, J	$1.88 \cdot 10^{-19}$	Adsorption energy, J	$2.76 \cdot 10^{-20}$
Co-adsorption energy*, J	$1.88 \cdot 10^{-19}$	Co-adsorption energy*, J	$2.90 \cdot 10^{-20}$
		General parameters	
Substrate temperature, K	373	Temperature in chamber, K	300
Ag–H_2O binding energy, J	$2.70 \cdot 10^{-20}$	Time step, s	$1.004 \cdot 10^{-12}$

*Adsorption energy on the homogeneous surface.

Table 8.2. Characteristics of the images shown in Fig. 8.2

Time	Layer													
	Silver							Water						
	1	2	3	4	5	N_{Ag}^*	N_{Ag+}^{**}	1	2	3	4	5	$N_{H_2O}^*$	$N_{H_2O}^{**}$
$4.02 \cdot 10^{-7}$	458	371	15	0	0	844	1837	1602	194	4	0	0	1800	247 361
$7.99 \cdot 10^{-7}$	355	1153	48	0	0	1556	3656	3870	1271	16	0	0	5157	492 250
$1.20 \cdot 10^{-6}$	309	1675	115	8	0	2107	5474	4518	2285	41	0	0	6844	737 139
$1.50 \cdot 10^{-6}$	294	1875	210	14	0	2393	6851	4591	2604	112	6	0	7313	922 660

Figures 8.2 b, c, d show that during the filling of the substrate the number of visible silver clusters. Cross sections of the adsorbate show that part of the silver clusters formed on the layer of the water adsorbate. This is confirmed by the data for the population of the layers by the silver adsorbate in the Table 8.2. It is seen that during the filling of the substrate the number of silver atoms in the second layer exceeds the number of silver atoms directly on the substrate. This means that in the emerging film of the silver condensate the bond of the silver atoms with the substrate – adhesion – is much less than the cohesion of the silver condensate. This creates prerequisites for detachment of the metal condensate film from the substrate, which is observed quite often in the experiments.

Another curious fact should be noted. Table 8.2 shows that when filling the substrate the number of silver atoms in the first layer decreases. This means that during the condensation of silver the water vapour adsorbate displaces the silver atoms from the first layer. Such 'aggressive' behaviour of the water vapour adsorbate vapour weakens even more appreciably the adhesion of the silver condensate to the substrate. The mechanism of displacement is based on the stochastic processes of surface diffusion of the adsorbate and desorption of particles of both components. The intensity of these processes can be judged by comparing the data in Table 8.2 for the total number of particles on the of the components on the substrate N_{Ag} and N_{H_2O} with the numbers of all particles of the components N_{Ag^+} and $N_{H_2O^+}$ which 'visited' the substrate. Comparison of these figures gives an idea of the rate and relative rate of adsorption of the silver and water vapours. At the time 1.5 μs the relative rate of silver vapour adsorption is on average of 35%, for water 0.8%.

8.1.6. Effect of contact angle

Figure 8.3 shows the filling of the substrate as a function of the wetting angle of silver by water. The calculations were performed with the following parameters:

The flow of water molecules on the substrate, m^{-2}·s^{-1}	$1.30 \cdot 10^{26}$
Saturation	0.0392
Binding energy of Ag–H$_2$O, J	0.5–$2.9 \cdot 10^{-20}$ J
Time, s	$1.5 \cdot 10^{-6}$

All other parameters are kept at the values listed in Table 8.1. In the absence of the silver vapour flow of silver the dynamic equilibrium of the water adsorbate 1.46 ± 0.88 molecules is established on the substrate. In the presence of a flux of silver atoms by varying the energy of the interaction of silver atoms with water molecules in this range the contact angle varies from 130 to 0 degrees. Moreover, the number of water molecules on the substrate varies from 4.36 ± 0.52 to 2072 ± 119. The first figure shows that even with poor wetting of silver by water in the adsorbate of the vapour mixtures the number of water molecules is much greater than in the pure water vapour adsorbate. This means that here we have the stimulated sorption of the water vapour. The second figure in comparison with the first means a strong dependence of the adsorption of water vapour, stimulated by silver, on the

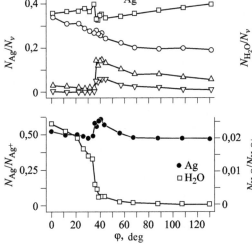

Fig. 8.3. Filling the substrate with silver atoms N_{Ag}/N_{Ag^+} and water $N_{H_2O}/N_{H_2O^+}$ as well as filling of layers N_{Ag}/N_v, N_{H_2O}/N_v depending on the angle of wetting. 1–4 – molecular layers.

wetting angle. The lower plot in Fig. 8.3 shows the relative rate of adsorption of vapour components, i.e. the fraction of the silver atoms and water molecules fixed to the substrate from all who 'visited' the substrate.

We see that the stimulation of adsorption of water by silver increases with improvement of wetting of silver by water. At an angle of wetting of about 35.9° we have the radical change. In a narrow range of contact angles the relative rate of adsorption the silver vapour and the population of the substrate with silver atoms are sharply reduced, and the amount of the water adsorbate and the relative rate of adsorption of water vapour rapidly increase. The upper graph in Fig. 8.3 shows layer-by-layer partial proportions of the adsorption vacancies occupied by water molecules and silver atoms. A detailed examination of these figures shows that a specified angle of wetting results in radical restructuring of the condensate structure on the substrate. The population of the third and fourth layers of silver atoms drops sharply, the population of the first layer grows. The population of the first layer by the water is also rising sharply. The water molecules occupy up to 40% of adsorption vacancies of the substrate, which suggests that the adsorption of water molecules is replaced by their capillary condensation, together with the condensation of silver atoms. In general, the silver and water condensates and on the substrate are compacted.

So, we are dealing with the mutual stimulation of adsorption of the components.

8.1.7. Adsorption isotherm

Figure 8.4 shows the dependence of stimulated absorption on the water vapour partial pressure in the chamber. The calculations were performed with the following parameters:

The flow of water molecules on the substrate, $m^{-2} \cdot s^{-1}$	$3.57 \cdot 10^{21} - 1.30 \cdot 10^{27}$
Saturation	$1.0792 \cdot 10^{-6} - 0.392$
Time, s	$1.5 \cdot 10^{-6}$

All other parameters corresponded to those in Table 8.1. Here we should clarify the term 'adsorption isotherm'. Calculations were carried out at constant temperature, varying the partial pressure of water vapour up to a constant time after supplying the vapour onto a substrate of 1.5 μm. During this time, the adsorbate of the vapour mixtures on the substrate does not reach equilibrium due to the continued condensation of silver vapour. Therefore, the term 'adsorption isotherm' has here a conditional, non-classical sense, and means that all calculations in this series were carried out at constant temperature.

In the absence of silver vapour the dynamic equilibrium of the water adsorbate water 15.6 ± 3.4 would be established on the substrate. The bottom graph in Fig. 8.4 shows the share of entrenched on the substrate silver atoms and molecules of water from all who visited on a substrate. In fact, these are relative rates of adsorption of vapour components. When the water vapour pressure is less than 100 Pa the rate of adsorption is of order 1, which suggests co-condensation of silver and water vapours. Since the saturation of water

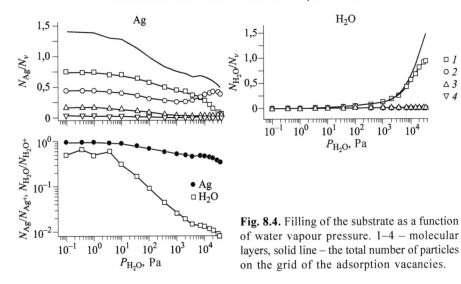

Fig. 8.4. Filling of the substrate as a function of water vapour pressure. 1–4 – molecular layers, solid line – the total number of particles on the grid of the adsorption vacancies.

vapour is insignificant, then for water this condensation means capture of the water molecules by the silver atoms. With the growth of the water vapour pressure the relative condensation rates of silver and water vapours decrease. For silver it is natural, since increasing pressure of water vapour increases the occupation of adsorption vacancies with water and this reduces the likelihood of the silver atoms bonding to them. This is not natural for the adsorption of water vapour because increasing water vapour pressure increases the likelihood of expected securing of the water molecules on the substrate. The upper right figure shows that, indeed, the filling of the substrate with the molecules of water vapour increases with increasing vapour pressure. The summary data for water at the bottom and top right graphs of Fig. 8.4 allow us to conclude that the probability of attachment to the substrate of each single molecule of water decreases with increasing water vapour pressure, and increase of the population of the substrate is due to an increase in the number of water molecules impinging on the substrate.

The upper left figure shows a feature in the population of the first and second layers with silver atoms. When the water vapour pressure is 7000 Pa the population of the second layer of silver atoms becomes larger than the population of the first layer. This means that for this and higher pressures, the water molecules displace silver from the first layer on the substrate. This phenomenon is reflected in the total amount of silver on the substrate in the form of a local perturbation in the curve.

8.1.8. Adsorption isobar

The term 'isobar' is used only to emphasize that the calculations were carried out at constant pressure. Figure 8.5 shows the dependence of stimulated adsorption on the substrate temperature. The calculations were performed with the following parameters:

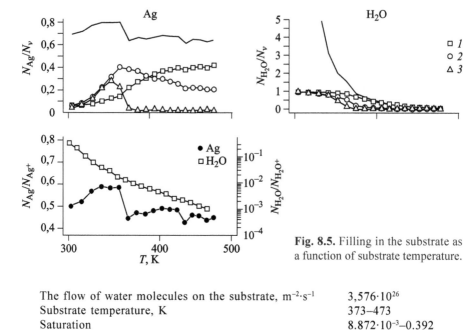

Fig. 8.5. Filling in the substrate as a function of substrate temperature.

The flow of water molecules on the substrate, m^{-2}·s^{-1}	$3{,}576 \cdot 10^{26}$
Substrate temperature, K	373–473
Saturation	$8.872 \cdot 10^{-3}$–0.392
Time	$1.5 \cdot 10^{-6}$

All other parameters are the same as those in Table 8.1. In the absence of silver vapour a dynamic equilibrium of the water adsorbate 15.6 ± 3.4 would be established on the substrate. The bottom graph in Fig. 8.5 shows the share of silver atoms and water molecules, fixed on the substrate, from all which 'visited' the substrate. The average relative rate of adsorption of silver in a time 1.5 µs is of order 1 in the entire temperature range. The top left graph of Fig. 8.5 shows that at temperatures around 370 K the population of the second layer of the silver condensate is higher than the population of the layer lying on the substrate, i.e. we observe the same effect of the restructuring of the silver condensate as the one which is visible in the isotherm in Fig. 8.4. At the same time the bottom graph shows a dramatic change in the relative rate of condensation of silver. The relative rate of adsorption of water is strongly dependent on temperature. When the temperature increases the relative rate of water adsorption decreases monotonically.

Figure 8.6 shows comparison of the relative rates of adsorption of pure water and silver vapours with the relative rates of the adsorption of these components in a mixture with the same parameters for 1.5 µs. It is seen that the adsorption of silver in the mixture is significantly hampered by the presence of water. In contrast, the adsorption of water vapour in the mixture is three orders of magnitude stronger than the adsorption of pure water vapour. Only when the substrate temperature is lowered there is a significant supersaturation of the water vapour which begins to condense, and the relative rate of condensation of pure water vapour becomes commensurate with the relative rate of capillary condensation in a mixture with silver vapours. It is ironic that the adsorption of water vapour together with silver vapour with

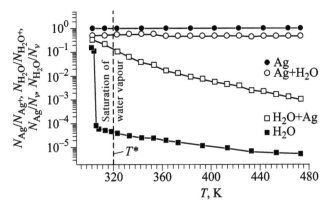

Fig. 8.6. Comparison of adsorption of pure water and silver and mixtures thereof.

increasing supersaturation does not lead to spontaneous nucleation of water and turns into condensation when the saturation is not reached.

8.1.9. Discussion of results

The fact that the presence of impurities in the cooled supersaturated vapours causes premature homogeneous condensation and reduces the supersaturation is widely known. Here we found on the substrate another previously not described phenomenon – adsorption and nucleation of silver vapour in the supersaturated state in the mixture with water vapour, stimulate the adsorption of water molecules which changes to the capillary condensation of water vapour whose concentration is significantly below saturation. The details of this phenomenon were also not previously known. Stimulated adsorption of water vapour has a counter effect on the condensation of silver vapour. First, the presence of water vapour decreases the rate of adsorption and condensation of silver vapour. Second, the influence of water vapour causes the restructuring of the silver condensate. The molecules of the water adsorbate displace silver atoms from the first layer on the substrate and move them to the upper layers of the adsorbate. Reducing the number of the silver atoms in the first layer on the substrate reduces the adhesion of condensation to the substrate. For the first time numerical experiments showed the mechanism of formation of adhesion and explained the well-known fact that the adhesion of the vacuum condensates depends on the presence of impurities in the atmosphere of the vacuum chamber where condensation takes place.

In the numerical experiments it was revealed that when the wetting angle of silver by water is changed the condensate restructuring occurs abruptly, with a little change in the contact angle. It also drastically changes the relative rates of condensation of both components. The reason for this critical change in the structure of the condensate is not completely clear. In the real process condensation leads to stacking of atoms and molecules in a configuration having the minimum free energy. Apparently, in the numerical experiment, in which statistical factors play a major role, condensation occurs by the same specific laws. And if some stacking of the molecules and atoms in the condensate ensures the minimum free energy at a certain ratio of the binding energies of atoms and molecules with the substrate and with each other, the statistical factors in the numerical implement this particular packing. Note that this phenomenon could not be observed in physical experiments, because in reality it is very difficult to achieve a specified ratio of binding energies, which determines the contact angle.

Analysis of the adsorption isotherm (see Fig. 8.4) shows that it consists of three characteristic regions.

1. On a site with a partial pressure of water vapour under 4 Pa the influence of the presence of water on the adsorption of silver is difficult to notice. Fluctuations in the curve for the relative rate of adsorption of water are caused by random deviations in the calculation of stochastic processes. The absence of any influence is understandable as the units of molecules on the 70×70 adsorption grid can not prevent the adsorption of silver atoms.

2. In the section with the partial pressure of water vapour of 4–5500 Pa the dependence of the relative rate of condensation of both components on the partial pressure of water vapour in the logarithmic coordinates is linear, indicating an exponential function with a negative exponent.

3. In the section with a partial pressure of water vapour of 5500–36000 Pa the dependence of the relative rate of condensation of both components of the partial pressure of water becomes linear in the linear coordinates, indicating a linear function with the negative coefficient at the linear term.

Note that the partial pressure of water vapour of 5500 Pa results in the restructuring of the condensate, which can be seen on the upper left graph of Fig. 8.4. We conclude that in the adsorption isotherm there are also critical condensation conditions and in transition through these conditions the structure of the condensate, its adhesion to the substrate and the growth law change.

The adsorption isobar (see Fig. 8.5) also suggests that the temperature range 360–380 K is characterized by the restructuring of the condensate: at the temperature below 380 K the number of silver atoms in the first molecular layer on the substrate is less than in the second layer. This also shows a drop in the condensate adhesion to the substrate. Generalizing this fact, we can conclude that in the field with the temperature – water vapour pressure there is a boundary between the zones with good and poor adhesion of the condensate to the substrate. Given the fact that a change in the wetting angle of silver with water also results in critical phenomena, we can say that in the space with the temperature–water vapour pressure–wetting angle there is a surface, separating the areas with good and poor adhesion and with different condensate structures.

Analysis of the graph for the relative rate of condensation of water vapour in Fig. 8.5 shows that the portion of the curve at $T > 360$ K is linear in the logarithmic coordinates, and at $T < 380$ K – in the linear coordinates. This indicates that the said boundary between the zones of good and poor adhesion shows also changes in the laws of growth of condensation.

Comparison of Figs. 8.4 and 8.5 revealed an unusual phenomenon. The solid lines on the top two graphs of Fig. 8.4 show that the growth of the population of the substrate by the water molecules causes a drop in the population by silver atoms. In Fig. 8.5 decrease in the population of water molecules decreases the condensation of silver. This apparent contradiction can be explained by the fact that the relative rate of adsorption of silver is reduced not because of rising populations of the substrate adsorbate of water, and by reducing the relative velocity of capillary condensation. Indeed, in both figures at the lower graphs reduce the relative velocity of capillary condensation of water corresponds to the reduction in the relative rate of condensation of silver. In other words, the components of the vapour mixture of mutually promote the adsorption and condensation of one another.

It is widely known that the presence of non-condensable gases in the condensing vapour reduces the rate of condensation. This is because the non-condensable gas creates an additional barrier to diffusion of vapour to the condensation surface. Figure 8.6 shows a decrease in the relative rate of condensation of silver in the presence of water vapour not as a result of the diffusion resistance but because of the occupation of some of the adsorption vacancies. This means that the common explanation of the effect of impurities on condensation must also be supplemented by these molecular processes on the condensation surface.

We should make one more clarification regarding Fig. 8.6. Water vapour in the presence of silver vapour condenses with a greater relative speed than in the pure form. This effect opens the possibility of intensification of condensation by the appropriate supplements to the condensing vapour. It is important to choose the right supplement. At first glance it seems that the criterion should be the ratio of the energies of mutual and dissimilar bonds of vapour molecules. This proposition may have important applications.

8.2. Non-stationary modes of formation of nanocomposites in vacuum deposition of two-component vapour

In order to explain the strange transients detected in the experiments to obtain nanosized composites by vacuum deposition of two-component vapour, a model based on the Langmuir adsorption model was constructed and software was developed for numerical simulation of condensation on a cooled substrate. The work of the model and programs was shown in the previous section for calculation of sorption in conditions when one of their vapour components was not in the saturated condition. In this section the model and the program are used to calculate the developed condensation in an environment where both components of the vapour mixture are supersaturated. Calculations have shown that there is a mutual stimulation of nucleation of condensed phases of the components with the leading role played the nucleation of silver. A number of exotic phenomena was also detected: increase in the relative rate of condensation of water vapour with a decrease in its partial pressure, lack of steady-state regimes of condensation, the existence of a border regime that separates the modes with higher and lower silver content in the condensate, the formation of porous frameworks from the condensation of one component with a filler of nanoparticles of the other component. The dispersion properties of the condensates, depending on the mode of condensation and condensate adhesion to the substrate, were calculated. The results of calculations help explain the previously unexplainable experimental results and develop technology for producing nanodispersed composites to a new level of understanding of the processes involved.

The model of the molecular processes at the surface, described in section 8.1, does not consider the dynamics of intermolecular interactions. It allows us to trace only the statistics behaviour of surface molecules. Therefore, we did not expect this model to obtain quantitative results, and hoped that the behaviour statistics of the surface molecules would allow us to illustrate a number of phenomena observed in the experiment which have not as yet been explained clearly enough. Expectations were based on the fact that, despite the 'coarseness' of the model, it is simple. This advantage allows us to describe the surface molecular processes at considerable intervals of time with a large number of molecules under consideration taking part.

8.2.1. The development of vapour condensation

The classical nucleation theory states that as a result of surface diffusion the adsorbate molecules collide, forming surface aggregates of molecules. At a sufficient supersaturation some of the aggregates are supercritical and stable. Therefore, islands of the condensate form on the surface of the crystal. Once a system of supercritical nuclei of the condensed phase forms on the crystal surface, vapour condensation begins to develop. Vapour molecules strike the surface, migrate by the mechanism of surface diffusion, fall into the adsorption vacancy where they bond with other molecules, and are fixed in these places, because their total binding energy significantly reduces the likelihood of migration and desorption. Nevertheless, desorption occurs. Part of the molecules break away from their neighbours, become single, and the probability of their desorption increases. But this process is less efficient than adsorption and condensation. Gradually, the entire surface of the substrate is filled with vapour molecules. New incoming vapour molecules fall on a monolayer of molecules already attached to the substrate. It is assumed that a monolayer of these molecules also forms the same lattice of the adsorption vacancies.

Equations (8.1) and (8.2) completely describe the mass transfer between the vapour and the substrate. Therefore, they can also be used to calculate the rate of condensation. The fraction of molecules, remaining on the surface, of all the molecules 'visiting' the surface in the physical sense means the relative rate of condensation:

$$\alpha_{C,i} = 1 - J_{V,i}/J_{K,i}.$$

When the program starts flows of dissimilar molecules are supplied to the original clean surface of the crystal–substrate. The first component of vapour is supplied from the heat source. The flow of vapour molecules on the substrate is calculated from the geometry of the source and the substrate assuming the spherical dispersal of molecules from the source nozzle. The second component is fed from the half-space filled with vapour at given values of pressure and temperature. We compute the dynamics of numbers of the molecules of the components in each layer on the condensation surface, as well as the total number of the components on the substrate.

8.2.2. The results of calculations

Calculations were performed for the condensation of silver and water vapour mixtures with the parameters listed in Table 8.3. The adsorption and co-adsorption energies were calculated from the latent heat of evaporation of the condensate of components and further refined by training the program under equilibrium conditions.

The calculations showed that the dynamic equilibrium of the adsorbed molecules is established during the period of the order of $1 \cdot 10^{-6}$ on the substrate.. Then the number of water molecules adsorbed on the surface of the crystal changes continuously over time, remaining at the mean value. There are local fluctuations in the number of molecules, leading to the formation of unstable aggregates of molecules in the adsorbed phase. These nuclei break up. Some of the nuclei do not disappear from the condensed phase but grow in the form of islands that are relatively close and fill the substrate. This

Table 8.3. Parameters for calculations

Ag		H_2O	
Defined parameters			
Evaporator temperature, K	2000	Chamber temperature, K	300
Evaporator diameter, m	0.12	Water vapour pressure, Pa	20–100
Substrate temperature, K	223	Lattice constant, m	$3.11 \cdot 10^{-10}$
Surface energy, N/m	0.8	Surface energy, N/m	0.074
Adsorption energy, J	$1.88 \cdot 10^{-19}$	Adsorption energy, J	$2.76 \cdot 10^{-20}$
Co-adsorption energy, J	$1.88 \cdot 10^{-19}$	Co-adsorption energy, J	$3.158 \cdot 10^{-20}$
Fraction of activation of diffusion	0.46	Binding energy with silver, J	$2.0 \cdot 10^{-20}$
Calculated parameters			
Flow on the substrate, $m^{-2}s^{-1}$	$1.795 \cdot 10^{+23}$	Flow on the substrate, $m^{-2}s^{-1}$	$(0.7–3.6) \cdot 10^{24}$
Flow from the substrate, $m^{-2}s^{-1}$	$-3.058 \cdot 10^{-13}$	Flow from the substrate, $m^{-2}s^{-1}$	$-1.623 \cdot 10^{23}$
Molecular volume, m^3	$1.716 \cdot 10^{-29}$	Molecular volume, m^3	$3.006 \cdot 10^{-29}$
Supersaturation	$5.869 \cdot 10^{35}$	Supersaturation	4.4–22
Crystal growth rate, Å/s	$3.079 \cdot 10^4$	Time step, s	$(0.6–3) \cdot 10^{-9}$
Incidence interval of atoms, s	$1.176 \cdot 10^{-8}$	Time, μs	0–353

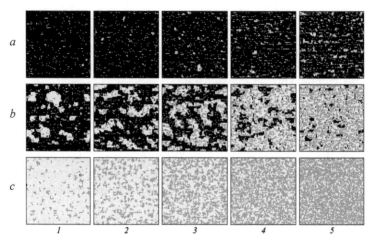

Fig. 8.7. Images of nanosized water–silver composite, obtained at the time of incidence on a substrate: a) 5000; b) 20 000; c) 50 000 water molecules. The composition of the vapour: 1) 5% Ag + 95% H_2O; 2) 8% Ag + 92% H_2O; 3) 10% Ag + 90% H_2O; 4) 15% Ag + 85% H_2O; 5) 20% Ag + 80 % H_2O.

is followed by the nucleation of the condensed phase on the surface of the islands, which indicates their growth in height. All this happens in the classical scenario described above.

Figure 8.7 shows images of the condensate on the substrate, obtained for five different compositions of the vapour supplied to the substrate. These compositions are calculated from the flows of incoming component molecules on the substrate. Black colour indicates a blank substrate, light gray color corresponds to the water condensate, dark gray to the silver condensate. The top row of images obtained after 5 000 water molecules fell on the substrate. This is the stage of nucleation of islands of the condensate. From left to right the images correspond to the reduction of water vapour pressure in the chamber where it condenses. There is a paradoxical phenomenon: along with a decrease in water vapour pressure and with decrease of vapour supersaturation the images show an increase in the number of islets of the water condensate. Table 8.4 presents the parameters corresponding to the images. It is seen that the number the water molecules adhering to the substrate increases with decreasing supersaturation, indicating the increase in the relative rate of condensation of water. The noticed fact requires a more detailed analysis of the results. Table 8.5 shows the relative rates of condensation of water and silver for the images as shown in Fig. 8.7. It was found that reducing the vapour pressure of water five times increases tenfold the relative rate of condensation. Table 8.5 shows that decreasing water vapour pressure also reduced the relative rate of condensation of silver. The observed phenomenon can be interpreted as the mutual stimulation of nucleation of components. The existence of this phenomenon was predicted by us in Refs. 13 and 14.

The second row of images was obtained when 20 000 molecules of water fell on the substrate – it is the filling phase of the substrate. The images show

Table 8.4. Calculation parameters and results of calculations of composites shown in Fig. 8.7

Number of H_2O molecules on substrate	Vapour composition	5 % Ag + 95 % H_2O	8 % Ag + 92 % H_2O	10 % Ag + 90 % H_2O	15 % Ag + 85 % H_2O	20 % Ag +80 % H_2O
	Supersaturation of H_2O vapour	22.03	15.42	9.91	6.61	4.41
5000	Time, μs	2.955	4.221	6.563	9.847	14.815
	Ag atoms	251	358	558	837	1259
	Ag adsorbate	217	295	428	651	956
	H_2O adsorbate	46	61	129	239	461
20000	Time, μs	11.801	16.869	26.22	39.36	59.01
	Ag atoms	1003	1434	2230	3347	5018
	Ag adsorbate	527	803	1302	2344	3718
	H_2O adsorbate	1220	1595	2030	4060	4808
50000	Time, μs	29.53	42.20	65.56	98.35	147.53
	Ag atoms	2509	3589	5576	8364	12548
	Ag adsorbate	1216	2015	3471	5896	9289
	H_2O adsorbate	13654	10118	8793	9623	8968

Table 8.5. Relative condensation rates α_{Ag} and α_{H_2O} for the images shown in Fig. 8.7

H_2O,	Number of H_2O molecules					
	5000		20 000		50 000	
	α_{Ag}	α_{H_2O}	α_{Ag}	α_{H_2O}	α_{Ag}	α_{H_2O}
100	0.865	0.00919	0.525	0.061	0.485	0.273
70	0.824	0.01217	0.5599	0.0797	0.561	0.202
45	0.767	0.0258	0.584	0.1015	0.622	0.1758
30	0.746	0.0477	0.7	0.203	0.705	0.192
20	0.759	0.0918	0.741	0.24	0.74	0.179
	Correlation coefficient					
1	0.969	−0.913	−0.913	−0.893	−0.987	0.882

an increase in the amount of the water condensate and increase the relative rate of condensation of water with a reduction of the vapour pressure and water supersaturation (fifth column in Table 8.5). However, at this stage the relative rate of condensation of silver also increases. Apparently, the decrease in vapour pressure and water supersaturation releases some of the adsorption vacancies for silver, which increases the relative rate of condensation of silver, and silver promotes the condensation of water by creating a larger number of stages of growth.

The third row of the images in Fig. 8.7 was obtained when 50 000 water molecules fell on the substrate. This is the stage of developed condensation.

Table 8.5 shows that at this stage a decrease in vapour pressure and supersaturation of water increases the relative condensation rate of silver, and the relative rate of condensation of water shows a downward trend. This is the expected result, due, apparently, by the competition of dissimilar molecules for the same adsorption vacancies [15, 16].

Table 8.5 also shows the correlation coefficients of the relative rate of condensation of silver and water with the partial pressure of water vapour. The absolute values of the coefficients are quite close to unity, which indicates a strong dependence. The sign of the correlation coefficient means a direct or inverse correlation. The stages of nucleation and filling of the substrate include 'exotic' phenomena – increasing the partial pressure of water vapour reduces the relative rate of condensation of water and at the nucleation stage increases the relative rate of nucleation of silver.

Our calculations for the same time after applying vapour flows to the substrate showed that the nucleation stage for silver is characterized by a high relative rate of condensation and at the same time by a low rate for water. At all three stages increasing water vapour pressure reduces the relative rate of condensation of silver. Apparently this is because part of the adsorption vacancies occupied by the water molecules are released for the sorption of silver atoms [15]. At the same time, increase of the water vapour pressure in the stages of nucleation and filling of the substrate reduces the relative rate of condensation of water. This fact clearly indicates the close relationship between the condensation of water and the condensation of silver. A more acceptable correlation of the condensation of silver and with the water vapour pressure is observed only in the stage of developed condensation.

The observed phenomena should radically affect the overall results of the condensation. Figure 8.8 a shows the change in the overall composition of the condensate in time. The composition of the condensate was calculated from the total amounts of the molecules of the components in the condensate. The peak at the beginning of the curves indicates that condensation begins primarily with the nucleation of silver. The concentration of silver in the concentrate is then dramatically reduced due to the increase of the relative rate of condensation of water. Further, the dynamics of the concentration of silver in the composite changes radically depending on the composition of the vapour mixture entering the substrate. The curves obtained for 5 and 8% of silver in the flows of the molecules reflect a monotonic decrease of the silver content in the condensate, and other lines a monotonic increase. The silver content in the condensate at the time of condensation of 350 µs is shown in Table 8.6. The table shows the discrepancy of the silver content in the incoming flows of vapour and in the condensate.

Figure 8.8 b and c show the dynamics of the relative rate of the condensation of water and silver. High values of the relative rate near zero time correspond to the adsorption on the still free vacancies. A sharp drop in the relative rate of condensation of silver is causes by an increase in the occupation of vacancies and also by desorption, nucleation and blocking of the growth stages by the water adsorbate. Filling of the substrate by the condensate occurs when the curves in Fig. 8.8 c take the form of a bend with a local minimum. The characteristic turning point in the curves corresponds to the onset of developed condensation of silver. Further the developed condensation of silver on the condensate takes place with 5 and 8% reduction of the relative rate curves for the Ag curves and with increasing relative rates for the other curves.

The dynamics of the relative rate of condensation of water vapour (see Fig. 8.8 b) is very different. Here, a sharp drop in the relative rate to ≈0 is due only to increase of desorption. The subsequent growth of the relative velocity is due to the increase in the number of silver nuclei. After filling the substrate with the condensate the curves are divided into monotonically increasing for the Ag content of 5 and 8%, and monotonically decreasing for higher silver content. This separation of the dynamic curves for the relative rate of condensation can be explained by competition between two components of the vapour in the sorption process on the same adsorption vacancies. This phenomenon is qualitatively discussed by us and other investigators [8, 9] and in this work is confirmed by numerical calculations.

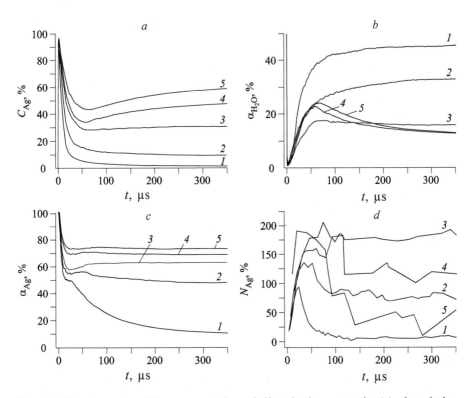

Fig. 8.8. The dynamics of the concentration of silver in the composite (a), the relative rate of condensation of water (b) and silver (c), the number of silver nanoparticles on the substrate (d) for different vapour compositions: 1) 5% Ag + 95% H_2O; 2) 8% Ag + 92% H_2O; 3) 10% Ag + 90% H_2O; 4) 15% Ag + 85% H_2O; 5) 20% Ag + 80% H_2O.

Table 8.6. Flux of silver atoms of a vapour mixture travelling to the substrate, and the Ag content of the condensate at a condensation time of 350 µs

Flux of silver atoms, %	Silver content of condensate, %	Flux of silver atoms, %	Silver content of condensate, %
5	1.14	15	47.38
8	9.44	20	58.94
10	30.73		

8.2.3. Histograms of particles

Images of the substrate with the condensate were analyzed using the Texture 2 program to calculate the dispersion composition of the condensate. The program detects only particles visible on the substrate as a solid, without inclusions of foreign atoms. It does not recognize the condensate as a particle in a matrix, with foreign atoms or particles visible on the background of the matrix. Therefore, the program provides complete information on the dispersion composition of the silver condensate for the silver content of 5, 8, and 10% and the water condensate for the silver content of 15 and 20%. In other cases, the program registers only truly individual particles. In view of this, the dynamics of the number of silver particles with a cross-section of more than one atom and of water can be seen in Fig. 8.8 d. It is seen that at the nucleation stage the number of silver particles increases rapidly for all silver contents. For the Ag content of 5 and 8% the increase in the number of particles followed by a decrease due to the fact that some of the particles are 'sheltered' by the water condensate. At 10% silver sheltering does not take place and the number of particles continues to grow. With silver contents of 15 and 20% the silver particles are combined into the matrix and can not be registered by the program. On the figure this is seen as a decrease of the number of particles detected with a large scatter.

A similar analysis of the dispersion composition of water in the condensate showed that at the nucleation stage the number of water particles is increasing rapidly. Then for the content of silver of 5 and 8% the number of detected particles of water decreases rapidly due to the formation of the matrix of the water condensate. For 10% silver content as a partial formation of the matrix of water ice. With silver contents of 15 and 20% the program records the stable number of water particles in the silver matrix.

8.2.4. Discussion of results

The above results help to explain the strange behaviour of condensation in the experiments to produce nanosized composites. Figure 8.8 a shows that the number of modes which can result in even approximately stable condensation is very limited. It is not possible to predict these modes in the experiments. If the selected mode is not accurate this will lead to a monotonic increase or decrease of the metal content in the composite. It is this fact which required in the experiments to continuously monitor and adjust the mode. Figure 8.8 a and Table 8.6 also helps explain why when creating a flow of vapour with the relatively poor content of the metal vapour the experimenters observed a high metal content in the composite. The obtained results may help researchers to create a more reliable technology for producing nanosized composites.

Figure 8.8 d clearly shows why it is difficult in the experiment to obtain a composite with a narrow fraction of the size of metal particles. A metal frame forms with a slight increase in the content of the metal vapour in the vapour mixture supplied to the substrate. When removing the condensate from the condenser, this frame is destroyed with the formation of metal particles. The resulting fraction of metal particles may be very different from the particles formed by nucleation.

Further analysis of the results of calculations leads to new useful results. The calculations determine the total number of silver atoms in each molecular layer of the condensate. These atoms belong to different silver nanoparticles whose shape is diverse – from single crystals, oriented perpendicular to the substrate, to a metallic porous frame. It is possible to calculate the generalized profile of the polycrystal, which will help one understand the general trend of growth of metal crystals form in the composite. The square root of the number of silver atoms in each molecular layer in the physical sense is the analogue of the radius of the generalized crystal. The result of this calculation is shown in Fig. 8.9 a. The figure shows that for all investigated silver contents of the vapour mixture supplied to the substrate the radius of the generalized crystal is smaller than the size of the substrate. During condensation at a silver content of 5 and 8% there is a temporary increase in the radius of the crystal, then the radius decreases with time and the silver content also decreases. With silver contents of 10, 15 and 20% the radius of the generalized crystal stabilizes. Increasing silver content in the mixture of vapours increases the radius increases which remains always smaller than the size of the substrate.

Studying Fig. 8.9 a it can be also concluded that in fact the quasi-stationary condensation modes still exist, and the non-stationary modes, as shown in Fig. 8.8 a, are caused by the strong non-stationarity in the stages of nucleation and filling of the substrate.

Figure 8.9 b shows the results of calculations of the content of silver atoms in each molecular layer. In fact, this is the distribution of silver across the condensate layer. Here we can see a unique feature. On the outer surface of the condensate in all cases there is a layer with a high silver content. This fact indicates that any new molecular layer is formed mainly by the nucleation of silver. Figure 8.9 b also shows that in the molecular layers adjacent to the substrate at a silver content of 5 and 8% there is a higher concentration of silver. The origin of these layers, rich in silver, is hard to explain. Apparently, this is a consequence of the benefits of nucleation of silver in the first molecular layer rather than in the subsequent layers.

The results of calculations allow us to calculate the adhesion of silver crystals to the substrate as the fraction of the adsorption vacancies of the substrate occupied by the silver atoms. Since it was assumed that the energy of adsorption of silver to the substrate is equal to the energy of mutual bonding

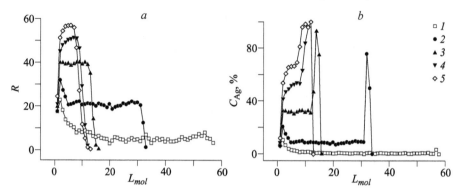

Fig. 8.9. Generalized profile of silver crystals in the composite (a) and the distribution of silver concentration in the height of the condensate layer (b) for different compositions of vapour entering the substrate. 1) 5% Ag + 95% H_2O; 2) 8% Ag + 92% H_2O; 3) 10% Ag + 90% H_2O; 4) 15% Ag + 85% H_2O; 5) 20% Ag + 80% H_2O. Condensation time $t = 352$ μs; R is the reduced radius of the crystals; L_{mol} are molecular layers.

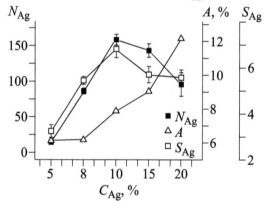

Fig. 8.10. Adhesion, the average number of nanoparticles N_{Ag} and average cross-sectional area S_{Ag} depending on the concentration of silver vapour of CAg in the initial mixture.

of the silver atoms, the calculated adhesion is expressed as a percentage of the cohesion of silver. The calculation results are shown in Fig. 8.10. The figure shows that if the silver content in the supplied vapour mixture is small, then the adhesion of silver to the substrate is negligible. This facilitates the removal of the condensate from the substrate in the manufacture of nanodispersed composites.

Figure 8.10 shows the practically useful information on the average number of silver nanoparticles and their mean area in the cross section. It is seen that these values correlate with each other, i.e. increase of the number of particles in the condensate on the substrate results in an increase in their size. This information is useful when selecting the condensation mode to produce a composite with the desired properties.

So, the modified Langmuir sorption model was used as a basis for constructing a model and a program for high-quality numerical simulation of nucleation and condensation in the supply of silver and water vapour flows on the cooled substrate. It is estimated that the nucleation stage is characterized by the mutual stimulation of nucleation of islands of condensed phases of the components. The leading role in the stage of developed condensation with the formation of the nanodisperse composite in the formation of a new molecular layer belongs to the nucleation of silver.

The stationary condensation regimes of the two-component vapour are available. The integral composition of the nanodisperse condensate varies continuously with increasing thickness of the condensate layer on a substrate. This is due to the fact that the stages of nucleation, the substrate and filling in the early development of the condensation relative rates of condensation of the components are significantly different and unstable. There is a borderline value of the ratio of vapour flows of components arriving at the substrate, separating regimes with a monotonic increase and decrease in silver content in the composite during the condensation.

Calculations showed that the modes with a high content of silver vapour degenerate into the condensation of the metal frame filled with particles of water ice, and the modes with low silver content degenerate to the formation of a frame of porous water ice containing silver particles. The results obtained can help in choosing the condensation mode in the development of production technology of nanosized composites with the desired compositions.

The generalized profiles of silver crystals in the condensate and their adhesion to the substrate were calculated. This is also useful information for developing the technology to a new level of understanding.

8.3. Conditions for the growth of columnar and fibrous crystals

The theory of the growth of columnar and fibrous crystals was developed in two main directions: the dislocation–diffusion model and the vapour–liquid-crystal (VLC mechanism). Almost all the known experimental data on the growth of whiskers are explained by these two models or their combination of different versions. In a review of experimental data and theoretical models E.I. Givargiz [17] identifies three key conditions for the implementation of any mechanism of the directional growth of crystals from the vapour:

a) preferential deposition of material at a point with the activity much higher than the activity of other parts of the crystal;

b) directional flow of matter under the influence of different factors;

c) the growth of crystals at strong anisotropy of crystallochemical bonds in the lattice.

In order to implement some mechanism it is necessary that to fulfill at least one of these conditions. The question arises whether the directional crystal growth is possible if none of these conditions exist? This issue will be addressed in the following sections, built on the numerical modelling of sorption, nucleation and condensation processes.

8.3.1. The model of molecular processes at the crystal surface

This section deals with the molecular effects, so modelling is based on a simple model of the molecular processes on the surface of an ideal crystal–substrate, successively developed by many researchers, described in [7] and in section 8.1. Here, this model is used as a variant of the single-component vapour above the substrate.

Figure 8.1 shows a portion of the condensate on the surface of the crystal when the surface molecules $A4$ and $B2$ have the maximum possible number of neighbours. The binding energy of the molecule $B2$ is calculated as the sum

$$E_{B2} = E_{BC}\, n_{BC} + E_{BB}\, n_{BB} + E_{BA}\, n_{BA} + E_{B2A2},$$

where, depending on the configuration of surrounding atoms

$$n_{BC} = 0 \div 4,\ n_{BB} = 0 \div 4,\ n_{BA} = 0 \div 4.$$

For the case when the molecules $A1$, $A2$, $A3$, $B1$, $B2$, $B3$, $C1$, $C3$ are homogeneous

$$E_{BB} = E_{B2A2} = \varepsilon_1,$$

then

$$E_{B2} = \varepsilon_1 (n_{BB} + L\,(n_{BC} + n_{BA}) + 1),$$

where $L = \exp\,(6 \cdot \ln\,(\mathrm{sqrt}\,(2) - 1))$ is a coefficient taking into account the interaction of molecules arranged approximately diagonally in the principal planes of the crystal lattice. The interaction of molecules in the diagonal planes is not considered.

The activation energy for surface diffusion is

$$U_{DB2} = \beta \cdot E_{B2}.$$

The characteristic time of desorption and the probability of desorption per time δt is

$$\tau_{B2} = v^{-1}\exp\,(E_{B2}/kT),\ P_{B2} = \delta t/\tau_{B2}.$$

The characteristic time to the start of the diffusive motion of the atom and its probability for time δt

$$\tau_{B2D} = v^{-1}\exp\,(U_{DB2}/kT),\ P_{B2D} = \delta t/\tau_{1D}.$$

The binding energy of the molecule is calculated as the sum of $A4$

$$E_{A4} = E_{AB}\, n_{AB} + E_{AA}\, n_{AA} + E_{A4S},$$

where, depending on the configuration of surrounding atoms

$$n_{AB} = 0 \div 4,\ n_{AA} = 0 \div 4.$$

For the case when the molecules $A3$, $A4$, $A5$, $B3$, B are homogeneous,

$$E_{AA} = \varepsilon_1,$$

then

$$E_{A4} = \varepsilon_1(n_{AA} + L\, n_{AB} + \varepsilon_s/\varepsilon_1).$$

The activation energy for surface diffusion of molecule $A4$

$$U_{DA4} = \beta \cdot E_{A4}.$$

The characteristic time of desorption and the probability of desorption per time δt

$$\tau_{A4} = v^{-1} \exp (E_{A4}/kT), \quad P_{A4} = \delta t/\tau_{A4}.$$

The characteristic time to the start of the diffusive motion of the atom and its probability for time δt -

$$\tau_{A4D} = v^{-1} \exp (U_{DA4}/kT), \quad P_{A4D} = \delta t/\tau_{A4D}.$$

The algorithm of the calculations is described in section 8.1.

8.3.2. The dynamics of sorption, nucleation and condensation

Calculations based on the algorithm described for the condensation of water were carried out with the following parameters:

Constant of the cubic crystal lattice, m	$3.11 \cdot 10^{-10}$
The size of the crystal surface in adsorption vacancies	70×70
Surface temperature of the substrate crystal, K	260
The energy of adsorption of water molecules on a water monolayer at 300 K, J	$3.05 \cdot 10^{-20}$
Water vapour pressure, Pa	3640
The temperature of water vapour, K	300
Supersaturation	16.75
The share of the activation energy of surface diffusion in adsorption energy	0.46
Adsorption energy of water molecules on the crystal–substrate, J	$(1–4) \cdot 10^{-20}$

It is known that the adsorption energy is a function of temperature. It is not possible to compute this function for the adsorption of water molecules on an arbitrary crystal. Therefore, the scale of the value of the adsorption energy of water molecules on the crystal–substrate was assumed to be ≈ 0.16 eV $= 2.56 \cdot 10^{-20}$ J, as it was used in [18] for the sorption of water molecules on the (001) face of the silicon crystal. The temperature dependence of the adsorption energy of the water molecules in the water vapour condensate was assumed to be similar to the temperature dependence of the latent heat of water vaporization. Finally, the energy of adsorption of water molecules on the water and the proportion of the activation energy of surface diffusion were refined on the basis pf the equilibrium condition of the water condensate at different temperature levels.

The initially clean surface of the crystal–substrate receives a flow of water molecules from the half space at the indicated values of pressure and temperature. The dynamics of water molecules in each layer on the surface of the crystal–substrate is computed. The total number of the water molecules on the substrate is also defined. An example of the calculation results of the dynamics of filling the first molecular layer and the characteristic images of the configuration of the water molecules on the surface of the crystal are shown in Fig. 8.11.

The calculations showed that a dynamic equilibrium of the adsorbed molecules is established on the substrate after a time of approximately $1 \cdot 10^{-7}$ s. The number of water molecules adsorbed on the surface of the crystal then changes continuously over time, focusing around the mean value. There are fluctuations in the number of molecules, leading to the formation of unstable aggregates of molecules in the adsorbed phase. As seen in Fig. 8.11 a the first stable nucleus capable of growth forms after 0.017 µs. More nuclei (see Fig. 8.11 b–f) then form and their area grows. The light background indicates a clean substrate. Gray dots represent single adsorbed molecules. Doubled and tripled points represent the subcritical assemblies of the molecules. The larger gray patches indicate supercritical nuclei of the condensed phase. The resulting nuclei of the condensed phase do not disappear, but grow in the form of islands which merge together and fill the substrate. When 12 000 water molecules fall on the surface the condensed phase starts to form on the surface of the islets which results in their growth in height. The nuclei of the second molecular layer on the islets of the first layer are shown in darker spots.

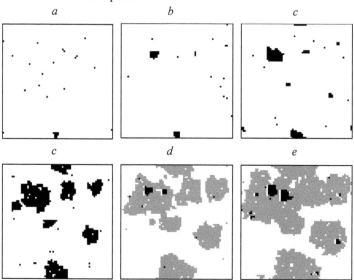

Fig. 8.11. The nucleation and dynamics of the configuration of nuclei on the substrate at a temperature of 272 K. Images a–f correspond to 1000, 2000, 4000, 8000, 12000 and 15000 molecules falling on the substrate, which is equivalent to 0.017, 0.033, 0.065, 0.130, 0.195 and 0.244 µs.

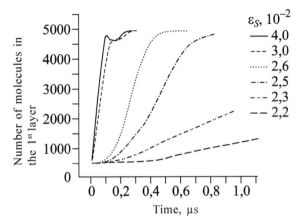

Fig. 8.12. The dynamics of the filling of the substrate at different adsorption energies. $T_S = 260°C$, $\varepsilon_1 = 3.05 \cdot 10^{-20}$ J.

The dynamics of filling the first layer with molecules for six different adsorption energies is shown in Fig. 8.12, the dynamics of filling of several molecular layers for three different adsorption energies is in Fig. 8.13.

Figure 8.12 shows the dynamics of filling the first molecular layer which significantly with decreasing energy of adsorption. When $\varepsilon_S = 4 \cdot 10^{-20}$ J the curve of the filling dynamics has a temporary dip in the number of molecules in the first layer. In Fig. 8.13 a the beginning of the dip coincides with the start of nucleation of islands of the condensate in the second molecular layer. This means that filling of the second molecular layer occurs by both the molecules coming from the vapour medium and the molecules migrating from the first molecular layer. With a decrease in the adsorption energy this effect becomes less noticeable and disappears probably due to reduction of the concentration of the molecules involved in surface diffusion on the surface of the crystal. With a decrease in the adsorption energy the rate of filling of the first molecular layer is reduced (see Fig. 8.12). The curves for small binding energies show that longer time is required for the start of nucleation in the first molecular layer.

Nucleation of the second molecular layer (see Fig. 8.13 a) starts when the filling of the first molecular layer is almost complete. At lower adsorption energies the filling of the next layer occurs long before the filling of the previous the molecular layer is completed (see Fig. 8.13 b, c). At the same time with a decrease in the adsorption energy the absolute times of all events after applying the vapour flow increases.

Figure 8.13 c show that new effects appear with a decrease in the adsorption energy. Shortly after the start of filling the first molecular layer the progressive filling rate decreases. The curve for filling dynamics shows a characteristic zigzag. This effect can be explained by the influence of surface diffusion. The progressive rate of filling of the molecular layer is ensured by the fact that an increase in the size of the islands their average radius of curvature and the likelihood of detachment of the peripheral molecules decreases. Reducing the

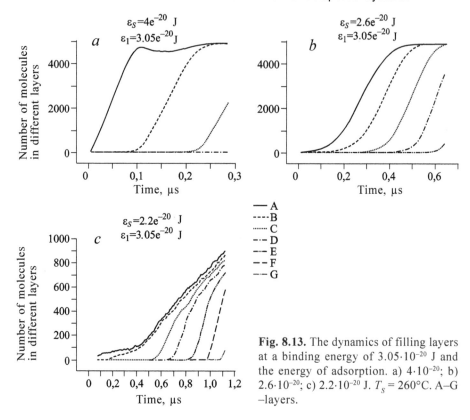

Fig. 8.13. The dynamics of filling layers at a binding energy of $3.05 \cdot 10^{-20}$ J and the energy of adsorption. a) $4 \cdot 10^{-20}$; b) $2.6 \cdot 10^{-20}$; c) $2.2 \cdot 10^{-20}$ J. $T_s = 260°C$. A–G –layers.

rate of filling occurs for two reasons. First, with increasing size of the islets their diffusion feed from the surface of the crystal changes from radial to plane-parallel form. This reduces the diffusion flux relative to the periphery of the islets. Secondly, with the nucleation of new molecular layers the same diffusion flux is expended on the growth of a larger number of layers. At the same time filling the second and subsequent molecular layers is preferred, since at the binding energy of the condensate greater than the adsorption energy, the likelihood of securing a migrating molecule at the periphery of the second and higher molecular layers is higher than at the periphery of the first molecular layer. It should be added that the growth of the second and subsequent molecular layers is limited by the area of the first molecular layer.

Thus, Fig. 8.13 c shows conditions for the growth of islands mainly in height rather than in the lateral direction.

8.3.3. Columnar growth of condensates

Apparently, the preferential island growth in height is a clear indication that we have found effects that can create conditions for the growth of columnar and fibrous crystals. To ensure more severe manifestations of these effects, calculations were carried with a further decrease in adsorption energy. The calculation results are illustrated in Figs. 8.14 and 8.15.

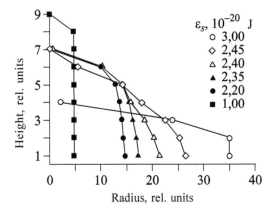

Fig. 8.14. The typical height and thickness of the columns.

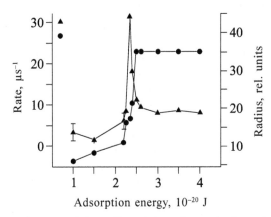

Fig. 8.15. The growth rate and the radius of the columns in dependence on adsorption energy.

Figure 8.14 shows the conditional generalized profiles of the columnar crystals. The conditional radius of the crystals is calculated by the formula

$$R_i = N_i^{0,5}/70/2,$$

where N_i is the number of molecules in the i-th layer. We see that when the adsorption energy is $3\cdot10^{-20}$ J the crystal radius is half the width of the working area of the crystal, adopted for the calculations. This means that the condensate surface is rough but distinctive columnar crystals do not form. A further decrease in the adsorption energy leads to the formation of the part of the crystal surface with no molecules. When the adsorption energy is $1\cdot10^{-20}$ J the pronounced growth of crystals of the condensate takes place only in height.

Figure 8.15 illustrates the dependence of the crystal growth rate of condensate in the height on the adsorption energy. In view of the stochastic nature of the processes, each running of the program for low adsorption energies leads to a new result. Therefore, the values of the growth rate at low adsorption energies are determined by averaging from several runs of the program for each value of energy. We see that when the adsorption energy is $2.6\cdot10^{-20}$ J and less the growth rate of the condensate starts to increase, while at the adsorption energy of $2.5\cdot10^{-20}$ J and less the filling of the crystal surface is incomplete, indicating that the nucleation of columnar crystals. When the adsorption energy is $2.35\cdot10^{-20}$ J the rate of growth of columnar crystals has a maximum value which is approximately four times the rate of growth of the solid condensate. At a further decrease in the adsorption energy there is a dramatic decrease in the rate of growth of columnar crystals with a decrease in their characteristic radius.

8.3.4. Discussion of results

Thus, in the process of calculating the adsorption of water molecules on the surface of a silicon crystal it became clear that after applying the flow of the molecules the adsorption equilibrium at the crystal surface is achieved within $1\cdot10^{-7}$ s. The flows incident on the surface of the molecules and desorbed as a result of thermal fluctuations are compared. The crystal surface is characterized by the establishment of an equilibrium concentration of adsorbed molecules moving by diffusion in the adsorption vacancies. In the process of surface diffusion the adsorbed molecules meet with random neighbours and form with them single-layer aggregates of molecules with lower mobility. These aggregates of molecules again break up to form single adsorbed molecules. When the vapour with respect to the crystal surface is in the supersaturated state, some the aggregates, accidentally reaching the critical size, are stable, they do not break down and begin to grow in the form of islands of the new phase. Then there is a gradual filling of the crystal surface with the monomolecular layer of the condensate. Similar sorption phenomena and the formation of aggregates of molecules take place on the surface of the condensate monolayer. The probability that nucleation occurs on the surface of the monolayer, i.e., supercritical nuclei are formed, depends on the supersaturation. If nucleation occurs, gradual filling of the second monolayer of the condensate takes place. At the same time filling of the second layer takes place by both the molecules, arriving from the vapour environment as well as by molecules that migrate over the surface of the first monolayer. If the binding energy of the condensate is equal to or greater than the adsorption energy, then part of the first layer of molecules also take place in building the second monolayer, which manifests itself in a temporary decrease in the occupancy of the first layer. Sequential filling of a growing number of molecular layers leads to the condensation process. If for some reason the adsorption energy is reduced, the time from starting the program to the start of nucleation increases and the rate of filling of monolayers of the condensate decreases. This leads to a situation where the formation of subsequent monolayers catches up with the filling of the previous ones and crystals of the condensate begin to grow in height with incomplete filling of the surface of the crystal–substrate. That is, we found that even in the absence of all three conditions for the growth of directional structures formulated by E.I. Givargizov and cited at the start of the chapter there is growth of the columnar crystals. However, this does not mean that the conditions formulated by this researcher are questioned. When the adsorption energies are smaller than the binding energy of the molecules, nucleation is followed immediately by the formation of nuclei of the condensed phase, whose surface is more favourable for condensation than the surface of the crystal–substrate. But this is in fact the first condition cited above. It can be concluded that even if there are no conditions for the growth of columnar and fibrous crystals on molecular deposition of the vapour, then these conditions may arise in the processes of adsorption and nucleation.

Thus, a simple model of molecular processes on the surface of the substrate–crystal in the presence of vapours gave a plausible picture of the molecular deposition and revealed new interesting results. First, it was found that the formation of the second monolayer of molecules on the previously formed monolayer on the crystal with the energy of adsorption equal to or greater than the binding energy of the deposited vapour molecules is accompanied by the 'grazing' of the first monolayer. Part of the first layer of molecules is used for filling the second layer. This is nothing more than a mechanism for the formation of roughness of the condensate layer, in addition to the fact that the nucleation itself is a source of subsequent monolayers of roughness. Secondly, it was found that decreasing adsorption energy slows down the filling of the first monolayer on the substrate–crystal, the nucleation rate of subsequent layers in contrast increases, which leads to a predominant increase in the height of the crystals. This means that in principle, the mechanism of growth of columnar and fibrous condensates is incorporated in the usual scheme of molecular deposition. But this mechanism is realized in a narrow range of deposition. Apparently, this is the reason why until now growing columnar and fibrous crystals is more art than science.

Conclusions

1. Using the modified Langmuir adsorption model we developed a model of co-adsorption of silver and water vapours on an ideal substrate. The numerical study of the model revealed the existence of a number of previously unknown phenomena. It was found that the adsorption and condensation of silver vapours significantly stimulate the adsorption of water vapour, which leads to the emergence of capillary condensation, even at low saturations. The rate of capillary condensation depends strongly on the contact wetting angle of water on silver. There is a characteristic contact angle at which the rate of capillary condensation changes drastically. It was also found that the water vapour has a significant effect on the adsorption of silver vapour, on the structure of its condensate on the substrate, and on the adhesion to the substrate. It is suggested that in the partial pressure of water vapour–the temperature of the substrate coordinates there are areas of good and bad adhesion of the condensate to the substrate and the structure of the condensate changes on the border between them. The presence of water vapour weakens adsorption and condensation of the silver atoms on the substrate as a result of

occupation of part of the adsorption vacancies and at the same time enhances the population of the silver condensate layer. The presence of the silver vapour, on the contrary, causes condensation of even unsaturated water vapour. It is therefore suggested that this fact opens up the possibility in principle to intensify condensation by adding foreign vapours, and a selection criterion was formulated for selecting the additives on the basis of the binding energies. The results are useful for understanding the details of two-component sorption, the development of technology of thin film growth in a gaseous environment, and improvement of the condensation process.

2. In order to explain the strange transients detected in the experiments to obtain nanosized composites by vacuum deposition of the two- component vapour, a model based on Langmuir adsorption and software for numerical simulation of condensation on a cooled substrate were developed. Calculations for the condensation of vapour mixtures of silver and water have shown that there is a mutual stimulation of nucleation of condensed phases of the components with the leading role played by the nucleation of silver. We calculate the dispersion properties of the condensates, depending on the mode of condensation and condensate adhesion to the substrate. The results of calculations help to understand the previously incomprehensible experimental results and develop technology for producing nanopowder composites on a new level of understanding of the processes involved.

3. The computer model of the initial stage of condensation processes was constructed. Stochastic processes of adsorption, surface diffusion and desorption were simulated on the 70×70 adsorption vacancy grid. The model was tested on an example approximating the sorption of water vapour on a conventional crystal face with a temperature of 260 K. The numerical results demonstrate the processes of nucleation, growth of nuclei and developed condensation. The calculations for different adsorption energies were carried out. It is shown that a decrease in adsorption energy changes the nature and rate of nucleation and subsequent condensation. By reducing the adsorption energy the probability of nucleation of islets on the substrate decreases, filling atomic layers on the islets occurs before the filling of the substrate, the rate of islet growth in height is increased in comparison with the condensation of a continuous film, and the growth rate of the area of the islands decreases. There is a characteristic adsorption energy at which the rate of islet growth in height reaches its maximum. A further decrease in the adsorption energy resulted in an increase in the height of the islets only, but the growth rate decreased. The phenomena observed in the calculations illustrate the mechanism of nucleation of columnar and fibrous whiskers.

List of Symbols

A - adhesion, calculated as a percentage of cohesion,%;

C_{Ag} - vapour concentration of silver;

E_{B2}, E_{A4} - the binding energy of the molecules with the nearest control neighbors, J;

E_{BC}, E_{BB}, E_{BA}, E_{B2A2}, E_{AB}, E_{AA}, E_{A4S} - binding energy control molecule with its nearest neighbors, J;

$J_{K,i}$ - flux of molecules arriving at the substrate from the chamber;

$J_{S,i}$, J_V - flux of molecules leaving the substrate, $m^{-2} \cdot s^{-1}$;

k - Boltzmann constant, $1.38 \cdot 10^{-23}$ J / deg;

L - coefficient taking into account the interaction of molecules;

L_{mol} - coordinate across the condensate layer, measured in molecular layers;

M_i - molecular mass, kg,

N_{Ag}, N_{H_2O} - the number of silver particles and water, respectively;

N_i - number of molecules in the i-th layer column;

n_M - number of molecules per unit area located surface, m^{-2};

n - number of nearest neighbors;

n_{BC}, n_{BB}, n_{BA}, n_{AB}, n_{AA} - the number of nearest neighbors of control molecules;

P_s, P_{SD} - the probability of events of desorption of surface migration, s^{-1};

p_{Ki} - partial vapour pressure in the chamber, N/m^2;

p_{B2}, P_{B2D}, P_{A4}, P_{A4D} - the probability of desorption and diffusion control molecules s^{-1};

R_i - radius of the conventional cross-section of the column in the i-th layer;

S_{Ag} - the average cross-sectional area;

T - temperature, K;

T_K - the temperature of a mixture of steam, K;

T_S - substrate temperature, K;

t - time, s;

U_D - the activation energy of surface diffusion, adsorption, $U_D = \beta \varepsilon_s$, J;

U_{DB2}, U_{DA4} - activation energy of surface diffusion control molecules, J;

a_{Ag}, α_{H_2O}, α_{Ci} - the relative rate of condensation;

β - fraction of the activation energy of surface diffusion in the adsorption energy;

Δt - time interval, s;

ε - the binding energy, J;

ε_S - the energy of adsorption, J;

ε_I - binding energy of the adsorbed molecules is homogeneous the first component vapour, J;

ν - frequency of thermal vibrations of the molecule, s^{-1};

τ - characteristic time, s;

τ_S - the lifetime of molecules in adsorbed state, s;

τ_{SD} - the lifetime of the molecule to move into the next position, s;

$\tau_{SD,I}$ and $\tau_{S,I}$ - the lifetime of each of the two interacting molecules in the neighboring vacancy adsorption to their diffusion or desorption, s;

τ_{B2}, τ_{A4} τ_{B2D}, τ_{A4D} - the characteristic time of desorption and diffusion control of the molecules, c;

φ - the contact angle, deg.

Indices:

Ag, H_2O - silver, water;
to - the chamber;
c - condensation;
D - diffusion;
S - substrate, sorption, desorption;
$S,1$ - surface, the first component;
$SD, 1$ - surface, the first component, the diffusion;
SD - surface diffusion;
1, 2, i - number of components;
V - evaporation, vacancies.

Literature

1.   Bochkarev A.A., Pukhovoy M.V., J. Vacuum, 1997, V. 48, No. 6, 579–584.
2.   Bochkarev A.A., et al., J. Colloid and Interface Sci., 1995, V. 175, 6–11.
3.   Bochkarev A.A., et al., Zavod. Lab., 1994, No. 4, 34–35.
4.   Brunauer S., et al., J. Amer. Chem. Soc., 1938, V. 60, No. 2, 309–319.
5.   Langmuir I., J. Amer. Chem. Soc., 1916, V. 38, No. 9, 2221–2295.
6.   Langmuir I., J. Amer. Chem. Soc., 1918, V. 40, No. 11, 1361–1403.
7.   Chernov A.A., et al., Modern crystallography, V. 3, The formation of crystals. Moscow, Nauka, 1980.
8.   Bochkarev A.A., Polyakova V.I., PMTF, 2009. V. 50, No. 5, 95–106.
9.   Bochkarev A.A., Polyakov V.I., Teplofizika i aeromekhanika, 2009, No 1, 103-114.
10. Bochkarev A.A., Polyakov V.I., Dokl. RAN, Fizika, 2009, V. 425, No. 5, 617-620.
11. Missol V., The surface energy of the interface in metals, Moscow, Metallurgiya, 1978.
12. Bochkarev A.A., et al., Sib. Fiz. Tekh. Zh., 1993, No. 2, 7–12.
13. Bochkarev A.A., Polyakova V.I., in: Thermal processes in the crystallization of substances, Proc., ed. A. Basin, V.S. Berdnikova, Novosibirsk, Publishing House, The Institute of Thermophysics, USSR Academy of Sciences, 1987, 128–135.
14. Bochkarev A.A., Polyakova V.I., in: Proc. Reports. II Conf. Modeling of crystal growth, Riga, November 2–5, 1987, 84–86.
15. Bochkarev A.A., et al., in: Thermophysics of crystallization and high-temperature treatment of materials, Proceedings, Institute of Thermophysics, Academy of Sciences, Novosibirsk, 1990, 98–117.
16. Held G., et al., in: Abstracts book 14th International Vacuum Congress, 31 August–4 September 1998, 253–254.
17. Givargizov E.I., The growth of filamentous and lamellar crystals from the vapour, Moscow, Nauka, 1977, 304.
18. Kazuto Akagi, Masaru Tsukada, Thin Solid Films, 1999, No. 343–344, 397–400.

9

Methods for direct conversion of kinetic energy to surface energy of the disperse system

Chapter 1 shows that the energy efficiency of the majority of the physical processes of dispersion used in technology is at the level of 10^{-2}–10^{-4}%. This is due to inefficient use of energy supplied to disperse the material. In this connection, search for ways to improve the supply of energy for dispersion is extremely important. The method of dispersing agents in the liquid state through a stage of its transformation into a thin film is attractive. The surface energy of the area of the free film with thickness δ and area S is

$$E_\sigma = 2S\sigma.$$

The specific surface energy of the liquid in the film can be determined by dividing E_σ by the volume of the fluid in the section of the film $S\delta$

$$\varepsilon_\sigma = 2\sigma/\delta. \tag{9.1}$$

Comparison of (9.1) with (1.3) in chapter 1 suggests that the specific surface energy of the fluid in the form of a film is equivalent to the specific surface energy of the fluid in the form of droplets with a radius

$$r = 3\delta/2. \tag{9.2}$$

Equation (9.2) offers the prospect of finding an effective process of pumping surface energy into the liquid in order to disperse in the direction of the processes of formation of thin liquid films and their subsequent fragmentation, preferably without the loss of surface energy.

For the liquids with relatively low viscosity, such as molten metal superheated above the liquidus temperature, a promising way of producing thin films is the use of certain hydrodynamic situations. The most studied and accessible scheme of the fluid flow with the formation of films is draining the liquid from the boundary of a solid stationary or moving surface. In this regard, this chapter discusses some issues associated with the hydrodynamic scheme and in particular surface phenomena taking place here. The next section discusses the simplest case of the flow in a flat film.

9.1. Liquid flow in a free thin film

The flow of a liquid in a free thin film, occurring in the collision of two jets of circular cross-section or impact of the jet on the barrier, is considered. At sufficiently high velocities of the jets the formed film is flat, indicating the insignificant effect of gravity. Therefore, in our analysis we neglect the influence of gravity and only the motion under the action of surface forces will be considered. Previously, the problem of liquid flow in a symmetric film was formulated by a number of authors [1, 2], who, based on the Bernoulli integral, believed that the flow velocity u in the film is constant; then the film thickness $\delta = Q/(2\pi r u_0)$, where Q is the volumetric flow rate; u_0 in the liquid velocity in the jet; r is the distance from the axis of the colliding jets. However, this solution does not satisfy the energy conservation law, since the surface energy per unit volume of liquid in the film is $\varepsilon_\sigma = 2\sigma/\delta$, where σ – surface tension, increases linearly with radius, and if the specific kinetic energy is constant, we do not understand the source of the surface energy. Figure 9.1 shows the variation of the specific kinetic and surface energy on the radius corresponding to the given solution; the energies are related to $\rho u_0^2/2$, where ρ – the liquid density, and the radius – to the value $R = \rho u_0 Q/(4\pi\sigma)$. In reality, the film exists from $r = d/2$, where d – the diameter of the jets, to the decay radius r_p, shown in Fig. 9.1 by the dashed line. The dependence of r_p on the parameters of the colliding jets is given in [3]. With increasing distance from the axis the specific energy density of the liquid in the film is more than doubled in comparison with the initial value. It can be concluded that the inclusion of surface energy in this problem is a necessary condition to obtain the correct solution.

9.1.1. Phenomenological model of the flow, taking into account surface tension

To obtain the equations of motion, we can write the surface energy of the circular plot of the film E with inner and outer radii r and $r + l$, respectively: $E = 2 \cdot 2\pi r l \sigma$. It was assumed that the width of the ring $l \ll r$, and the factor

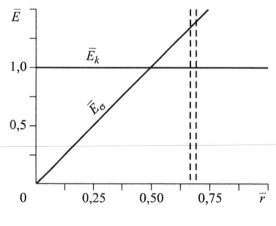

Fig. 9.1. Change of the relative kinetic and surface energy along the radius of the film.

2 indicates that the film has two surfaces. If we assume that the motion is potential, its equation $mdu/dt = -\partial E/\partial r$, where m – mass of the circular section; u – the velocity of the mass centre of the liquid ring in the film; t – time. In our case,

$$\frac{\partial E}{\partial r} = 2 \cdot 2\pi\sigma \left[l + r\frac{dl}{dr} \right]. \qquad (9.3)$$

The first term on the right side of (9.3) is the sum of surface forces acting on the considered part of the film in the radial direction, and the second is associated with the deformation of this ring section when the radius changes. If we assume that this term is due to the inhomogeneity of the velocity across the film thickness, i.e. with irreversible loss of energy, it should not be included in the equation of motion.

Then

$$m\frac{du}{dt} = -4\pi\sigma l. \qquad (9.4)$$

Introducing the film thickness δ, l can be eliminated from the equation (9.4), which leads to the formula

$$\rho r\delta\frac{du}{dt} = -2\sigma. \qquad (9.5)$$

Using the continuity equation $Q = 2\pi ru\delta$, equation (9.5) takes the form

$$du = -\frac{4\pi\sigma}{\rho Q}dr. \qquad (9.6)$$

Introducing the dimensionless variables $\bar{u} = u/u_0, \bar{r} = r/R$ from (9.6) we obtain

$$d\bar{u} = -d\bar{r}. \qquad (9.7)$$

The solution (9.7) with initial condition $\bar{u} = 1$ at $\bar{r} = \bar{r}_0$ will be

$$\bar{u} = 1 - (\bar{r} - \bar{r}_0). \qquad (9.8)$$

Now from the continuity equation, taking into account the fact that $Q = u_0\pi d^2/2$, we can obtain the following dependence of film thickness on the radius:

$$\delta = \frac{2d}{We_d}\frac{1}{\bar{r}\left[1 - (\bar{r} - \bar{r}_0)\right]}, \qquad (9.9)$$

where $We_d = \rho u^2/\sigma$ – the Weber number, defined by the diameter of the jets. It is seen that at some $\bar{r}$ the film has a certain minimum thickness δ_{min}. Exploring the function (9.9) by conventional methods, we find that the equation

$$\delta_{min} = \frac{8d}{We_d(1 - r_0^2)} \qquad (9.10)$$

is implemented with $\bar{r}=(1+\bar{r}_0)/2$.

Another possible solution is obtained if we take into account both terms in the right-hand side of (9.3). It is easier to obtain this solution directly from the energy conservation law, written for two circular sections of the film:

$$\frac{2\sigma}{\delta_1}+\frac{\rho u^2}{2}=\frac{2\sigma}{\delta_2}+\frac{\rho u_2^2}{2},\tag{9.11}$$

Making equation (9.11) dimensionless and taking the same initial condition $\bar{u}=1$ when $\bar{r}=\bar{r}_0$ we obtained

$$\bar{u}^2+2\bar{r}\bar{u}-(1+2\bar{r}_0)=0,\tag{9.12}$$

whence

$$\bar{u}^2=\sqrt{\bar{r}^2+2\bar{r}_0+1}-\bar{r}.\tag{9.13}$$

The dependence of film thickness on the radius is determined as in (9.9):

$$\delta=\frac{2d}{We_d}\frac{1}{\bar{r}\left(\sqrt{\bar{r}^2+2\bar{r}_0+1}-\bar{r}\right)}.\tag{9.14}$$

The change in film thickness along the radius, calculated by formulas (9.9), (9.14), is shown in Figure 9.2 by the curves *1* and *2*, respectively. Curve *3* – the result of the calculation by the formula $\delta=\dfrac{2d}{We_d}\dfrac{1}{\bar{r}}$, corresponding to the flow with a constant radial velocity. The vertical dashed line shows the radius of the spontaneous decay of the film.

Graphs of change of velocity, obtained by (9.8) and (9.13) in the approximation $\bar{r}_0\ll\bar{r}$ are shown in Fig. 9.3 in the form of continuous curves. Both solutions show a significant change in the flow velocity in the film.

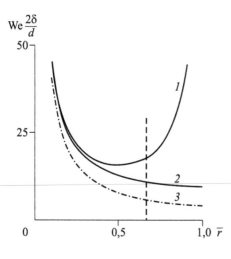

$We\,\dfrac{2\delta}{d}$

Fig. 9.2. Change of the reduced film thickness along the film radius. *1* – calculation by (9.9); *2* – calculation by (9.14); *3* – the potential flow $u=1$.

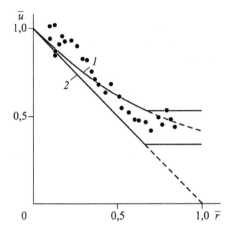

Fig. 9.3. Speed change in radius. *1* – calculation by formula (9.8); *2* – calculation by formula (9.13); points – experiments, dashed lines – the speed in the film without its collapse

9.1.2. Experimental verification of models

To test the models, the liquid velocity in the film was measured. The experimental setup is shown in Fig. 9.4. Water is supplied through a distributor *1* in a calibrated glass tube *2* used as nozzles. The nozzles are mounted on holders, allowing for alignment. To measure the flow a flow meter is placed in front of the distributor *3*. The length of pipes and nozzles is selected the same, which ensures the equal flow through the nozzle and provides an opportunity to produce a horizontal film. The rate was determined by the dynamic pressure on the plate *4*, soldered to one arm of the balance *5*. The balance is equipped with hydraulic damping of oscillations *6*. The force acting on the plate is equal to $F = \rho u Q_p$ where Q_r is the share of consumption attributable to the plate, which is determined by the geometry of the system. The relative error in measuring the velocity, associated with the instability of the flow and accuracy when measuring distances and weighing, is not more than 18%. The speed measurements at a flow rate $Q = 26 \cdot 10^{-5}$ m³/s and the nozzle diameter

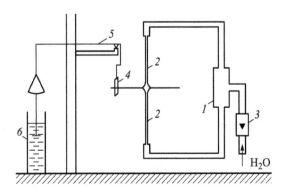

Fig. 9.4. Schematic of the experiment. *1* – distributor; *2* – nozzle; *3* – meter; *4* – measuring plate; *5* – balance; *6* – damper.

$d = 2.45$ mm are shown in Fig. 9.3. These measurements show a significant change in flow velocity along the radius of the film. A possible reason for the reduction of the velocity is air friction, but in [2] it is shown that due to this the velocity does not decrease by more than 7%. This suggests that the influence of friction can be taken into account by a small correction.

It is interesting to note that the solution of (9.9) can be used to predict the radius of collapse of the film, if, as a condition that determines the collapse of the film, we accept to the condition of equality of the surface energy ε_σ and the energy of chaotic motion ε_χ, occurring during the deformation of the film. The energy of random motion of a unit volume ε_χ can be determined from the energy conservation law:

$$\varepsilon_\chi = \frac{\rho}{2}\left(u_0^2 - u^2\right) - \frac{2\sigma}{\delta}.$$ (9.15)

The condition $\varepsilon_k = \varepsilon_\sigma = 2\sigma/\delta$ and (9.8), (9.9) and (9.15) can be used to determine the radius r_p at which the film disintegrates. Assuming that $r_p \gg r_0$, we obtain

$$r_p = 2R/3.$$ (9.16)

For the case of films formed during the collision of two jets, the radius of the collapse can be expressed through the Weber number We_d:

$$\frac{r_p}{d/2} = \frac{1}{6}\mathrm{We}_d.$$ (9.17)

This result agrees well with experimental data [3] for the numbers $\mathrm{We}_d < 800$. At high Weber numbers, the decay of the film is due to another mechanism. Note also that the local Weber number $\mathrm{We}_\delta = \rho u^2 \delta/\sigma = 1$ at $r = r_p$.

The dependence of the local Weber number of the radius can be obtained from (9.9):

$$\mathrm{We}_\delta = \frac{2\left(1 - \bar{r}\right)}{\bar{r}}.$$ (9.18)

By setting the condition $\mathrm{We}_\delta = 100$, meaning that the specific surface energy of the liquid is 1/25 of the specific kinetic energy of (9.18), we obtain $\bar{r} = 0.02$. For the radii $\bar{r} < 0.02$ the influence of surface forces can be neglected with an error of about 4%. The range of radii $\bar{r} = 0.02–/67$ is the range of applicability of the model.

9.1.3. Derivation of the equation of motion from the Navier–Stokes equations

Finally, we discuss the question of how to obtain a solution of (9.8), using the traditional methods of hydrodynamics. First, we note that in the derivation of (9.5), the surface forces are considered to be attached to the entire volume of the liquid, i.e. essentially a viscous problem is considered. This can be explained by the fact that the film thickness is very small. Unfortunately, we do not know the data about the thickness of the boundary layer near the free surface, so as a rough estimate we take the formula for a thin plate: $\Delta/x \approx 1/\sqrt{Re_x}$, here Δ – the boundary layer thickness, and $Re_x = \rho u_0$. For our typical mode at $x = 10$ mm, we have $d = 6 \cdot 10^{-5}$ m and the thickness of the film at such a distance does not exceed 10^{-4} m. Thus, it is plausible that the flow in the film is viscous.

We obtain for our case an analog of the Bernoulli integral, integrating the Navier–Stokes equations for the radial velocity component u in the film thickness. We write this equation in the form

$$\rho u \frac{\partial u}{\partial r} + \rho w \frac{\partial u}{\partial z} + \frac{\partial p}{\partial r} = \mu \frac{\partial^2 u}{\partial z^2} + \mu \Delta_r u, \tag{9.19}$$

where $\Delta_r u$ identifies the components of the Laplacian, containing derivatives with respect to r; z is the coordinate along the symmetry axis of the film. Assuming that the flow velocity across the film is $w \ll u$, we omit the second term on the left side of the equation (9.19), and integrating this equation over the thickness of the film, we get

$$\int_{-\delta/2}^{\delta/2} \rho u \frac{\partial u}{\partial r} dz + \int_{-\delta/2}^{\delta/2} \frac{\partial p}{\partial r} dz = \int_{-\delta/2}^{\delta/2} \mu \frac{\partial^2 u}{\partial z^2} dz + \int_{-\delta/2}^{\delta/2} \mu \Delta_r u\, dz.$$

Applying the mean value theorem, we proceed to the equation

$$\left\langle \rho u \frac{\partial u}{\partial r} \right\rangle \delta + \left\langle \frac{\partial p}{\partial r} \right\rangle \delta = \mu \frac{\partial u}{\partial z} \bigg|_{-\delta/2}^{\delta/2} + \mu \langle \Delta_r u \rangle \delta, \tag{9.20}$$

where $\langle X \rangle$ denotes the average over the film thickness.

Assuming that the radial velocity profile is symmetric about $z = 0$, we obtain

$$\mu \frac{\partial u}{\partial z} \bigg|_{-\delta/2}^{\delta/2} = 2\mu \frac{\partial u}{\partial z} \bigg|_{\delta/2} = 2\tau_{rz} \big|_{\delta/2},$$

where $\tau_{rz}\big|_{\delta/2}$ is a component of the stress tensor on the boundary of the film. Suppose that at the boundary of the film there are the shear stresses generated by surface tension and equal to the derivative of the surface tension forces on the area: $\tau_{rz} = \partial F/\partial S$. Then, considering only the first term in equation (9.9), we have $\tau_{rz} = -\sigma/r$. Equation (9.20) takes the form

$$\left\langle \rho u \frac{\partial u}{\partial r} \right\rangle \delta + \left\langle \frac{\partial p}{\partial r} \right\rangle \delta = -\frac{2\sigma}{r} + \mu \langle \Delta_r u \rangle \delta.$$

We pass to dimensionless form of this equation, introducing the notation $\bar{u} = u/u_0$, $\bar{r} = r/R$, $\bar{p} = p/\rho u_0^2$. Ignoring the difference between $\langle X \rangle$ and X, we write the formula

$$\bar{\delta} \bar{u} \frac{\partial \bar{u}}{\partial \bar{r}} + \bar{\delta} \frac{\partial \bar{p}}{\partial \bar{r}} = -\frac{2}{r} \frac{1}{We_d} \frac{d}{R} + \bar{\delta} \Delta_r \bar{u} \frac{d}{R} \frac{1}{Re}, \tag{9.21}$$

where we have introduced the Reynolds number $Re = \rho u_0 d/\mu$.

We estimate the order of the members of (9.21), assuming $1/We_d$ is a small parameter:

$$\frac{d}{R} = \frac{4\pi\sigma d}{\rho(1/2)\pi d^2 u_0^2} = \frac{8}{We_d} = O\left(We_d^{-1}\right).$$

$$\delta = \frac{Q}{2\pi r u R} = \frac{1}{4}\frac{d^2}{R^2}\frac{1}{ur} = O\left(We_d^{-2}\right).$$

Here O is the formal function of a small parameter.

For a free film the term is determined only by Laplace pressure, which can be approximately written as

$$p_L = \frac{\sigma \partial^2/\partial r^2}{\left[1+\left(\partial\delta/\partial r\right)^2\right]^{3/2}}.$$

Then

$$\overline{p}_L = \frac{\sigma \partial^2/\partial r^2}{\rho u_0^2 \left[1+\left(\partial\delta/\partial r\right)^2\right]^{3/2}} = \frac{\partial^2/\partial r^2}{\left[1+\left(\partial\delta/\partial r\right)^2\right]^{3/2}}\frac{1}{We_d}\frac{d}{R}.$$

Since differentiation with respect to r does not change the order of smallness of We_d, we can now write down the orders of all terms in equation (9.21):

$$\overline{\delta u}\frac{\partial \overline{u}}{\partial r} = O\left(We_d^{-2}\right),$$

$$\overline{\delta u}\frac{\partial \overline{p}_L}{\partial r} = O\left(We_d^{-6}\right),$$

$$\frac{2}{r}\frac{1}{We_d}\frac{d}{R} = O\left(We_d^{-2}\right),$$

$$\overline{\delta \Delta_r u}\frac{d}{R}\frac{1}{Re} = O\left(We_d^{-3} Re^{-1}\right).$$

Thus, up to the higher-order terms of smallness with respect to We_d equation (9.21) reduces to the expression

$$\rho u \frac{\partial u}{\partial r}\delta = -\frac{2\sigma}{r}, \tag{9.22}$$

which is equivalent to the equation of motion (9.5) obtained earlier.

In [4] it is shown that the shear stress tensor components at the interface with a constant coefficient of interfacial tension are equal to zero. Assuming $\tau_{zz} = 0$ in (9.21) up to the higher-order terms of smallness we obtain to the equation

$$\rho \delta u \frac{\partial u}{\partial r} = 0,$$

whose solution $u = $ const does not hold at $0.02 < \bar{r} < 0.67$ (see section 9.1). This casts doubt on the universal validity of the claims in [4]

Thus, in problems of hydrodynamics in which the flow occurs during the change of the free surface, it is necessary to take into account the shear stresses that arise due to surface tension forces. The specific form of the tangential component of the stress tensor on the boundary depends on the geometry of the problem.

It should be noted that the results in this section are largely questionable and require further study.

The next section presents the results of the study ofa little more complicated case of a film flowing from a rotating barrier.

9.2. Formation of droplets in during the flow of liquid from a rotating barrier and in melting of a rotating rod

At the time of the first publication of the results of this section, the literature did not contain sufficiently detailed analysis of the formation of droplets falling from the edge of a melted core. The generalized experimental information obtained at that time in the form of the dependence on a single criterion is of particular interest. Calculations of the size distribution function of the resulting droplets are original to the present time.

9.2.1. The similarity in the centrifugal atomization process

A model of the process of spraying a rotating rod with a melting end is considered. The resulting droplet size dependence on the parameters of the process is confirmed by comparison with experimental data. Described an attempt to describe the distribution function of droplet size.

There is a description [6, 7] of the method for producing metal powders by melt a spinning rod, the end of which is heated by the plasma arc. The theoretical description of this process would facilitate the application of this method for the production of powders of different materials.

To determine the general laws of the sputtering process of a melting rotating rod with a radius R_0 we consider a simple model. We believe that the material of the rod melted by heating flows in the form of film or separate jets over the end surface of the rod. The separation of droplets occurs from the edge of radius R_0. The mechanism of separation of droplets is to overcome the surface tension forces of the melt by centrifugal forces. The condition for separation of the droplet can be written as the equation of balance of forces

$$F_c = F_\sigma, \tag{9.23}$$

where F_c is the centrifugal force acting on the separated amount of the melt; F_σ is the surface tension force holding the melt at the edge of the end of the rod.

Under the assumption that the melt separated from the edge of the end of the rod in the form of spherical droplets, the centrifugal force can be written as

$$F_c = \frac{4}{3}\pi r_m^3 \rho \frac{v^2}{R_0}. \tag{9.24}$$

Here r_m is the radius of the drop; ρ is the density of the melt; V is the peripheral speed at the edge of the end of the rod. If we imagine that the droplet separates from the end of the melt stream flowing from the end of the rod, the surface tension forces can be approximately written as

$$F_\sigma = 2\pi r_m \sigma. \tag{9.25}$$

Substitution of (9.24) and (9.25) into equation (9.23) gives

$$\frac{2\rho r_m^2 v^2}{3R_0} = \sigma. \tag{9.26}$$

From equation (9.26) we obtain the formula

$$\frac{r_m}{R_0} = \sqrt{3/2}\, \mathrm{We}_R^{-1/2}, \tag{9.27}$$

where the Weber number $\mathrm{We}_R = \rho V^2 R_0 / \sigma$ is defined by the outer radius of the rod. The physical meaning of We_R is the ratio of the centripetal force, acting on the unit volume of the rod material, to the surface force of the melt acting on the film flowing down from the end of the rod.

Formula (9.27) can be written as

$$\frac{r_m}{R_0} = \frac{3}{2}\mathrm{We}_r^{-1}, \tag{9.28}$$

where the Weber number $\mathrm{We}_r = \rho V^2 r_m / \sigma$ is defined by the radius of the droplet. The physical meaning of We_r is the ratio of the kinetic energy of the detached drop to its surface energy. Equation (9.28) shows that the radius of the droplets formed is determined by this ratio and the radius of the rod.

For use in practical calculations, formula (9.27) leads to a more convenient form

$$r_m = \frac{A}{\omega\sqrt{R_0}}, \tag{9.29}$$

where $A = \sqrt{3\sigma/2\rho}$ is an analog of the capillary constant; ω is the angular velocity of rotation of the rod. Here are the values of A for several metals, calculated in the SI system of units:

Material	Zn	Sn	Pb	Al	Cu	Ti	Fe
A, m	0.0137	0.0110	0.0082	0.0183	0.0157	0.215	0.0187

A was calculated using the reference data [8].

The results of calculations by formula (9.27) are shown in Fig. 9.5 in the form of a continuous line. The same graph shows by the points the data [7] for the average mass particle size obtained in experiments on the dispersion of the rods made of titanium alloys. In processing the experimental data we used

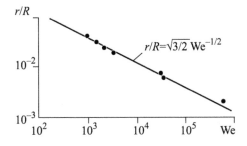

Fig. 9.5. Generalized dependence of the radius of droplets formed on the parameters of a rotating rod.

the surface tension of the melt at the melting point of titanium. It is seen that the experimental data agree well with the model under consideration process.

9.2.2. The distribution function of particle size by centrifugal spraying a rotating rod

The question of the size distribution function of droplets formed by the centrifugal atomization is interesting because it determines the fractional composition of the powder produced in this process. In order to describe the distribution function of droplets, it is assumed that the release of droplets of the melt during melting of the rod occurs locally, i.e. in the area of melting of the material. The radius of the droplets r, formed by melting the material at the radius R, is defined by

$$r \frac{A}{\omega \sqrt{R}}.$$ (9.30)

This assumption seems plausible for a convex or flat end of the rod shape at high rotation speeds, when the centrifugal forces are large, and the melt flow in the film on the end of the rod is made difficult by viscosity.

The amount of material m consumed from the layer of the rod with a radius R in unit time is proportional to R:

$$m = B \cdot R.$$ (9.31)

Here B – the coefficient of proportionality depending on the power supplied to the melted surface. Having determined R from (9.30) and substituting in (9.31), we obtain

$$m = \frac{A^2 B}{\omega^2 r^2}.$$ (9.32)

The droplets with radii in the range $r_m - r$ formed in unit time have the total mass

$$G = \int_{r_m}^{r} m \, dr.$$ (9.33)

Substituting (9.32) into (9.33) and integrating, we obtain

$$G = \frac{A_2 B}{\omega^2}\left(\frac{1}{r_m} - \frac{1}{r}\right).$$
(9.34)

From (9.34) at $r \to \infty$ it follows that the mass of droplets formed per unit time at the steady-state form of the rod end is

$$G_{\Sigma} = \frac{A^2 B}{\omega^2 r_m}.$$
(9.35)

The fractional composition of droplets f, i.e. the mass fraction of droplets, the size of which lies in the range r_m–r, is defined as

$$F = G/G_{\Sigma}.$$
(9.36)

Substitution of (9.34) and (9.35) into (9.36) gives

$$f = 1 - r_m / r.$$
(9.37)

The results of calculations using formula (9.37) as the dependence of $f - f_m$ on r/r_m are shown in Fig. 9.6. The value of f_m is defined by (9.37) at $r = r_m$. It is seen that for the model $f_m = 0$, which means no drops with radii less than r_m form. However, experimental data [7] show that nearly half the mass of the resulting powders have particle sizes less than r_m. This is due to the presence of an additional mechanism of fragmentation of droplets detached from the melted surface. One can point to a possible refinement of droplets as a result of their interaction with the oncoming gas in the air after separation and the size reduction effects of the arc. If we assume that generally the largest droplets are refine, and the product of refining are droplets with sizes less than r_m, the comparison of (9.37) with the experimental data is possible. In processing the data [7], the value f_m was determined from the measured curves of the fractional composition under the condition $f > f_m$. The results of such data processing [7] for $f > f_m$, obtained in the range $We_R = 10^3 - 6 \cdot 10^5$, are plotted in Fig. 9.5.

On the basis of the comparison of the calculated curve and the experimental data we can not conclude that they are in agreement. However, the correlation

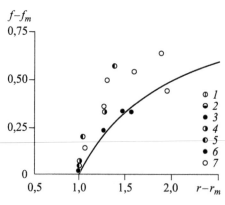

Fig. 9.6. Comparison of the calculated distribution function of the droplet size with experimental data for different Weber numbers. Solid curve – calculation. For points 1–7 We_R = 960; 1496; 2159; 3380; $3.08 \cdot 10^4$; $3.46 \cdot 10^4$; $6 \cdot 10^5$, respectively.

between them is, of course, observed. This means that the model gives a correct description of the function of the size distribution of droplets that are not subjected to additional refinement. Description of the distribution function in the whole range of droplet size requires consideration of additional refinement mechanisms.

Thus, the resultant theoretical model allows us to calculate the mean mass size of the particles produced by melting of the rotating shaft of various materials, and assess the size distribution of the powder in the particle size range large than the mean mass range.

9.2.3. Deformation of liquid melt droplets on impact on a hard surface

Consideration of the deformation of liquid droplets upon impact on a solid surface is of interest for the development of technologies for producing granules in the form of thin flakes. The scheme of the process shown in Fig. 9.7.

A spherical droplet with radius r and velocity V strikes a hard surface. We assume that during deformation the droplet has the shape of a spherical segment. During deformation of the droplet the kinetic energy of movement of the droplet is converted to surface energy. Consider the case of a solid surface wetted with liquid droplets, so the calculations can take into account only the spherical surface of a segment.

The energy conservation equation can be written as

$$E_k = E_\sigma + E_\sigma^0. \tag{9.38}$$

Here E_k is the kinetic energy of the droplet before impact, E_σ is its surface energy, E_σ^0 is the surface energy of a spherical segment with its size maximum in the deformation process. The equation of conservation of energy in the form (9.38) is valid in the case of hitting the cold surface when in the process of deformation the droplet rapidly solidifies. For the case of impact on the surface whose temperature is above the freezing point of the liquid, the droplet rebounds from the surface with the initial kinetic energy minus the energy of viscous dissipation during deformation of the liquid. Here we consider the case of complete solidification of the liquid at the point of its maximum deformation.

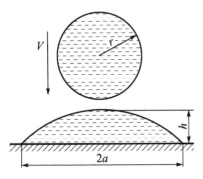

Fig. 9.7. Scheme of the deformation process of a droplet on a solid surface. r is the radius of the droplet prior to deformation; h is the height of the droplet after deformation in a spherical segment; $2a$ – chord.

Using the known formula for the volume of the sphere and the spherical surface of a spherical segment, we can write

$$E_k = \frac{4}{6}\pi r^3 \rho v^2, E_\sigma = 4\pi r^2 \sigma, E_\sigma^0 = \pi\left(h^2 + a^2\right)\sigma. \tag{9.39}$$

Substitution of (9.39) into (9.38) gives

$$\frac{4}{6}r^3\rho v^2 + 4r^2\sigma = \left(h^2 + a^2\right)\sigma. \tag{9.40}$$

In equation (9.40) the unknowns are h and a. To eliminate one of the unknown conditions, we used the condition of conservation of matter:

$$V_k = V_c. \tag{9.41}$$

V_k and V_c is volume of the droplet and the spherical segment defined by the known formulas

$$V_c = \frac{1}{6}\pi h\left(3a^2 + h^2\right), \ V_k = \frac{4}{3}\pi r^3 \tag{9.42}$$

With the help of (9.42), condition (9.41) gives the equation

$$h^3 + 3a^2 h - 8r^3 = 0. \tag{9.43}$$

The discriminant of equation (9.43) $D = (-4r^3)^2 + (a^2)^3 > 0$, so the equation has two imaginary and one real solution:

$$h = \sqrt[3]{4r^3 + \sqrt{\left(4r^3\right)^2 + a^6}} + \sqrt[3]{4r^3 - \sqrt{\left(4r^3\right)^2 + a^6}}, \tag{9.44}$$

by substituting (9.44) into (9.40) we obtain

$$\left(\frac{4}{6}\rho r^3 v^2 + 4r^2\sigma\right) = \left(\sqrt[3]{4r^3 + \sqrt{\left(4r^3\right)^2 + a^6}} + \sqrt[3]{4r^3 - \sqrt{\left(4r^3\right)^2 + a^6}}\right)^2 + a^2. \tag{9.45}$$

The solution of equation (9.45) with respect to a is difficult, so when calculating the required flight speed of the droplet with radius r for producing a circular flake with radius a it *is* appropriate to use it in the form

$$V = \sqrt{\frac{\left[\left(\sqrt[3]{4r^3 + \sqrt{\left(4r^3\right)^2 + a^6}} + \sqrt[3]{4r^3 - \sqrt{\left(4r^3\right)^2 + a^6}}\right)^2 + a^2 - 4r^2\right]\sigma}{2\rho r^3}}. \tag{9.46}$$

At high impact velocities, when $a \gg r$, (9.45) takes the form

$$\frac{4}{6}\rho r^3 v^2 + 4r^2\sigma = a^2\sigma. \tag{9.47}$$

From (9.47) we can obtain the formula

$$\frac{a}{r} = \left(\frac{2}{3}We_r + 4\right)^{1/2}, \tag{9.48}$$

where the Weber number is defined by the radius of the drop: $We = \dfrac{\rho V 2r}{\sigma}$.

At low impact velocities, when $a \ll r$, from (9.45) we obtained similarly

$$\frac{a}{r} = \left(\frac{2}{3} We_r\right)^{\frac{1}{2}}. \tag{9.49}$$

Equations (9.48) and (9.49) can be used for assessment calculations, the formula (9.46) – to calculate the size of flakes produced at $We_r < 10^3$. When $We_r > 10^3$, in our opinion, the presented model of the process and the resulting formulas are unfair, because the droplet will spread on impact.

9.3. Using Taylor bells for dispersing liquids

Studies of water film bells produced by discharge of water from an annular nozzle were started by Enteneuer [9]. The experimental data, obtained in these papers, were described by a theoretical model that takes into account the interaction of the inertial forces with the Laplace gradient in the curvature of the film. Enteneuer was the first scientist who noticed a paradox of increasing surface energy of the film with a non-obvious source of this energy. The author [9] was satisfied with the explanation that most of his experiments lie in the range of small changes in velocity of the liquid in a relatively thick film.

A more extensive study of water bells, produced by a collision of a water jet with a conical barrier with different apex angles, was conducted by Taylor [2]. The disagreement of the experimental data for the form of bells with the calculations using the model, taking into account the inertial forces, the Laplace pressure gradient and gravity, was attributed by Taylor to the influence of friction of the film on the surrounding air.

The films, flowing down from the edge of a solid surface in different situations, were studied by Dombrovsky et al [10, 11].

The aim of the research, described in this section, was to establish the similarity in the forms of water bells and determine the conditions for producing the thinnest film of the liquid that is suitable for use in the dispersion processes.

9.3.1. Derivation of the general equation for calculating the shape of a bell

The problem of the form of a bell-shaped liquid film, formed by the leakage of a non-viscous liquid from the edge of a rotating axisymmetric conical surface, is studied. The calculation scheme is shown in Fig. 9.8.

The condition of local equilibrium of forces for a film projected on the normal to it is the equation

$$2\sigma\left(\frac{1}{R_1} + \frac{1}{R_2}\right) = p_b - p_a + \delta\rho\frac{v^2}{R_v} - \delta\rho g \sin(\alpha), \tag{9.50}$$

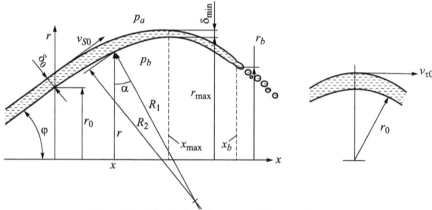

Fig. 9.8. Calculation diagram of the Taylor bell.

where σ is the surface tension; R_1 and R_2 are two principal radii of curvature; δ is the local film thickness; R_V is the local radius of curvature of the stream lines of the liquid; ρ is the density of the liquid; g is gravitational acceleration; $V = (V_\tau^2 + V_S^2)^{1/2}$ is the local total velocity of the liquid in the film, consisting of the tangential motion and the motion in the plane r–x.

Using Meunier's theorem for bodies of revolution can be defined as the radius of curvature

$$\frac{1}{R_1} = \frac{\cos\alpha}{r} = \frac{1}{r\left(1+r'^2\right)^{1/2}}, \frac{1}{R_2} = \frac{r''}{\left(1+r'^2\right)^{3/2}}. \tag{9.51}$$

Here, ' denotes differentiation with respect to x. The local radius of curvature of the line of flow is determined by Euler's theorem:

$$\frac{1}{R_V} = \frac{1}{R_1}\cos^2\Theta + \frac{1}{R_2}\sin^2\Theta,$$

where Θ – angle between the full speed vector and its tangential component. In this regard,

$$\cos^2\Theta = \frac{V_\tau^2}{V_\tau^2+V_S^2}; \sin^2\Theta = \frac{V_S^2}{V_\tau^2+V_S^2}; \frac{1}{R_V} = \frac{\left(1+r'^2\right)V_\tau^2+r''rV_S^2}{V^2r\left(1+r'^2\right)^{3/2}}. \tag{9.52}$$

The local angle of the film is expressed as

$$\sin\alpha = \frac{\mathrm{tg}\alpha}{\left(1+\mathrm{tg}^2\alpha\right)^{1/2}} = \frac{r'}{\left(1+r'^2\right)^{1/2}}. \tag{9.53}$$

Substitution of (9.51) (9.52) (9.53) into (9.50) leads to a dimensional equation

$$rr''+1+r'^2 = \left(p_b - p_a\right)\frac{r\left(1+r'^2\right)^{3/2}}{2\sigma} + \frac{\delta\rho}{2\sigma}\left(1+r'^2\right)V_\tau^2 + \frac{\delta\rho}{2\sigma}r''rV_S^2 + \frac{\delta\rho g}{2\sigma}rr'\left(1+r'^2\right),$$

$$\tag{9.54}$$

inconvenient for calculations in connection with the dependence of V_s on x and V_τ on r.

From the solution of the energy equation in a gravitational field

$$V_s = \left(V_{s0}^2 + 2gx\right)^{1/2}, \tag{9.55}$$

from the equation of conservation of the angular momentum –

$$V_\tau = V_{\tau 0}\, r_0/r, \tag{9.56}$$

from the continuity equation $V_{s0}r_0\delta_0 = V_s r\delta$ –

$$\delta = \frac{V_{s0}r_0\delta_0}{rV_s}. \tag{9.57}$$

After substituting (9.55), (9.56) and (9.57) into (9.54), introducing the notation

$$\mathrm{We}_\tau = \frac{\rho\delta_0 V_{\tau 0}^2}{2\sigma}, \ \mathrm{We}_s = \frac{\rho\delta_0 V_{s0}^2}{2\sigma}, G = \frac{\rho g r_0 \delta_0}{2\sigma}, \ p_a = \frac{(p_b - p_a)r_0}{2\sigma}$$

and making the linear dimensions with respect to r_0 and azimuthal velocity V_{s0} dimensionless

$$\bar{V}_s = \sqrt{1 + \frac{2gr_0}{V_{s0}^2}\bar{x}},$$

we obtain the equation

$$\bar{r}''\left(\bar{r} - \mathrm{We}_s\bar{V}_s\right) + \left(1 + \bar{r}'^2\right)\left(1 - \frac{\mathrm{We}_\tau}{\bar{r}^3\bar{V}_s} - G\frac{\bar{r}'}{\bar{V}_s}\right) - p_a\bar{r}\left(1 + \bar{r}'^2\right)^{3/2} = 0, \tag{9.58}$$

suitable for calculating the form of the liquid bells at any initial parameters r_0, V_{s0}, $V_{\tau 0}$, φ_0, p_a, p_b. Note that as the coefficients equation (9.58) contains similarity criteria. The first two are the Weber numbers, defined by the tangential and meridional velocity, the criterion G is a relation between the Laplace and hydrostatic pressures. Criterion p_a is the relation between the pressure drop of the gas on the film and the Laplace difference.

9.3.2. Particular cases of analytical solutions

In the absence of rotation ($V_{\tau,0} = 0$), gravity ($g = 0$) and azimuthal flow, equation (9.58) degenerates into the equation of the meniscus with constant curvature:

$$\bar{r}\bar{r}'' + \left(1 + \bar{r}'^2\right) - P_a\bar{r}\left(1 + \bar{r}'^2\right)^{3/2} = 0,$$

suitable for calculations of the form of films produced, for example, between two hollow cylinders, wetted with a liquid with the addition of surfactants.

Another special case of the analytical solution of equation (9.58) is obtained neglecting gravity ($G = 0$, $V_s = 1$), in the absence of pressure drop inside the bell and in the surrounding area (p_a = 0), and in the absence of twisting (We_τ = 0). Equation (9.58) in this case is simplified to the form

$$\bar{r}''\left(\bar{r} - We_s\right) + 1 + \bar{r}'^2 = 0. \tag{9.59}$$

Replacing $q = \bar{r}'$, $qq' = \bar{r}''$ gives

$$\left(1 + q^2\right)\left(\bar{r} - We_s\right)^2 = C.$$

The integration constant C can be determined using the boundary condition

$$\bar{r} = 1, \ q = tg\varphi_0,$$

$$q = \pm\left[\frac{\left(1 + tg^2\varphi_0\right)\left(1 - We_s\right)^2}{\left(\bar{r} - We_s\right)^2} - 1\right]. \tag{9.60}$$

Given that $q = d\bar{r}/d\bar{x}$, equation (9.60) with the boundary condition

$$\bar{x} = 0, \bar{r} = 1$$

gives a solution which is suitable for easy calculation of the shape of the liquid bell:

$$\bar{x} = \pm(1 + tg^2\varphi_0)^{1/2}\left(1 - We_s\right)\left(\sin\varphi_0 \pm \sqrt{1 - \left(\frac{\bar{r} - We_s}{1 - We_s}\right)\cos^2\varphi_0}\right). \tag{9.61}$$

The maximum diameter of the bell can be obtained from (9.60) if q = 0:

$$\bar{r}_{max} = We_s + \frac{1 - We_s}{\cos\varphi_0}. \tag{9.62}$$

Substitution of r_m in (9.62) gives the position of the cross section with a maximum diameter

$$\bar{x}_{max} = \pm tg\varphi_0\left(1 - We_s\right). \tag{9.63}$$

Equations (9.62) and (9.63) are useful for evaluation of liquid bell calculations in practical use.

9.3.3. Collapse of the Taylor bells. The minimum film thickness

In [2, 3] is it shown that the collapse of the liquid film can occur at

$$We_\delta = \frac{\rho\delta V_{So}^2}{2\sigma} = 1. \tag{9.64}$$

In view of (9.57) from (9.64) we obtain the dimensionless radius of the collapse of the bell

$$\bar{r}_b = We_s. \tag{9.65}$$

Substituting (9.65) into (9.61), we can get the length of the bell before the collapse:

$$\bar{x}_b = \pm \frac{1 - \text{We}_S}{\cos \varphi_0} (\sin \varphi_0 \mp 1). \tag{9.66}$$

The $\pm$ sign in the beginning of the formula (9.66) implies the symmetry of solutions with respect to the origin of the coordinates, inside the formula – the possibility of the collapse of the bell before and after reaching its maximum diameter. Depending on what transitional flow regimes were obtained by the bell of a given size, the collapse in both sections can be implemented in the experiments.

From the practical reasons outlined at the beginning of this chapter, of the greatest interest are the sections of the bell having a minimum thickness of the film, because they have a maximum specific surface energy. Formula (9.57) shows that the minimum thickness corresponds to the sections with the maximum diameter of the bell. Therefore, together with (9.62), this gives

$$\delta_{\min} = \frac{\delta_0}{\bar{r}_m} = \frac{\delta_0}{\text{We}_S \pm \dfrac{1 - \text{We}_S}{\cos \varphi_0}}. \tag{9.67}$$

If the collapse of the bell to droplets occurs at $x < x_{\max}$, it should be borne in mind the thinnest film forms in sections of the bell prior to collapse.

In conclusion, it should be noted that in terms of using the Taylor bells dispersion of liquids there is no urgent need for experimental verification of the above calculations. From Taylor's work [2] it is known that the observed differences between the experiments and particular cases of the calculations relate to the Coanda effect and the friction of the film on the surrounding atmosphere. Against the background of these effects it is difficult to experimentally identify the more fundamental question about the need to consider changes in the surface energy of the liquid in the film in the original equations.

9.4. Practical use of the results

Chapter 1 shows that in terms of saving energy in dispersing substances to obtain powders with a particle size of $2 \cdot 10^{-7} - 4 \cdot 10^{-5}$ m one can recommend a two-stage process: the first stage is 'pumping' of the surface energy into matter by obtaining a thin film of melt, in the second stage the melt film is dispersed with the gas.

The method of dispersion of a thin film of the melt may be implemented by various methods. For example, in [12] a thin film was produced by centrifugal forces of a spinning melt flow in a nozzle. The authors of [13] proposed a device for production of powders, in which a thin film of the melt is formed by the collision of a jet with a barrier. Industrial implementation of this method is described in the author's certificate [14] for producing tin powders. In this equipment, the melt film is produced by the discharge of a metal stream from a pressurized metal tube, which creates excess, and by collision of the jet with a flat circular barrier. Molten metal spreads into a thin film and is easily dispersed by a stream of air or inert gas supplied from the edges of the barrier. A similar method of creating a film of the melt was used in the author's certificate [15], in which the atomizing gas is fed from the top of the barrier.

Theoretical fundamentals of the description of axisymmetric flows with the formation of thin liquid films are set out in sections 9.1 and 9.3. According to the results described in these sections, we can conclude that the simplest solution for creating a thin film of melt in a continuously operating facility is the method of collision of the metal with a flat barrier. Moreover, the necessary pressure for the creation of the jet can be obtained, for example, using a pressure tube of the corresponding height.

9.4.1. Technology of obtaining powders of low-melting solders

As a first step in using the results, described in sections 9.1 and 9.3, a technology of producing powders of low-melting solders was developed. The technology is based on the sputtering of a thin film of the melt formed in a collision of a melt jet with a barrier. The scheme of the system, protected by the author certificate No. 928723, 1980, is shown in Fig. 9.9. The source of the melt was a heated crucible made of ferrous metals and hermetically sealed to increase the pressure and velocity of the metal flow in the jet impinging on a barrier. Tests on the technology with tin to produce a powder with an average size of 45 µm with a productivity of 83 g/s showed that the specific consumption of the dispersion gas for 1 kg of powder was 0.105 Nm^3/kg. The low specific consumption of the dispersion gas, unobtainable in other known dispersion methods, allowed to use this technology to be used with argon as the dispersion gas. The shape of the powder particles and the particle size distribution are shown in Figs. 9.10

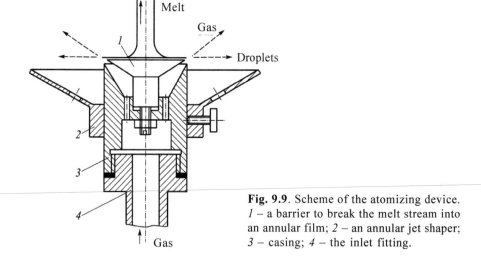

Fig. 9.9. Scheme of the atomizing device.
1 – a barrier to break the melt stream into an annular film; *2* – an annular jet shaper; *3* – casing; *4* – the inlet fitting.

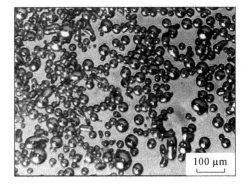

Fig. 9.10. The shape of tin powder particles dispersed with argon, ×500.

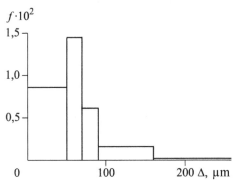

Fig. 9.11. Particle size distribution of tin powder.

and 9.11. The technology is used at the Novosibirsk Tin Concern [14] on the pilot-industrial scale for the production of solder pastes for tinning car bodies and automatic soldering of electronic circuit boards in the electronics industry. The technology using the same principle of dispersion is introduced on one of the plants for producing dispersed materials for special applications according to the Author certificate No. 176251, 1980.

Long-term experience in the use of this technology has not shown any major difficulties associated with the physical processes involved and the scale of production.

9.4.2. The process of obtaining zinc powder

As mentioned above, if the metal stream is pre-converted into a thin film of the melt, atomization process is much more efficient, with a markedly lower consumption of the atomizing gas. Therefore, the method of dispersion of the thin film melt was chosen as the method for producing the zinc powder. However, because zinc is more refractory and aggressive towards structural materials in comparison with the solder, extensive additional research is needed to create the technology for producing zinc powders.

As already mentioned, the easiest way to obtain a thin melt film is that the jet of molten metal, flowing from a small hole under high pressure, hits the flat barrier and flows from it in the form of a thin film. In the case of circular barriers placed in the horizontal plane, the stream of metal, flowing from the edges of the barrier, forms the so-called thin-wall Taylor bell. If the initial part of the bell is then subjected to the effect of radial flow of gas, it disintegrates into small droplets which crystallize and form a metallic powder.

The jet of the zinc melt was produced using the hydrostatic pressure of the liquid column. At the height of the liquid h the pressure at the bottom of the pressure tube where there is a hole for producing the metal jet, is determined from the simple expression

$$P = \rho_M g h, \tag{9.68}$$

where ρ_M – the density of the liquid metal. On the other hand, the excess pressure of the inhibited liquid is determined by the equation

$$p_{\text{liq}} = \rho_M v^2 / 2. \tag{9.69}$$

Using (9.68) and (9.69), we obtain

$$v = \sqrt{2gh}. \tag{9.70}$$

The flow of the liquid through the hole with diameter d, neglecting the viscous effects, can be determined from the expression

$$G = \rho_M v \pi d^2 / 4. \tag{9.71}$$

Thus, from the equations (9.70) and (9.71) we obtain an expression for determining the performance of the installation

$$G = \frac{\rho_M v \pi d^2}{4} \sqrt{2 \rho_M h}. \tag{9.72}$$

9.4.3. Durability of structural materials in molten zinc

In the preparation of zinc powder by gas atomization of a thin film of melt formed from the jet, the molten metal is held for a long time in some container, and, flowing through the nozzle, hits the flat barrier. These conditions impose stringent requirements on the materials from which the container, the pressure tube, the nozzle and the flat barrier are produced. This material must have sufficient heat resistance, mechanical strength, processability and mechanical processing and should not interact with the molten zinc. This set of properties is found only in metals. However, the refractory metals such as tungsten, molybdenum, are not suitable due to their low thermal stability in air. Materials based on iron are easily wetted by the molten zinc and dissolve quickly. 12Cr18Ni10Ti stainless steel in contact with the molten zinc at a temperature of 500°C becomes brittle and cracks form after about half an hour of contact. Thus, almost all pure metals and alloys do not possess sufficient stability in contact with the molten zinc and air atmosphere at a temperature of 500°C.

A composite material, based on asbestos and cement, the so-called TsIIT, showed a fairly high thermal stability in molten zinc and satisfactory mechanical properties. Thus, components made of this material showed evidence of erosion after their stay in molten zinc for 24 hours. In addition, they are not wetted with liquid zinc and do not form cracks during multiple thermal shocks. However, when heated to about 700°C the dehydration of the cement takes places and the components are gradually destroyed. Commercially produced TsIIT components represent plates with the thickness not exceeding 40 mm, which does not allow vessels of the desired size to be produced. In addition, plates made of TsIIT showed enough high mechanical strength in the transverse direction, i.e. weak bonds between layers. Therefore, this material can be used to make only small dimensions that do not require high mechanical strength.

Electrode graphite in tests in molten zinc showed a fairly high thermal stability and weak interaction with zinc. In addition, commercially available graphite blocks allow components of virtually any size to be produced. Therefore, the basic units of laboratory experimental setup that are in contact with the molten zinc are made of graphite. The general view of the laboratory experimental setup is shown in Fig. 9.12. It consists of the melting crucible 1, the pressure tube 2, the check valve 3, the nozzle for the liquid metal 4, diaphragm 5 and nozzle 6. All elements of the experimental setup are placed in the tank 7, which has channels for the release of dispersion gas 8, closed with fabric filters 9.

The melting crucible is a steel vessel, with the heating elements in the side walls and in the bottom. The inner surface of the vessel is lined with a one-piece graphite cup placed in a steel container on a soft pad of silica sand. The lateral clearance between the vessel walls and the cup were filled with sand.

The pressure tube 2 with the length of 1.6 m is made of individual graphite sections, held together by a threaded connection. On top of the pressure tube there is a heating element and a heat-insulating coating. The pressure tube is connected to the graphite cup through a conical seal. At the lower end of the pressure tube with threaded connections there is a nozzle for the formation of the jet of liquid metal made of graphite or TsIIT composite.

Shut-off valve 3 is designed to eliminate the spilling of metal during the melting process and to stop dispersion when the crucible is not completely

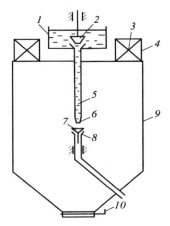

Fig. 9.12. The experimental setup. 1 – crucible; 2 – shut-off valve; 3 – fabric filters; 4 – output channels for spraying gas; 5 – pressurized tube; 6 – nozzle for liquid metal; 7 – diaphragm; 8 – the injector; 9 – bunker; 10 – shutter.

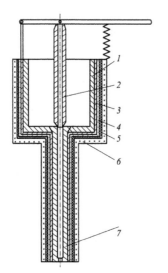
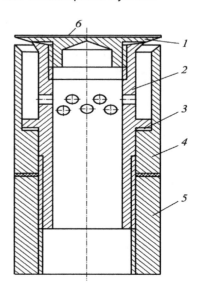

Fig. 9.13. Scheme of the melting crucible with a pressure pipe. *1* – graphite liner; *2* – shut-off valve; *3* – steel body; *4* – insulation; *5* – resistive heater; *6* – thermal insulation; *7* – graphite tube with a nozzle.
Fig. 9.14. (right) Spray nozzle. *1* – barrier; *2* – pin; *3* – shims; *4* – body; *5* – nut; *6* – enamel coating.

empty. The shut-off valve has a graphite rod with a through hole which is sealed in a graphite cup with a conical seal and pressed to the cup by a spring. The schematic view of the crucible, the pressure tube and the shut-off valve is shown in Fig. 9.13.

Several different materials were tested as a material of the barrier, separating the jet of metal into a thin film. When using a steel sheet as the barrier, the zinc jet wets the metal the melt film is rather thin, and the melt flow rate in the film is small. For this reason, the dispersion process becomes ineffective, requiring a high flow rate of gas approaching the rate used in the method of 'focusing' the jets. In addition, the resistance of the steel plate in contact with the molten zinc was inadequate. The coating of the steel plate by plasma spraying of aluminum dioxide powder significantly increased the resistance of the barrier. However, due to poor adhesion the film of aluminium dioxide at high temperature peeled from the steel substrate and the barrier broke down.

The best results in manufacturing were obtained for a barrier made of steel 08KP (A 622, USA) coated with a crockery enamel by conventional technology. This enamel, without wetting and not reacting with the molten zinc, withstood multiple thermal shocks, i.e. rapid heating to a temperature of 500°C and cooling to room temperature. Under the barrier, separating the zinc jet into a thin film, there was an annular nozzle supplying the flow of dispersion gas at an angle to the melt film. This section was provisionally named the nozzle. The schematic view of the nozzle is shown in Fig. 9.14.

9.4.4. Design of the spray nozzles

The process of spraying a thin film of the melt, obtained in a collision of a jet of molten metal with a barrier, requires that the gas flows around the film at a large angle. However, as was established experimentally, for an axisymmetric nozzle the increase of the exit angle of the flow by more than 20° from the horizontal resulted in reversal of the flow up to the pressure tube. As a result, the end face of the pressure tube and the nozzle was covered with a powder, which led to the breakdown of the dispersion mode. Furthermore, it was found that the outer diameter of the pressure tube with a layer of insulation, the diameter of the annular nozzle of the injector, and the length of the metal are inter-related. By increasing the diameter of the pressure tube to maintain a radial flow of gas it was necessary to increase the length of the metal stream. However, the length of the stream is limited by the hydrodynamic instability of the free liquid jet and, to a greater extent, by its spatial variations which lead to the fact that at the small diameter of the nozzle the jet went beyond the barriers.

In [13] it is proposed to install a ring collar to reverse the flow in the radial direction. However, in this case, the zinc powder rather quickly, within 5–10 seconds, covered the surface of the nozzle and overlapped the slit gap of the nozzle. It was also shown that the gas stream flowing from the annular nozzle, in the presence of the collar, has only two stable positions – either the current flows along the radius of the collar or it 'jumps' to the pressure tube. At a high ratio of the diameter of the pressure pipe and the nozzle barrier (~2), as well as the small distance between the pressure tube and the nozzle (~d_n – the diameter of the nozzle), the slightest disturbance is sufficient for the transition of the flow regime from radial to bell-shaped. For this reason, the angle of exit of the radial flow from the slit nozzle should not exceed 15° from the horizontal.

The value of the diameter of the nozzle plays a dual role. On the one hand, in the spreading of a jet of metal in the radial direction on a large–diameter nozzle the moving film of molten metal is inhibited and thickens due to the influence of friction forces. This increases the size of the particles of the powder and the required dispersion powder increases. On the other hand, at the small diameter of the nozzle and a constant diameter of the pressure tube the instability of gas flow in the radial direction becomes greater. This makes it necessary to increase the distance between the pressure tube and the cup. Thus, when the nozzle diameter is 40 mm and the diameter of the pressure tube 60 mm the minimum (in terms of stability of the radial gas flow) length of the metal stream will be 40 mm. To halve the diameter of the nozzle it would be necessary to double the length of the jet. However, such a length of the jet begins to affect the spatial instability of a jet of molten metal due to its interaction with a turbulent gas flow.

The reason for these phenomena is the fact that the influence of the radial gas flow results in the formation of a low pressure zone near the pressure tube. The larger the ratio of the diameter of the pressure tube to the nozzle diameter and the smaller the length of the metal jet, the greater the degree of dilution due to injecting action of the radial flow. Under the influence of the resulting pressure gradient the radial gas turns towards the side of the pressure tube. For this reason, in order to preserve the stability of the radial flow, the exit angle of the flow from the slit nozzle was reduced to 10° from the horizontal.

9.4.5. Experimental studies of the process of zinc spraying

In studies of dispersion in a laboratory setting a one-off portion of zinc of about 20 kg loaded into the crucible, heating to the melting point and superheated to a temperature of 500°C. At the same time, the pressure tube was closed from the crucible by a graphite valve. After bringing the temperature of the melt and the temperature of the pressure tube to the desired value (500°C), the valve was opened and the metal was fed to the pressure tube from which it was discharge in the form of a jet with a diameter of 3 mm and hit the barrier of the nozzle. The duration of establishment of the mode, associated with the need of accommodation of the temperature of the barrier, was 3–4 s. The duration of spraying in laboratory equipment depended on the mass of the charge and did not exceed 90 s. Figure 9.15 shows a photograph of the resulting powder particles. As can be seen in the photograph, most particles are needle-shaped. When spraying a thin film of the melt flowing down from the barrier, the separated droplets can not form a sphere due to the influence of the oxide film. The melt temperature is also important in this process. For a small superheating of the melt with respect to the melting point the droplet solidifies in the form in which it is separated from the film. With increasing superheating and using an inert gas for dispersion the form of the particles is close to spherical as shown by the photograph in Fig. 9.16.

At a small superheating of the metal the melt film, flowing on the plane of the barrier of the nozzle, manages to partially crystallize before its descent from the surface. As a result, as the temperature of the melt decreases, the dispersion process becomes unstable, the powder particles become larger and, as a rule, the process of dispersion is spontaneously terminated due to an increase in the film thickness of the skull on the surface of the barrier. The failure of the dispersion mode begins with a decrease in melt temperature

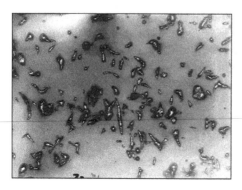

Fig. 9.15. The particle shape of the powder in dispersion with air.

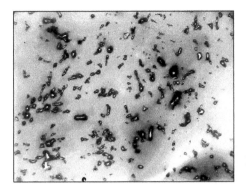

Fig. 9.16. The shape of the powder particles in dispersion with an inert gas.

below 450 °C. For this reason, when spraying zinc the metal temperature in most experiments was not maintained below 500 °C.

9.4.6. Testing and development of the operating modes of the laboratory setup for obtaining zinc powder in production conditions

In the laboratory conditions, a portion of the metal was melted and brought to the desired temperature directly in the crucible. This process is periodic and can not be applied on an industrial scale, requiring high performance. In addition, in an industrial environment the metal is melted in special furnaces. Therefore, an additional task of feeding the metal from the furnace into the dispersion unit must be tackled.

To verify the efficiency of the system in producing metallic powders in the long-term regime where it is required to process large quantities of metal, the installation was tested under industrial conditions. For this purpose, the Belovo Zinc Plant constructed in the workshop for production of zinc powder by the gas phase method a system which is a modification of the laboratory setup described above. Here the experience gained in the test of the installation in industrial environments is described.

A pressure tube 1.6 m long and a crucible 0.3 m were made of graphite. The crucible was inserted in a resistance furnace, made of metal in the form of a cylinder, on the side of which there was a nichrome wire heater. The bottom of the furnace was heated by tubular heaters, soldered to the bottom. The total power of the furnace was 4 kW, the heater power of the pressure tube 2 kW. The diameter of the nozzle for the jet of liquid metal was 3 mm. Spraying was carried out with nozzles with a diameter of the barrier of 20, 30 and 40 mm, coated with enamel crockery. The equipment was located in close proximity to an industrial gas reverberatory furnace for smelting zinc.

In preliminary experiments with this apparatus liquid metal heated to 500°C was poured into a crucible at a temperature of 500°C. At the same, the pressure tube was also warmed to a temperature of 500°C. The dispersion process began with the supply of gas to the nozzle at a pressure of 0.2 MPa at the inlet and pouring of metal on the barrier. During emptying of the crucible a new portion of the metal was poured from the furnace using a ladle. The duration of the

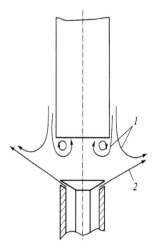

Fig. 9.17. Flow pattern of gas flows. *1* – injected gas flow; *2* – direction of the main gas flow.

experiment was about 15 min. The process of spraying was stable, without significant disruption of the regime. At the end of the experiment it was found that on the front surface of the pressure pipe there is a powder coating which is prolonged use can lead to failure of the dispersion mode. The analysis of the nature of the appearance of the powder coating on the front surface of the pressure tube showed that the influence of the radial gas flow forms an injected flow directed along the pressure tube. The air suspension of the fine fraction of powder in the chamber is carried away by the axial flow and falls in the area of the end of the pressure tube and the drain. Figure 9.17 shows a diagram of the gas flows in the presence of injecting suction. Due to the fact that the end of the pressure tube has a temperature close to the melting point of the metal, the suspended fine fraction sticks to the hot surface. At the same time no sticking of the powder to the heat insulating coating of the pressure tube was observed.

To organize a fully continuous operation of the equipment, the flow of metal in it was carried out on an inclined open tray 1.2 m long, on which the metal from the furnace in a stream flowed into a graphite crucible. In these experiments it was found that about 5 min after the start of spraying the build-up of metal appeared on the edges of the nozzle and at the end of the pressure tube, which soon closed up, blocking the exit of the metal from the pressure tube. The cause of failure of the dispersion mode in this case could be cooling of the metal during its motion on the tray. Thus, control of temperature of the melt in the crucible and in the pressure tube showed that it amounts to 420–440°C. The power of the heaters of the crucible and of the pressure tube is not enough to heat the metal to the desired temperature. Therefore, the formation of the crust at the nozzle is due to the low temperature of the melt, just as it happened during testing of the laboratory setup.

9.4.7. Pilot plant for zinc powder production

Based on the results of laboratory and industrial experiments, along with the laboratory setting a variant of the pilot plant in relation to the specific conditions of obtaining the molten zinc was also developed at the Belovo Zinc Plant. The main difference between this installation from the laboratory one is the supply system liquid metal from the furnace. A schematic view of the pilot plant is shown in Fig. 9.18. In this design, the flow of metal from the furnace is realized from a separate pocket *1* built in an industrial furnace for smelting zinc, *2,* and the level of metal in this pocket and in the furnace was the same due to the organization of supply of the metal into the pocket with the tray located below the melt in the furnace. To maintain the temperature of the metal in the pocket within the prescribed limits, the latter was heated by an additional gas burner. The temperature of the metal in the pocket and the furnace was controlled by chromel–alumel thermocouples. The pocket and the pressure tube *3* were connected using a conical seal. The brick wall of the pocked contained a graphite block *4* with a tapered bore. The pressure tube 2.5 m long with a conical thickening at the tip was inserted into the tapered bore of the graphite block and compacted under its own weight.

To maintain the temperature at a given level the graphite tube was inserted into a tubular electric resistance furnace *5* with a length of 2.5 m. The lateral surface of the resistance furnace was covered with asbestos insulation, and the lower end of the furnace was made of an insulating conical tip using TsIIT composite. The pressure tube ended in a conical nozzle for forming a metal jet.

The conical nozzle and the conical heat-insulating tip of the resistance furnace reduced twisting the injected gas flow and prevented sticking of the fine fraction of the powder to the end of the pressure tube. The spray nozzle in the pilot plant was identical to the nozzle in the laboratory setting.

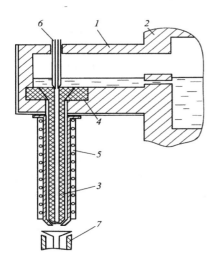

Fig. 9.18. Scheme of the pilot plant. *1* – pocket; *2* – furnace; *3* – pressure tube; *4* – graphite block; *5* – tubular oven; *6* – shut-off valve; *7* – injector.

To interrupt the spraying process, the pocket contains a shut-off valve 6, which consists of a graphite conical tip and a metal rod. If it is necessary to interrupt the process of spraying, the graphite tip is inserted into the discharge orifice of the pocket and seals it by its weight. For long- term shutdown the pocket can be cut off from the furnace volume by covering the feeding tract with refractory cement. At the same time, the molten metal remaining in the pocket can be fully utilized during the spraying process.

Unfortunately, the pilot plant could not be tested in long-term modes because it was destroyed in a night shift by shop workers, who feared that the new high-productivity technology will deprive them of well-paid jobs at the gas-phase production of zinc powder.

So, as a result of studies in the laboratory conditions and tests of the laboratory setup at the factory, with the creation of the pilot plant it was possible to reveal several features of the process of gas spraying of molten zinc with an intermediate stage of transformation of the melt jet into a thin free film. In the studies thin films were obtained as a result of impingement of the melt with a flat barrier. In the process of laboratory studies of the method the results various features of the behaviour of structural materials in molten zinc and powder samples were obtained with a fraction (2–200) μm. Steps have been taken to establish the industrial variant of the implementation of the method of spraying thin films, which required a solution of the supply of the metal jet from the furnace to the spray unit.

Conclusions

1. The radial free liquid film was used as an example to shows the need to consider changes in the surface energy in the hydrodynamic calculations and experiments under conditions where the free surface area of the film increases due to the decrease of the dynamic pressure of the flowing fluid. This is a direct way of converting the kinetic energy to the free surface energy.

2. It is shown that the processes of separation of liquid droplets from the edge of a rotating hard surface or from the edge of a melted bar are similar. For the melting of the convex end of the rotating shaft the size distribution function of the separated droplets was determined. The kinetic energy of rotation is partially transferred to the free surface energy of the droplets.

3. A general equation describing the shape of Taylor bells with a number of similarity criteria as coefficient was derived. The partial analytical solution of this equation, neglecting gravity, pressure drop and twisting, was shown. A simple method of calculating the basic parameters of the Taylor bells was proposed for this case.

4. The transition of par of kinetic energy to surface energy during the formation of a free liquid film was used to develop a radically new cost-effective technologies for producing powder materials. The technology for producing solder powders is used in the industry. Principal technical problems were solved for the technology of producing zinc powders and recommendation for use in the industrial conditions presented.

List of symbols

A, B	– coefficients;
d	– diameter, m;
E	– energy, J;
F	– force, H;
g	– acceleration due to gravity;
h	– height of the column of liquid;
l	– length, m;
m	– mass, kg;
p	– pressure;
Q	– volumetric flow rate, m3 / s;
Re	– the Reynolds number;
R, r	– radius, m;
S	– area, m2;
t	– time, s;
V, u	– velocity, m / s;
We	– Weber number;
x, z	– coordinate on the axis of symmetry, m;
Δ	– the boundary layer thickness;
δ	– thickness, m;
ε	– the specific energy J/m3;
μ	– dynamic viscosity, kg / (m · s);
ρ	– density, kg/m3;
σ	– surface tension, N / m;
ω	– angular velocity, s–1.

Indices:

a – a drop;
s – segment;
σ – the surface;
χ – a chaotic;
p – decay;
k – kinetic;
τ – a tangent,
d – diameter

δ – thickness;
R – radius;
c – a centrifugal;
S – meridian;
max – the maximum;
min – the minimum;
0 – initial;
– dimensionless.

Literature

1.   Lavrentiev M.A., Shabat B.V., Problems of hydrodynamics and their mathematical models, Moscow, Nauka, 1973.
2.   Taylor G.I., Proc. Roy. Soc. London, Ser. A, 1959, V. A253, No. 1274, 289–295; 296–312; 313–331.
3.   Huang J.C.P., J. Fluid Mech., 1970, V. 43, N 2. P. 305-319.
4.   Batchelor J.C., Introduction to fluid dynamics, Springer-Verlag, 1973.
6.   Powder metallurgy of materials for special purposes, Ed. J. Bark, V. Weiss,

Moscow: Metallurgiya, 1977.

7.   Altunin Yu.F., et al., Tekhnol. Legkikh Splavov, 1970, No. 2.
8.   Tables of Physical Quantities, Ed. I.K. Kikoin, Moscow, Atomizdat, 1976.
9.   Von Enteneuer G.A., Einfluß der Oberflächenspannung auf die Ausbildung von Flüssigkeits-Hohlstrahlen, Forschung auf dem Gebiete des Ingenieurwesens, 1956. Bd. 22, No. 4. 109–122. doi: 10.1007/BF02592597.
10.   Dombrowski N., et al., Chem. Eng. Sci., 1960, No. 12, 35–50.
11.   Dombrowski N., Hooper P.C., J. Fluid Mech., 1963, V. 18, 392–400.
12.   Zholob V.N., Koval' V.G., Poroshk. Metall, 1979, No. 6, 13–16.
13.   Author Cert. 928723 USSR, MKI B 22 F 9/06. A device for spraying, A.A. Bochkarev, et al.
14.   Author Cert. 1110036 USSR, MKI B 22 F 9/08. Device for spraying powders of low-melting metals, V.E. Dyakov, et al., Otkrytiya. Izobreteniya, 1986, No. 21.
15   Author Cert. 1002096 USSR, MKI B 22 F 9/08. A device for obtaining metal powders from melts, N.A. Raspopov, et al., Otkrytiya. Izobreteniya, 1983, No. 3.

10

Production of metallic powders of the required composition by melting a rotating rod

Despite the abundance of publications describing various methods of producing metal powders by rotating a molten rod, the production of powders of a given chemical composition can be difficult, as the technology of their production usually involves some materials or media that introduce impurities [1, 2]. In producing powders of high purity metals and alloys, such as materials for chemical power sources, the issue of protection from contaminants is extremely important. Among the known methods for producing powders, contributing a minimum amount of impurities, we note the method of exploding wires [3]. However, the difficulties of obtaining a powder with a given fractional composition, low productivity, and the complexity of the process equipment do not allow us to consider this method promising at present.

The method of arc melting a rotating electrode in an inert atmosphere [4–8] makes it possible to obtain powders (granules) of metals with a minimal change in chemical composition compared to the original material, to control the average size of particles, and is quite productive. The method can be applied to a broad class of materials. In the process of obtaining a powder the admixture may be introduced during heating and partial destruction of the fixed electrode made of a refractory metal. However, this method cannot be used to obtain powders with the chemical composition that does not differ from the source material.

When the material of rotating and stationary electrodes are identical, it may be possible to produce of high-purity powders of materials with particles of a given chemical composition. Modifications of methods that enhance its performance are available. At the heart of technology development there should be a model of a physical process, which allows to calculate the main parameters of the process. Section 9.2 shows the methods of calculating the average particle size and the size distribution function of particles, and the cooling of the particles is also calculated. Laboratory setup is described, which was used for experiments aimed at obtaining a zinc powder of a given initial composition. Different versions of the centrifugal method are compared to find a method with minimum contamination of the source material with impurities.

10.1. A brief review of the literature

A centrifugal method for manufacturing metal granules, described by S. Abkowitz [5, 6], proved to be promising in obtaining high-temperature titanium alloys of high purity. Later, the method was developed in [4, 7–9], aimed at increasing its productivity and lowering the impurity content in the formed granules (powder particles). Yu.F. Altunin et al. [4] described equipment with electric arc melting of a rotating rod. The authors addressed issues such as selection of electric drives, bearings, seals for the vacuum input shaft rotating at a speed of 15 000 rpm, the current lead system, etc. The influence of the basic parameters of the equipment (the speed of rotation of the consumable electrode arc current, the diameter of the electrode) was studied. The chamber of the equipment was evacuated by a vacuum pump to a pressure of $1.33 \cdot 10^{-1}$ Pa and then filled with an inert gas (argon, helium) to a pressure above atmospheric. The strength of the arc current was varied within wide limits – from 100 to 2000 A. Productivity was 1.5–2.0 kg of granules of titanium alloy per shift. The authors of [4] reported that the granulation process should be performed at the minimum current strength that ensures the stability of the arc. Doubling the diameter of the electrode (from 15 to 30 mm) allows to increase the productivity of the process four times (from 0.5 to 2.0 kg), replacing argon with helium in the chamber has a positive effect on the product quality. The dependence of the fractional composition of the granules on the speed of rotation of the consumable electrode, arc current and the diameter of the rotating electrode was investigated.

The installation with the automated feed of consumable electrodes was proposed by Japanese authors [7]. The casing of the system was vacuum-tight, filled with an inert gas which is saved due to the continuity of the prepared electrodes inside the unit. A similar device is described in a German patent [10]. A similar installation with higher performance and producing higher quality powder was proposed I.A. Kononov et al [11]. It is equipped with special chambers for the accumulation of blanks and stubs. Since the chambers are separated from the melting section of the installation by vacuum-tight gates, it is possible to supply additional blanks, without interrupting the spraying process. The authors of the patent [12] suggest using a long rod as a consumable electrode. This increases the sputtering time of a rod and reduces the load on the feeder mechanism for feeding the blanks. However, using a long rod, it is difficult to solve the problem of stability of the consumable electrode at high rotation speeds.

The patent [13] describes a method for producing ultrafine powders by a scheme similar to [4]. It is believed that one of the electrodes, made of a refractory metal, is not destroyed, the other one works in the droplet formation mode. The temperature of the end of the melting electrode is equal to the melting point, and the electrode rotates at a frequency $n = 20\ 000$–$30\ 000$ rpm.

The process of obtaining a powder is usually performed in a controlled atmosphere of argon or helium (the gas pressure varies from atmospheric pressure to 10^3 Pa). Droplets are scattered from a rotating electrode under the influence of centrifugal forces and by heat exchange with the gas they cool down becoming spherical particles of the powder.

A.S. Buffed and P.U. Gummeson set out a method to obtain high-quality powders with low oxygen content by melting a spinning rod. The size range of the powder particles is relatively narrow (30–500 μm), and the particles are completely spherical particles with a smooth surface [14]. The experimental setup is similar to that described in [4].

Improved equipment for the production of powders by melting the spinning electrode is described in a French patent [15]. It is intended to produce powders of materials with high melting points. The form of the powder can be adjusted during production. The non-consumable electrode is moved over the surface of the end of the spinning electrode for uniform melting. The non-consumable electrode erosion products can bring in impurities unacceptable in the case of powders of high-purity materials.

Equipment for spraying a powder by sputtering a rotating workpiece was proposed by N.F. Anoshkin et al [16]. In order to regulate the form of powder particles its chamber is equipped with a water-cooled conical screen positioned coaxially with the workpiece and configured for rotation and axial movement. The preform to be sprayed is placed in the chamber which is then sealed, evacuated and filled with helium. The workpiece rotates at a rate of 10–20 thousand rpm. The plasma generator (plasmatron) is activated and the end face of the workpiece is melted. The molten metal is separated by the centrifugal force from the end of the workpiece and falls on a rotating water-cooled screen. By changing the distance from the end of the sprayed workpiece to the screen it is possible to adjust the flight time of molten particles before crystallization, and hence the shape of the powder particles. If the screen is located in the immediate vicinity of the spray area, we obtain powders of lamellar-flaky form or mixed with a powder of spherical shape. The disadvantage of this method for obtaining powders of high-purity materials is the use of a heat source which introduces impurities. The material of the electrodes of the plasmatron comes into the powder in the form of the vapour phase or in the microdroplet state.

A method of producing spherical nickel particles was proposed by French authors [8]. The method consists in the fact that the end surface of the cylindrical ingot from the molten metal is brought in vacuum to the melting temperature by the heat source. The ingot is set in rotation around its axis, and the spherical particles obtained during the solidification of liquid droplets ejected from the ingot by centrifugal force are collected. In the method, the surface of the ingot is brought to the melting point by moving the zone of influence of the heat source above it. During the time of rotation of the ingot the zone of the effect describes curves uniformly distributed around the axis of the ingot from the periphery to the centre. Thus, it is possible to obtain particles (powder) of ultrapure materials. This technique is effective for materials with a high melting point, since the energy losses in radiation significantly increase at high temperature. Cooling of the droplets takes place only due to the radiation of particles. However, to obtain a powder of low-melting materials, this method would require a significant increase in the device dimensions.

The original plant for producing granules from pure metals and alloys was described in [9]. A characteristic feature is its lack of rotating parts. The the area of the molten material is affected by the electromagnetic field, resulting in the metal film being subjected to centrifugal atomization. The authors of former West Germany [9] made a device for a continuous supply of material into the area of granulation using a system of vacuum locks. A water-cooled non-consumable electrode provides the necessary purity of the powder material with minimal destruction.

There are a number of patents [13, 16–18] for different versions of the method.

The review [19] is devoted to the study of the formation of particles of metals and non-metals in rotating a liquid in a crucible. Arc-melted drops of a fixed electrode are placed in a rotating cooled copper crucible. The liquid metal spreads over the surface of the crucible and is sprayed from the edge of the crucible by the centrifugal force to form granules. The power released by an arc is about 50 kW. The voltage on the electrode is 80 V, arc current is 2500 A. The arc is ignited by an ignition device, operating at 2 kV and a frequency of 3 mHz. Before spraying the plant is pumped with vacuum pumps to a pressure of 0.13 Pa and then filled with high-purity argon. The density of the sputtered metal differs – from 2.4 to 16 g/cm^3, melting point is 673–3773 K. The review notes that one of the advantages of the method lies in the fact that the sprayed metal remains superheated up to and including the process of spheroidization, resulting in high output of spherical particles. The method can be implemented in the mode of the consumable crucible (anode) and the non-consumable stationary electrode (cathode) made of refractory material such as tungsten. In this variant, the described method does not differ from the method of spraying a rotating electrode [4]. It can be realized in the mode of consumable electrodes, made from materials such as nickel or steel. The arc is ignited and its power rises to the melting of the cathode. It is noted that the shortcoming of such a method is the difficulty of monitoring and process control.

In the installation with non-consumable electrodes the material in the form of granules (pieces) falls into the arc region through a special channel in the cathode, liquid droplets spread out on the rotating crucible and the resulting film material is sprayed in a controlled atmosphere of the installation. The method was used to obtain spheroids carbides, oxides and other compounds (UC, UO_2, ZrO_2, Al_2O_3, B_4C, ZrB_2). The pressure in the installation is maintained below atmospheric pressure and is determined by the stability of arcing and cooling of the resultant particles. The powder of the spray material contains the ultrafine fraction – the spheroids with the size of ≈ 0.02 μm in the amount of $\approx 0.1\%$. Given the extremely high chemical activity of the particles with a size of less than a micron, one should pay attention to the problem of evacuating the powder and its storage.

In the survey [19] noted that several authors [18–23] investigated the formation of particles during centrifugal atomization, and the relationship of the particle diameter with the parameters of the process has the form:

$d = \dfrac{A}{\omega}\left(\dfrac{\sigma}{D\rho}\right)^{1/2}$, where d is the diameter of the particle, A is a constant; ω is the
rotational speed; D is the diameter of the rod; ρ is the density of the liquid
metal; σ – surface tension coefficient.

The process of formation of the so-called 'bundles' – needle-like formations on the edge of the crucible or consumable anode was described. The
formation of a liquid metal film on the surface of the crucible was investigated.
The trajectories of the particles of metal, leaving the edge of the crucible, were
calculated. It is believed that the particle in flight is affected by the drag force
of the environment and gravitational force. The calculation is made assuming
the spherical shape of the particles. Calculations were carries out to determine
the trajectories of particles of the size of 50–2000 µm, cut off from a crucible
76 mm in diameter, rotating at a frequency of 3000 rpm. The particles were
cooled in argon at atmospheric pressure and a temperature of 323 K. The
dependence of the cooling time of the particles of iron in the temperature range
T_m = +50... –50 K on the diameter of the particles at different speeds of the
crucible was also calculated taking into account the forced convection and
radiation. It was assumed that the initial temperature of the particles and their
emissivity are known and that the volume of the particles is heated uniformly.
Experimental results are presented in the form of the size distribution of
spheroids in dependence on the different methods of melting the spray material.
The particles were spherical or in the form of flakes. The flakes are the result
of plastic deformation of powder particles on impact with a hard surface.

Study [24] describes a system for the production of powders of titanium
alloys by spraying a molten rotating rod. It is known that the production
of titanium powder is associated with the high activity of its particles, so
the control of the atmosphere is given special attention. The installation is
evacuated with vacuum pumps and filled with helium. The fixed electrode
is made of tungsten. The rotational speed of the sprayed electrode is 10–25
thousand rpm and is determined depending on the diameter of the particles
to be collected. The arc current is 400–800 A. A characteristic feature is the
spherical shape of the powder particles and the small difference in size. The
size distributions of particles, obtained at different rotation speeds of rods of
different diameters, are determined.

In [25], concerned with the powder metallurgy of titanium alloys, the
authors describe the advantages of powders obtained by spraying a rotating
rod. The installation is pumped by vacuum pumps and filled with helium. The
neutral atmosphere is necessary for the production of powders with controlled
chemical composition and preventing the oxidation of powders in the process
of accumulation in the chamber. The particle size is controlled by the speed of
rotation of the rod and by its diameter. The resulting powders have a specific
surface area of 0.009 m²/g with an average particle size of 200–250 µm. The
bulk density of the powder is 64–66% of the theoretical value.

In [26] a sprayed rod was fixed eccentrically in a holder, which rotates
at a frequency determined by the size of the particles. The installation is
made on the basis of a patent [13]. The unusual arrangement of the atomizing

electrode produces particles with a high degree of sphericity, but only with the size of 50–500 µm and smaller. The rotational speed of the sprayed core is 600–20 000 rpm.

In the equipment described in the patent [27] for the production of powders and granules of metals by spraying a rotating electrode the atomization chamber is evacuated by pumps and filled with a neutral gas. The equipment has a feeder for supplying blanks running in the automatic mode. If necessary, the chamber can be used for simple operations, using gloves attached to the chamber.

The characteristics of the process and equipment for the production of powders by centrifugal atomization of the rotating workpiece are described in the work by I.A. Kononov and V.T. Musienko [28], in which the mechanism of formation of granules in centrifugal spraying of the rotating installation is outlines. The dependence of the size of granules on the process parameters of spraying, the sprayed material and the dimensions of the workpiece is determined. The principal possibility of using the method of centrifugal sputtering of a rotating workpiece to produce powders of materials not containing in their structure large precipitates of refractory phases is shown.

Spraying of a spinning disc with a skull was studied by V.A. Kozlov et al [29]. It is noted that the method has high performance, the powder particles do not have cavities, the range of the fractional composition of the powder is smaller in comparison with the gas jet method. The authors obtained powders of lead, tin, copper, aluminium and bronze. During spraying, needles, scales, and flakes flew from the disc. It is established that the viscosity of the melt affects the mean diameter of the particles:

$$d_{mean} = \frac{C}{(\omega r_{sc})^{1/2}} (\zeta_m v)^{0.25} \left(\frac{\sigma}{\rho g} \right)^{0.125} .$$

Here C is an experimental constant, r_{sc} is the radius of the skull, ζ_m is the specific heat of the melt; v is viscosity; g is the acceleration due to gravity. The disc was made of quartz glass. The powder particles were cooled with water.

In his monograph, V.D. Johnson [30] discusses the various methods of obtaining powders, including the method of centrifugal sputtering, in which the liquid is fed into the centre of a rotating bowl. Under the influence of centrifugal force the liquid travels up the wall of the bowl, reaches the edge, and breaks away from it as a film. At sufficiently high speeds of rotation the film can be very thin. Studies of this method have shown that the centrifugal atomization is characterized by high homogeneity of the powder particle size. The diameter of the particles of the liquid metal is given by

$$d\omega \left(\frac{D\rho}{\sigma} \right)^{1/2} = A.$$

The productivity of the method is a few pounds per hour. It is indicated that the method of centrifugal spraying can be useful if it is required to produced the powder with a uniform particle size. The design of systems for centrifugal atomization of magnesium and aluminium is shown in a British patent [31]. Some of the problems that arise in the implementation of the method are discussed in a U.S. patent [32]. Production of a fine lead powder is described in [33]. The first patent, describing a centrifugal method for producing particles, issued in 1883 to W. Cross [34].

The above discussed technical solutions of the centrifugal method of producing powders (granules) by melting a spinning rod are designed to increase the productivity of the method, control the shape of the powder particles, and obtain particles with minimal impurities. However, in all plants, either a heat source, introducing small amounts of impurities, is present, or the process takes place in a vacuum, making it difficult to obtain powders of volatile metals. Replacing the non-consumable electrode material by the material of the sprayed rod, we can produce powders without impurities from the heat source in an inert gas. The prospect of obtaining powders of high purity materials is here promising. It is expected that it will be possible to zinc powder iron powders with the iron impurity content not greater than 0.0005%, zinc oxides – not more than 1%. In the form after screening the powder particle size should be in the range of 0.1–0.3 mm. Such a powder is needed for the production of chemical current sources at the Novosibirsk Condenser Plant. Further experiments in this direction will be carried out.

10.2. Calculation of cooling of droplets

It is assumed that a spherical droplet of the melt with a melting point T_m separated from the melting rotating crucible and flies in the gas cooling it (argon), which has a temperature of T_k, some distance l with velocity v, defined by the rotational speed ω of the rod and its radius R. For some time τ the droplet transfers from the liquid into solid state, releasing the heat of the phase transition into the surrounding gas:

$$l = v\tau = \frac{2\pi R}{\omega} \cdot \frac{r_1 \rho_2 \frac{4}{3}\pi r^3}{\alpha(T_m - T_k)4\pi r^2}. \tag{10.1}$$

Using (9.29)

$$r = \frac{A}{\omega\sqrt{R_0}}$$

and substituting in A the values of surface tension and density for molten zinc $\sigma_r = 0.810$ N/m, $\rho_r = 6.42 \cdot 10^3$ kg/m³, we obtain an expression for the radius of the detached zinc droplet

$$r = 0.1308\frac{1}{n\sqrt{R}}. \tag{10.2}$$

Substituting (10.2) in (10.1) and introducing the dimensionless heat transfer coefficient (Nusselt number – Nu), we can write

$$l = 112778 \frac{1}{n\mathrm{Nu}}. \tag{10.3}$$

In the calculation were used: $T_m = 419.5°C$, $T_k = 20°C$, $\lambda = 0.0174$ W/(m·deg), $r_1 = 102\,236$ J/kg, $\rho_r = 6.42 \cdot 10^3$ kg/m³. To determine the number Nu, we use data on heat transfer of a sphere as described in [35], where Nu is represented as a function of the value of the product of the Reynolds and Prandtl numbers:

$$\mathrm{Nu} = \mathrm{Re}^{0.54}\,\mathrm{Pr}^{0.33}. \tag{10.4}$$

In calculating the Reynolds number, the characteristic size is the droplet diameter, in the calculation of the Prandtl number we use data [36] for argon:

$$\mathrm{Re} = \frac{\rho_3 v 2r}{\eta_3} = 2211\sqrt{R},\ \mathrm{Pr} = 0.662,$$

$$\rho_r = 17873 \text{ kg/m}^3, \tag{10.5}$$

$$\eta = 2210 \cdot 10^{-8}\,\mathrm{Pa \cdot s}.$$

Substituting (10.5) (10.4), we compute the product

$$\mathrm{Re}^{0.54}\,\mathrm{Pr}^{0.33} = 55.6 R^{0.27}.$$

For different values of the radius of the rod R we can find the value of Nu from the graph in [35]. Substituting Nu into (10.3), we find the minimum distance at which the phase transformation of the droplet occurs. The result of calculation is shown in Fig. 10.1 as the dependence of the distance on the speed of the rod. These data can be used in the calculation of the parameters of the equipment for producing powders by melting the spinning rod.

10.3. Description of the experimental setup

Figure 10.2 shows a diagram of the apparatus. In the casing *1* there are: the receiver of the powder *2*, disc *3* of the starting material spinning in the horizontal plane, the heat source *4*, set at an angle to the vertical ($\alpha \approx 40°$). The powder *5* is collected in the receiver. The disc is rotated by a drive motor *6* with the speed meter *7*.

The heat source is an electric arc ignited between the rotating disc and the stationary electrode set at a distance providing steady burning of the arc. A plasma torch from the PVP-1 semi-automatic air–plasma cutting equipment and induction heating were also used.

Figure 10.3 shows a diagram of the device for heating the disc, where the fixed electrode *4* on node *2* is mounted on a holder *1* connected to a power source. In the body of the electrode there is a hole *3* to supply the plasma-forming neutral gas (argon, helium). The arc *5* burns between the rotating disc

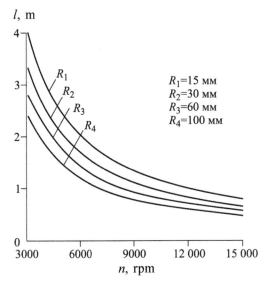

Fig. 10.1. The minimum flight distance of the particles required for the crystallization of the liquid zinc droplet in dependence on the number of revolutions of the rod. Calculated from (10.3).

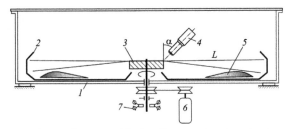

Fig. 10.2. The experimental setup with the disc rotating in a horizontal plane. *1* – the casing body; *2* – powder receiver; *3* – disc, layer of sputtered particles; *4* – heating source; *5* – powder; *6* – DC motor; *7* – the device counting the number of revolutions.

6 and the electrode *4*. The photograph in Fig. 10.3 shows the appearance of the sprayed disc and the stationary electrode.

Figure 10.4 shows a diagram of a stationary electrode with a circular flow of neutral gas. The electric arc *7* burns between the electrode *5* and the rotating disc *6*. The central part of the electrode is surrounded by an annular flow of inert gas pumped through the tube *3* in the space between the ceramic tube *4* and the water-cooled electrode *5*. The fixed electrode *5* is the cathode in the circuit of the electric arc. The appearance of the described device is shown on the right.

Figure 10.5 shows a similar device with a fixed electrode, in which a tungsten insert is pressed into the copper electrode *1*. The neutral plasma-forming gas is fed through the hole *3*. The electrode is cooled by water flowing through the tubes *2*. The photo at right shows the appearance of the stationary electrode with a tungsten insert.

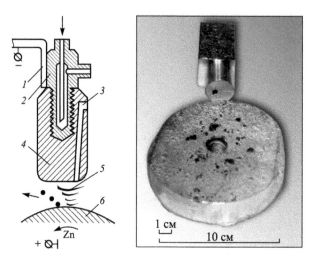

Fig. 10.3. Diagram and appearance of the heater for the disc. *1* – the holder; *2* – cooling unit; *3* – the hole for input of the plasma-forming gas; *4* – a zinc electrode; *5* – electric arc; *6* – the rotating sprayed.

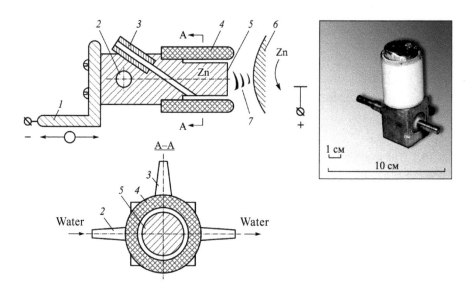

Fig. 10.4. Diagram and appearance of the stationary electrode with the annular neutral gas flow. *1* – motion input; *2* – hole with cooling tubes; *3* – neutral gas input tube; *4* – ceramic tube; *5* – electrode; *6* – sprayed disc; *7* – the area of the electric arc

In the experiments to obtain trhe zinc powder the heating source was also a plasma torch (plasmatron), whose scheme is shown in Fig. 10.6. The gas, supplied through the pipe *2*, passes through the flow twisting device *3* into the zone of the electric arc burning between the cathode *4* and the anode *6*, is heated, through a nozzle in the anode reaches the surface of the spining

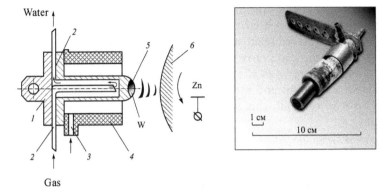

Fig. 10.5. Diagram and appearance of the stationary electrode with a tungsten insert. *1* – zinc electrode; *2* – cooling pipe; *3* – the plasma gas input; *4* – a ceramic tube; *5* – tungsten insert; *6* – sprayed disc.

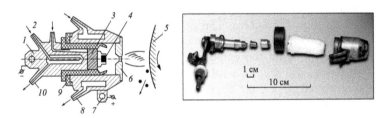

Fig. 10.6. Diagram and appearance of the plasma torch PVP-1. *1, 7* – electrodes connected to a current source; *2* – gas inlet; *3* – flow twisting device; *4* – zirconium insert; *5* – rotating disc; *6* – anode; *8, 10* – cooling water tube; *9* – insulators.

disc *5*, heats it and melts. Pipes *8* and *10* are used for cooling, the electrodes *1, 7* – to connect to a power source, aand washer *9* – for insulation. On the right there is the appearance of parts of the plasma torch. Main technical characteristics of the PVP-1:

The highest rated current, A	250
The open circuit voltage, V	180
Maximum arc power, kW	30
Air flow, m^3/h	2–5
Operating air pressure, kPa (kgf/cm^3)	9.8–294 (0.1–3.0)
Cooling water pressure, not less, kPa (kgf/cm^2)	245 (2.5)
Water consumption, not less, m^3/h (l/min).	0.15 (2.5)
The diameter of the cathode insert, mm	1.5

10.4. The processes at the electrodes

When using an electric arc for heating and melting the spinning disc the electrode processes largely determine the quality of the powder. In order to obtain a powder with a given fractional and chemical composition it is necessary to control the parameters of the arc in terms of erosion of electrodes with high chemical activity. The peculiarity of the process in that the surface of one electrode (anode) is moved relative to each other. Travel speed $v' = 2\pi R\omega$, where R is the radius of the disc; ω is the frequency of rotation of the disc. When $R = 0.1$ m, $\omega = 3000$ rev/min $v' = 31.4$ m/s. The experiments were determined by the current in the arc and the operating voltage on the electrodes at the frequency of disc rotation, calculated for zinc particles with the size of $d = 0.1$–0.3 mm.

A. The experiments were conducted in air at atmospheric pressure. The fixed electrode, made of ultrapure zinc (see Fig. 10.3), was installed at a distance $l_d \approx 20$ mm from the disc. Once the disc rotation speed was stabilized, the electrode was moved up to the spinning disc by using a coordinate device and by touching the electrodes the electric arc was ignited. Then, the distance between the electrodes was increased, keeping the arc steady burning. A bright glow was observed and a characteristic crack was heard. Arc power was increased until the start of erosion of the spinning electrode. Droplets of liquid zinc scattered in the form of a fan on the disc. The scattering process was observed through the filters. The operating voltage $U_a = 50$–60 V, arc current $I_a = 120$–150 A, power $P_a = 6$–9 kW were measured. For stable arcing it is necessary to maintain the cathode–anode distance equal to 2–2.5 mm. At greater distances the arc, as a rule, goes out. It was found that the fixed electrode is destroyed during the operation, with one electrode dispersing 3–5 discs. After that, the electrode should be replaced. Cooling of the electrode has a positive effect on the duration of its work, but as a result of the destruction of the electrode, water can get into the working chamber. The water, circulating directly in the body of the electrode, sometimes seeps through the cracks in the material (zinc). Figure 10.7 shows a stationary electrode after spraying. Traces of erosion of the cathode can be seen, and in the end part there is a hole for cooling. Figure 10.8 shows the sprayed discs. The disc on the left was obtained in the experiment when the electrode was fixed above the plane of the disc at an angle $\alpha \approx 40°$; the disc on the right – $\alpha = 0°$. The thread used to mount the disc to the rotation shaft almost completely disappeared.

1 см

10 см

Fig. 10.7. Stationary electrode after spraying.

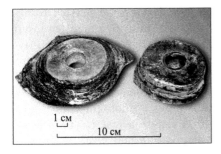

Fig. 10.8. Rotating electrodes (discs) after dispersion.

Fig. 10.9. Discs dispersed in different atmospheric conditions. Left – air with water steam, right – the air with an excess of nitrogen.

Probably the disc temperature was sufficient for the disc material to move into the area of plastic deformation.

The uneven dispersion of the first disc can be explained by 'wobbling' due to a lack of strong attachment of the disc to the rotation shaft. The surface of the discs in places of the effect of the arc is covered with zinc oxide – white patches (sometimes yellow), zinc nitrides – black film having good adhesion to the disc. In some places, the shiny surface of a pure metal is visible.

Figure 10.9 shows two discs sputtered at different atmospheric conditions. Left – in air with water vapour, with the zinc particles falling into the receiver with de-ionized water, right – spraying with an excess of nitrogen. The first disc is covered mostly with oxides, the second – nitrides.

Given that spraying in the air is accompanied by chemical reactions with the formation of undesirable compounds such as oxides and nitrides, spraying was carried out in argon. Experiments have shown that with an increase in the difference between the voltage across the electrodes the arc current increases more rapidly than in the experiments in air. Upon reaching the power necessary to spray the disc the current was increased to 200–300 A which resulted in the start of a catastrophic destruction of the cathode. Operating voltage was $U_a \approx 30$ V.

B. To increase the voltage drop in the arc, a plasma torch was used in which the cathode is made of ultrapure zinc with the gas blown onto it to form the channel of the arc (see Fig. 10.4), while the anode is the spinning disc. It was found that the ceramic tube *4* (see Fig. 10.4) is rapidly destroyed. The arc 'washes' disc from all sides and the disc can not be dispersed thoroughly. The chosen system works but the design needs to be improved. The cathode is exposed to erosion. Figure 10.4 shows the destruction of the cathode after the first few seconds of arcing.

The use of ultrapure zinc as an electrode (cathode) in the centrifugal method for obtaining zinc powders prevents the introduction of metal impurities (subject to other conditions), but the erosion of the zinc in the arc is extensive. The resulting zinc powder meets the requirements of the chemical composition and size.

C. Experiments with spraying were conducted using the plasma torch, where the fixed electrode is made of a cooled copper rod with a tungsten insert – cathode (see Fig. 10.5). In operation, the device showed high reliability: the arc is characterized by a steady burning when the cathode–anode distance is changes and the cathode material is damaged only slightly. The surface of the disc, treated with the arc, is clean, slightly dull, sometimes shiny. Black zinc nitride films do not form. Figure 10.10 shows a photograph of a zinc disc after spraying with a plasma torch with a tungsten insert. The powder meets the requirements of the chemical composition and size. Arc power P_a = 6.3 kW, arc current I_a = 180 A, the voltage drop in the arc U_a = 35 V, speed 4000 rpm.

D. In order to simplify the technology for producing the ultrapure zinc powder, the heating source for the spinning disc was a plasma torch of a PVP-1 semi-automatic machine for the air-plasma cutting of metals (see Fig. 10.6). The air was replaced with argon, but the zirconium insert quickly failed. The discs treated with the plasma torch are shown in Fig. 10.11. The resulting zinc powder has an admixture of copper, which comes into the powder from the partial destruction of the copper anode. The disc surface treated with the plasma torch is clean, free from black crusts of nitrides and white or yellow particles of zinc oxide.

Of the four heating devices (A, B, C, D) the most suitable system for the production of the powder turned out to be the scheme (B) of the plasma torch with the Zn cathode and the scheme (C) of the plasma torch with a tungsten insert. The quality of the powders, obtained by using these devices, meets the requirements of technical specifications.

10.5. Analysis of the powder. Size, shape, chemical composition

The main characteristics of the zinc powder are the size distribution and chemical composition. The particle size prescribed by the technical conditions should be 0.1–0.3 mm, the percentage of iron in the powder – not more than 0.0005%.

The particle size was determined using a microscope with an object micrometer (MBI-15, MBS-9) and using a measuring microscope. The size distribution was determined by sieving the powder through a calibrated screen.

1 см
10 см

Fig. 10.10. Drive after spraying with a plasma torch with a tungsten insert.

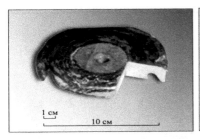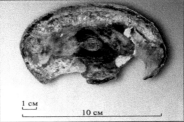

Fig. 10.11. A disc after spraying with a PVP-1 plasma torch.

Fig. 10.12. Zinc powder particles obtained in the experiment with zinc electrode into the spray air. *a* – the needle–shaped particles; *b* – teardrop–shaped particles close to the spheres; *c* – plate deformed drops.

The chemical composition was determined by the colorimetric method as described in [37, 38].

Figure 10.12 shows the powder obtained in the experiment with a zinc electrode with dispersion of the zinc disc in an air atmosphere. The powder contains zinc particles of characteristic shape: needle-like droplet-like particles that are close to spheres, elongated droplet-like particles, plates, deformed droplets. The length of the particles reaches 1 mm, the thickness (diameter) 0.5 mm. In the powder there are zinc particles in the form of zinc oxide in the form of thin films, plates, spheroids. The colour of these particles is predominantly white, rarely – yellow. There are also black particles, presumably coated with zinc. The most common black films cover a plate of zinc nitrides. The chemical composition of the powder meets the requirements (Fe $\leq$ 0.0005%) in cases where the experiments were performed in a chamber with an atmosphere of air passing through a cloth filter (cotton, gauze). In cases where the experiments were performed without a filter, in an atmosphere of room air, the powder had a total iron concentration higher than normal. The particles, shown in Fig. 10.12, were obtained by spraying a spinning disc with a diameter of 130 mm, speed 3500 rpm, arc current 200 A.

Figure 10.13 shows the zinc powder particles obtained in the experiment with a zinc electrode. The particles, detached from the melting spinning disc, fell into the receiver of the powder with deionized water. The distance between the axis of the disc and the water surface was 250 mm. Impact of liquid Zn drops on the water surface produced particles in the form of flakes.

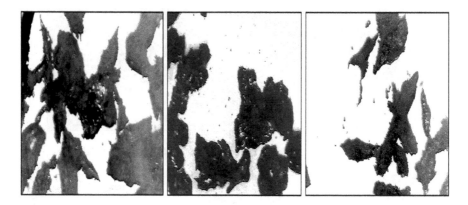

Fig. 10.13. Zinc powder particles obtained in the experiment with the zinc electrode. Zinc particles after falling in the receiver with deionized water.

The particles were coated with zinc oxide, which in the form of white, yellow, films and lumps coated the zinc particles. The length of the particles varied up to 1 mm, the thickness of the particles was 0.1–0.2 mm. Many of the particles were in the form of broken shells – concave plates with torn edges. Flat particles were found only rarely. The particles in Fig. 10.13 were obtained at a frequency of rotation of the disc n = 5000 rpm, arc current 100 A. The particles which hit the surface of the chamber walls were in the form of flat circular flakes. The deformation of the particles decreased with increasing length of their flight. When collecting the particles in the vessel with deionized water the chemical analysis of iron was satisfactory. When the deionized water was replaced by tap water, the iron concentration increased dramatically.

Figure 10.14 shows the particles of zinc obtained in the experiment with a zinc electrode in the setup with the plasma torch (see Fig. 10.4). Argon was blown on the zinc electrode. The arc was stable, the voltage drop in the arc was up to 70 V, current 150 A. When the frequency of spinning of the disc was ω = 3000–3500 rpm the powder contained particles with a size of 0.1–1.0 mm, with droplet, needle and complex shapes. In the atmosphere of filtered air the powder contained iron in the amount not exceeding the norm. The content of zinc oxide was less than 1%.

Figure 10.15 shows the particles obtained by spraying a spinning disc when it is heated by the plasma torch with a tungsten insert (see Fig. 10.5). The plasma-forming gas was argon or helium. The zinc particles had a shiny surface.

The particle size decreases with increasing frequency of rotation of the disc and this is accompanied by an increase of the number of particles whose shape is nearly spherical. The particle images were obtained at a frequency of rotation of 4000 rpm, the voltage drop in the arc 60 V, the current – 180 A The iron content in the powder 0.0003–0.0004%, zinc oxide no more than 1.1%.

Figure 10.16 shows a zinc powder produced during by spraying a spinning disc is heated by the plasma torch of the semiautomatic equipment for air-

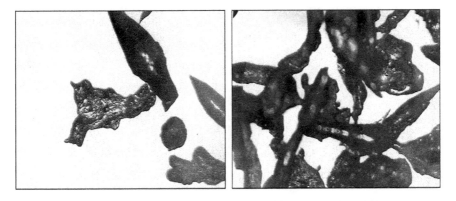

Fig. 10.14. Zinc particles obtained in experiments with zinc electrode (Figure plasmatron)

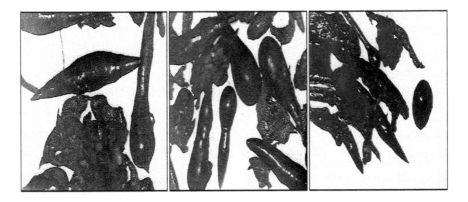

Fig. 10.15. Particles of zinc obtained by the plasma torch with a tungsten insert.

plasma cutting of metals. Different particle shapes were found: needle-shaped, droplet-shaped, spherical, greatly deformed particles, plates. The content of the spherical or almost spherical particles did not exceed 10%. The majority particles had a shiny surface. The content of the zinc oxide particles was 0.8%. The maximum particle size was 1.5 mm, the minimum – 0.1 mm. The plasma torch worked in the air as the plasma-forming gas. The zinc particles were collected at various distances from the disc. The particles collected at a distance of 1.5 m showed a developed surface, at a distance of 2 m the particles were droplet-shaped. The frequency of rotation was 3000 rpm. The admixture of iron was not more than 0.0005%.

Figure 10.17 shows the size distribution of the zinc powder obtained by spraying the disc heated with a plasma torch with a tungsten insert. It is seen that 85% of the powder particles have a size of 0.1–0.3 mm. The rotation frequency was calculated in advance and was equal to 4000 rpm. The arc current was 180 A, the voltage drop in the arc 60 V, the power 10.8 kW.

Comparing the results, obtained in different experiments, one can argue that any variant of the tested method of producing powders and granules, based

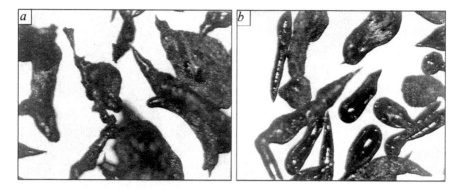

Fig. 10.16. Particles of zinc obtained using the plasma torch PVP-1. Particles collected at different distances: a – 1.5 m, b – 2 m

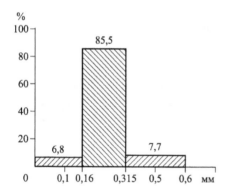

Fig. 10.17. The size distribution of zinc powder obtained by spraying the disc heated with the plasma torch with a tungsten insert.

on spraying a spinning melting rod or disc, can be used to produce the zinc powder with the particles size of 0.1–0.3 mm, with an iron content of not more than 0.0005%. The disc can be heated with the electric arc ignited between the spinning disc and a stationary electrode, made of ultrapure zinc. Heating can also be carried out using a plasma torch, including industrial torches, such as PVP-1. The particle size is regulated by the number of revolutions of the sprayed disc with a constant radius. The chemical composition of the zinc powder meets the requirements in the case when the spraying takes place in a controlled (filtered air) or inert atmosphere.

10.6. Preparation of powders and granules of metals using induction heating

To reduce contamination of the sprayed metal, it is possible to obtain metal powders and granules by spraying a rotating rod, melted by a high-frequency heat source. The end surface of a rotating sprayed rod is heated with high-frequency electromagnetic field (HEMF). The frequency of HEMF must be such that the depth of heating rod δ_Δ (skin depth) was significantly less than its diameter and should be about 0.1–0.001 of the diameter of the rod, preferably

0.01 diameter, to ensure that the rotating shaft does not lose its mechanical strength. If the frequency of HEMF is be such that the magnitude of the depth of heating is close to the diameter of the rod, a significant portion of the volume of the rotating rod during heating to the melting point can become unstable and collapse prior to centrifugal atomization. The depth of penetration of HEMF into the metal depends on its physical properties [39]:

$$\delta_\Delta = A\left(\frac{\rho'}{\mu f''}\right)^{1/2},$$

where A is a constant; ρ' is electrical resistivity; μ is magnetic permeability; f' is the frequency of the electromagnetic field. The power of the electromagnetic field should be sufficient to melt the rotating rod and to form of a liquid metal film on its front surface. In the process of producing the powder the centrifugal force acts on the liquid metal and leads to the formation of droplets which cool down in flight.

In the process of heating the material of the rod and producing the powder the heat source does not introduce any impurities into the powder. Heating of the sprayed material is such that the temperature in the zone of formation of the powder does not exceed greatly the melting point of sprayed material, which reduces the probability of formation of impurities such as oxides, nitrides, carbides, taking place in processes of production of the powders, using the electric arc or plasma torch for heating [13, 16–18].

Figure 10.18 shows the various devices that implement the method. Figure 10.18 b shows the equipment for obtaining a powder of a high purity metal by melting a spinning rod. The rod 1 is melted on the end surface 2 by the HEMF emitted by the inductor 3. Under the influence of centrifugal force, a droplet of liquid metal 4 leaves the spinning rod and hardens in flight in the air. Figure 10.18 c shows simultaneously rotating melting rods 1. The high-frequency inductor 3 transfers energy to the rods in such a way that the end faces 2 facing each other, are heated and fused. Under the effect of the centrifugal force the droplets 4 detach from the rods, cool down and solidify in flight in a space filled, for example, with the neutral gas. By varying the frequency of rotation of the rod with a constant radius, one can obtain a powder with the desired dimensions.

The device shown in Fig. 10.18 a was used to produce the powder of ultrapure zinc. A rod with a diameter of 70 mm, made of ultrapure zinc, was placed at a distance of 3–5 mm from the high-frequency single-turn coil 3 with an inner diameter of 80 mm. The frequency of the high-frequency field $f' = 440$ kHz, the power of high-frequency equipment 10 kW. Rod 1 was rotated at a frequency of 5000 rpm by the power supplied to the inductor. summed output to the inductor. After a time $\tau \approx 20$ s droplets of molten zinc started to separate from the rod and solidified in flight. The particle size was 100–500 μm. The chemical composition of the powder corresponds to the chemical composition of the starting material. The shape of the particles is shown in the photograph (Fig. 10.19). It is seen that the shape of the particle is close to the spheroids. There are no particles in the form of hard deformed drops, crusts

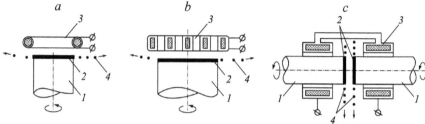

Fig. 10.18. Devices that implement the method for producing metal powders and granules, with the use of the high-frequency electromagnetic field. *1*– sprayed material (disc, rod); *2* – melted part; *3* – inductor; *4, 5* – metal particles.

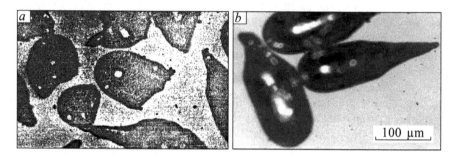

Fig. 10.19. Particles of zinc powder obtained by spraying a rotating shaft.

and needles. In Fig. 10.19 *a* the particles resemble the shape of cylinders, whose length is not significantly greater than the diameter. Among the particles there are pieces of zinc, separated from the disc during rotation (not shown in the photographs). The dimensions of the pieces are substantially (10 times) greater than those of the zinc particles. Their presence can be explained by the non-optimal regime of heating of the sprayed material. Figure 10.20 shows the disc after the sputtering process.

The method of induction heating the melting disc has an advantage compared with the above methods, described in [1, 2, 40], since the contact of the sprayed material with the material of the device involved in spraying is excluded. In comparison with the centrifugal method of production of powders and granules [2, 13], where the heat source is the electric arc or plasma torch, the induction method allows to obtain powders of metals with the chemical composition only slightly different from the chemical composition of the starting material.

10.7. Discussion of the results

In comparative experiments various schemes of heat sources were tested: A – the electric arc burning between the stationary electrode, made of the sprayed material, and a rotating disc, B – a plasma torch with an electrode of the ultrapure zinc, C – a plasma torch with a tungsten insert, D – a plasma torch of the semi-automatic equipment for air–plasma cutting. All the schemes have shown the suitability for the production of zinc powders alloy with the iron

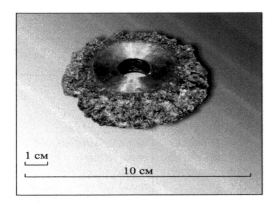

1 см

10 см

Fig. 10.20. Disc of ultrapure zinc after the spraying process.

admixture of no more than 0.0005%. For reasons of convenience in practical realization, the best results were obtained for the plasma torches with a zinc electrode, the index B, and with a tungsten insert at the anode, the index C. The atomized disc was also heated used the high-frequency electromagnetic field, the index D. A summary of the test results is shown in Table 10.1.

The average particle size is mainly determined by the frequency of rotation and the sprayed disc diameter. It was established experimentally that the relationship of these parameters obtained in [13] satisfactorily explains the results. In order to obtain the desired particle size range, the frequency of rotation of the disc should be n = 3000–5000 rpm, with a radius of the disc R_0 = 60 mm. Regardless of the method of heating the size of the resulting powder can be controlled relatively easily. However, the shape of the particles in this case will depend on the cooling conditions. To obtain non-deformed particles, the minimum distance of flight of particles needed for the crystallization of a liquid zinc droplet should depend to a certain extent on the speed of the discs at a known radius. It was established that the resulting connection allows to choose correctly the distance at which the deformation of the cooled particles is insignificant.

The presence of needle-like particles in the powder is determined by the mechanism of formation of a particle in separation in the conditions when there is no superheating of the molten metal above the melting point, and the melt surface is covered with oxide films significantly altering its surface properties. The details of the separation of droplets are not considered here. In order to increase the proportion of spheroidal particles in the resulting powder, it is necessary to find ways to increase the superheating of the zinc melt on the surface of the melted electrode to the formation of particles and reduce the concentration of oxidants in the atmosphere in which melting and spraying are carried out.

The chemical composition of the powder with a specified minimum content of impurities is much more difficult to obtain, given the many factors that contribute to the introduction of impurities in the sprayed material. Metals, zinc in particular, at high temperature have a high chemical activity, so that it is necessary to create certain conditions for dispersion, which would take

Table 10.1 Summary table of test different types of heating melts the rod or disc

Index	Figure number	Name	Cleanliness	Particle Shape
A	10.3	Fixed homogeneous electrode, electric arc	Oxides and nitrides; Fe ≤ 0.0005% when particles fall in deionized water, filtering the atmosphere	0.1–1.0 mm, needle-shaped droplet particles, droplet elongated, plates, deformed droplets, concave plates with ragged edges, flat round flakes
B	10.4	The plasma torch with a zinc electrode, argon	Zinc oxide less than 1%, Fe ≤ 0.0005%	0.1–1.0 mm droplet and needle-shaped particles
C	10.5	The plasma torch with a tungsten insert, argon, helium	Fe = 0.0003–0.0004%, zinc oxide no more than 1.1%	0.1–1.0 mm, needles, tailed drops, particle of complex forms with a shiny surface
D	10.6	PVP-1 plasma torch for air plasma cutting of metals	Admixture of copper, iron, not more than 0.0005%, 0.8% zinc oxide	85% of the particles have a size of 0.1–0.3 mm, needle-shaped, droplet-shaped, spherical, not more than 10%, heavily deformed particles, plates
E	10.18	Inductor	No impurities	0.1–0.5 mm, cylindrical droplets with tails, spheroids, pieces of zinc

this into account. Spraying should be conducted in a neutral atmosphere with preliminary evacuation of the installation. Impurities introduced by the heat source material should be kept to a minimum. The experiments showed that satisfactory results (by traces of iron) can be obtained even with a heat source using the scheme of the plasma torch with air as the plasma-forming gas at moderate electric arc currents.

Analyzing the obtained powders of zinc, one should pay attention to its main characteristics: the average size and chemical composition. No less important properties of the powder, such as bulk density, particle shape, flow, were not central to this stage. The measurement of these properties is also required to further improve the production technology of powders by centrifugal methods.

To summarize this phase of research, it is possible to formulate a conclusion that, to obtain highly pure zinc powders, it is desirable to eliminate the plasma as an environment conducive for high-temperature chemical reactions. The process of particle formation should be performed at a certain degree of overheating of the melt. Table 10.1 shows that the powders obtained using the induction heating method have the most spherical particles. In addition, the induction heating method eliminates the introduction of impurities of materials of equipment. By these criteria, the use of the high-frequency electromagnetic field is the most promising and attractive. In addition to the melting of the sprayed material, the high-frequency electromagnetic field can also heat the particles of the melt at the time of separation, or in flight, which contributes to spheroidization of the powder particles.

10.8. The recommendation to create industrial plant

The test result allow us to formulate the requirements that must be met for the installation for producing the ultrapure zinc powder:

1) the size of the installation must be such that the melt drops could crystallize in the air, to cool down so that after hitting the surface of the chamber the deformation of the particles was negligible, the length of the path of zinc particles should not be less than 2 m;

2) the atmosphere in the system should be neutral, with a high degree of purification to remove oxidation impurities and dust inclusions;

3) the chamber should be vacuum-tight to allow preliminary pumping of of atmospheric gases and prevention of leakage of harmful gases and aerosols;

4) to increase efficiency and productivity it is necessary to provide a bunker for blanks in the chamber and a device for quick reinstallation of blanks on the rotating shaft;

5) to increase the range of possible modes of operation, the installation must be equipped with several method of heating melts of the material and variations in the composition of the neutral atmosphere in the chamber;

6) the system must be cooled in accordance with the power of the methods used for heating capacity the melted material;

7) due to the high rotational speed of the disc (3000–5000 rpm) the rotation must be reliably fixed; protection from emergency situations with a rotating rod or disc must be provided.

Conclusion

The task of obtaining ultrapure zinc powders was solved by analysis of the existing technologies to obtain powders of the controlled chemical composition and dispersion. Mathematical models of the mechanism of formation, cooling, deformation of the particles allowed to satisfactorily describe the experimental results, plan the work ahead and carry out it according to the task. The fundamental question regarding the possibility of using the method of spraying a rotating molten rod (disc) for producing the zinc particles with a size of 100–300 μm and the iron content in the powder not higher than 0.00055%, can be considered solved. Attention should then be given to the control of tyhe dispersion and shape of the particles, lowering the concentration of zinc oxides and nitrides in the powder.

List of symbols

A	– an experimental constant;
d	– diameter of the particle, m;
f	– the frequency of the electromagnetic field, s^{-1};
I_d	– the arc current, A;
l	– length, m;
l_d	– the distance the rod electrode, m;
n	– rotation speed, rev./min;
P_d	– arc power, W;
R, D	– the radius and diameter of the rod, m;
r	– radius of the drop, m;
r_1	– the latent heat of fusion, J/kg;
T	– temperature, K. °C;
U_d	– electric arc voltage, V;
v	– velocity, m/s;
α	– the average heat transfer coefficient, $W/(m^2 \cdot K)$;
δ	– thickness, m;
δ_Δ	– skin–layer thickness, m;
η	– dynamic viscosity, Pa $\cdot$ s;
λ	– thermal conductivity, $W/(m \cdot K)$;
μ	– magnetic permeability, H/m;
ρ	– density, kg/m^3;
ρ'	– electrical resistivity, Ohm$\cdot$m;
σ	– the surface tension, N/m;
τ	– time, s;
ω	– frequency of rotation, s^{-1}.

Indices:

g – gas. scull.
d – electric arc
a – room;
mp – melting.

p – melt;
cp – average;
r – radius of the drop;
0 – initial.

Similarity criteria:

$Nu = \alpha l/\lambda$ – Nusselt number;
$Pr = v/a$ – Prandtl number;
$Re = \rho v l/\eta$ – Reynolds.

Literature

1. Powder metallurgy of materials for special purposes, Ed. J. Bark, V. Weiss, Moscow, Metallurgiya, 1977.
2. Nichiporenko O.S., et al., Dispersed metal powders, Naukova Dumka, 1980.
3. Kace Kaopy, et al., J. Jap. Soc. Powder Met., 1970, V. 16, No. 8. 338–344.
4. Altunin Yu.F., et al., Tekhnol. Legkikh Splavov, 1970, No. 2, 120–126.
5. Abkowitz S., Metals Progress, 1966, V. 89, No. 4, 62–65.
6. Abkowitz S., J. Metals, 1966, V. 18, No. 4, 458–464.
7. Pat. 52–25826 Japan. Publ. 07/09/1977, Bulletin, No. 2–646.
8. Pat. 2401723 France, Procede et dispositif pour la fabrication de particules sferiques, Devillard Jacques, Publ. 05/04/1979, Bulletin No. 18.
9. Pat. 2532875 Germany, Verfahren und vorrichtung zum tiegellosen granulieren von metallen und metallegierungen, Winter Heinrich, Ruckdeschel Walter, Publ. 04/23/1978.
10. Pat. 2210451 Germany, Einrichtung zur fienpulverherstellung aus fluessigem metal, Glasunow S.G., Publ. 07/31/1975.
11. Author Cert. 534.304, USSR, Installation for a powder, I.A., et al., Publ. 05.11.1976, Bull. No. 41.
12. Pat. 2.101.657 France, Method for production of powders and device for its implementation, Publ. 05.05.1972, Bull. No. 18.
13. Pat. 3099041 USA, Method and Apparatus for Making Powder, A.R. Kaufmann, July 30 1963.
14. Buffed A.S., Gummeson P.U., Metal Prog., 1971, V. 99, 68.
15. Pat. 2253591 France, Improved device for the production of powders by the melted electrode. Publ. 08.08.1975, Bull. No. 32.
16. Author Cert. 497097 USSR, N.F. Anoshkin, et al., Publ. 12/30/1975, Bulletin No. 48.
17. Pat. 2253591 France. Perfectionnement des machines á fabriquer des poudres par fusion d'une électrode fournanre. 07/12/1973.
18. Author Cert. 534.304m USSR. Publ. 11/05/1976, Bulletin, No. 41.
19. Hodkin D.J., Powder Metall., 1973, V. 16, No. 32, 277–313.
20. P. Bär, Doct. Dissert. Tech. Coll. Karlsruhe, 1935.
21. Walton W.H., Prewett W.C., Proc. Phys. Soc. London, B, 1949, V. 62, 341–350.
22. Muraszew A., Engineering, 1948, V. 166, 316.
23. Fraser R.P., Eisenklam E.P.. Trans. Inst. Chem. Eng., 1956, V. 34, 294–307.
24. Wilson B.T., Precision Metals, February, 1968, 31–35.
25. Friedman G., Metal Progress, 1975, V. 107, P. 81–84.
26. Pat. 3784656 USA. Method of Producing Spherical Powder by Excentric Electrode Rotation, A.R. Kaufmann, 12/03/1971.
27. Pat. 5038074 Japanese. Power supply for automobile heater parts, Iwata Toshio, Kaneyuki Kazutoshi, 12/02/1993.
28. Kononov I.A., Musienko V.T., Treatment processes of light and heat resistant alloys, Moscow, 1981, 205–212.
29. Kozlov V.A., Golubkov V.G., Poroshk. Metall., 1981, No. 3, 1–6.
30. Jones V.D., Fundamentals of Powder Metallurgy. Production of metal powders. Springer Verlag, 1965. 200 s.
31. Pat. 746 301 England. Atomizing magnesium. 14.03.1956; Pat. 754 180 England. Atomizing aluminium or aluminium alloys. 01/08/1956.
32. Pat. 2825108 USA. Metallic Filaments and Method of Making Same, R.B. Pond, 04/03/1958.

33.   Pat. 0702736 USA. Apparatus for reducing fusible materials to dust, A.F. Madden, 17/06/1902.
34.   Pat. 1883/633 England, W. Cross.
35.   Kutateladze S.S., Fundamentals of the theory of heat transfer, Nauka, Novosibirsk, 1970.
36.   Kutateladze S.S., Borishanskii V.M., Handbook of heat transfer, Moscow, Leningrad, Gosenergoizdat 1959.
37.   Bulatov M.I., Kalinkin I.P., Practical guide to photometric and spectrophotometric methods of analysis, 4th ed., Leningrad, Khimiya, 1976.
38.   Charlo G., Methods of Analytical Chemistry. Quantitative Inorganic Analysis Connects, Translated from French, ed. Yu.Yu. Lurie, Moscow, Khimiya, 1966.
39.   Brokmayer K., Induction melting furnaces Moscow, Energiya, 1972.
40.   Powder metallurgy of superalloys and refractory metals, V.S. Rakovski, et al., Moscow, Metallurgiya, 1974.

11

Some aspects of coalescence processes in aerosol systems

The metastable state of aerosol dispersion systems, consisting of excess of the surface energy of the particles, is supported by the disunity of the particles in space. When the particles collide with each other they may merger, which is an elementary act of relaxation of metastable states to stable states. In the aerosol dispersed systems, including solid particles, the result of their merger may be the formation of conglomerates, for example, as shown in Fig. 4.7 in chapter 4. In systems with liquid aerosol droplets leads to an enlargement of the latter. In the processes of producing powders by spraying the substance in the liquid form the coalescence reduces the efficiency of the process and leads to a distortion of the fractional composition of the obtained powders.

Investigations of processes of coalescence, due to their complexity, was reduced until recently mostly to the introduction of the coefficient of efficiency of coalescence in collision of the particles. The physical content of this factor remains unclear. In this paper, we present the results of studies that can provide practical guidance for determining the coefficients of the efficiency of collisions, and highlight some of the physical phenomena involved in the process of coalescence. We investigate the role of the vapour layer that forms in the collision of particles of a disperse system with each other or the surface of a solid or liquid phase. Along the way, we present some results that represent physical interest.

The possible existence of a vapour layer between a hot body and a spheroidal liquid at temperatures lower than the boiling point, at low body–liquid temperature gradients, is shown in [1], which also describes the mechanism of this effect. In the case of experiments in an atmosphere saturated with vapours of the liquid, a high pressure associated with the evaporation of the liquid is generated in the layer. At surface temperatures of the solid above the boiling point, higher pressure is generated in the layer associated with evaporation of the liquid. At temperatures of the surface of the solid higher than the boiling point the layer is characterized by the flow of the vapour which creates a pressure gradient to support the weight of the droplet. For this temperature range the theories developed in [1, 2] agree with the experiment. At temperatures of the solid surface and droplets below the boiling point the excess pressure in the vapour layer is associated with the diffusion of vapour into the environment. The diffusion process in the vapour layer and the non-isothermal state of the solid surface and the droplet are affected by the saturation of the environment with liquid vapours [1]. In [1] the authors also predict the minimum temperature below which the spheroidal state of liquids can not be obtained. However, in experiments with the spheroidal state of the liquid on the body surface the realization of the predicted relatively small gradients, in which the liquid is separated from the body by the vapour layer, meets with considerable difficulties. As shown in [1], the difficulty in the experiments lies in defining the influence of vibration of the liquid droplet and the surface roughness of the body. The practically achievable purity of polished surfaces is not sufficient for the stationary existence of a vapour layer at low temperature gradients due to its thickness being commensurate with the size of the roughness.

The next section discusses the possibility of the existence of a vapour layer in a wide temperature range and interacting surfaces.

11.1. Experimental studies of the interaction of the liquid with hot solids

When heating a solid to a sufficiently high temperature, a vapour interlayer, preventing wetting, forms between its surface and the liquid bordering with it. When lowering the solid temperature the destruction of the vapour layer is due to the loss of its hydrodynamic stability. Surface roughness contributes to the destruction of the vapour layer and wetting. In this regard, it is interesting to consider the question about the wetting of a heated solid in the absence of destabilizing factors – significant hydrodynamic pulsations and surface roughness.

The aim of the study is to investigate experimentally the existence of a vapour layer and its destruction in the conditions of gravitational stability and the absence of pulsations associated with the evacuation of vapour and surface roughness.

11.1.1. Experimental procedure

The experimental setup, shown in Fig. 11.1, was selected A copper rod *1*, equipped with a heater *2*, thermocouples *4*, vertically immersed in a liquid, was placed in a glass cuvette *6*. The rod was equipped with thermal insulation *5* and a copper jacket *3* to reduce heat leakage through the lateral surface. For visual observation of the vapour layer the cuvette was placed on the lens of an optical microscope *7*. The liquid level in the cuvette was maintained in the focal plane of the microscope. The heated rod was moved relative to the cuvette with a micrometer screw of the microscope table with a scale division of 2 μm. The distance between the working end of the rod and the surface of the liquid was determined with a probe *8*, taking into account thermal expansion, the absolute error did not exceed ±5 μm. The diameter of the working part of the copper rod was $d = 2 \cdot 10^{-3}$ m. The heat flow was measured

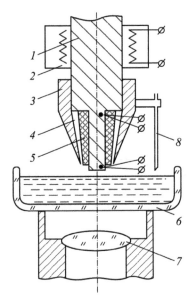

Fig. 11.1. Schematic of the experiment. *1* – copper rod, *2* – heater, *3* – screen *4* – thermocouple, *5* – insulation, *6* – cuvette, *7* – microscope objective, *8* – feeler.

on the basis of the readings of the thermocouples and the known thermal conductivity of copper. The error of temperature measurements was ±0.25 K.

The end of the copper rod was spherically blunted with a radius $r = 2 \cdot 10^{-3}$ m and polished by chemical methods. In the experiments measurements were taken of the average heat flux q through the end surface of the rod depending on the depth of its immersion h in a liquid. At the same time visual observations were carried out of the end of the rod surface in monochromatic light with a wavelength of 561 nm. The temperature of the liquid was maintained at room temperature, $T_l = 291–293$ K. To reduce the effect of vibration and thermocapillary convection, the experiments used water with artificially increased viscosity by the addition of 0.1% polyethylene oxide.

11.1.2. The experimental results

Figure 11.2 shows examples of the results of measurements of the average specific heat flux for different temperatures of the rod end T_w. The measurements were performed in the range $h = 100–170$ μm, $h = 170$ μm corresponded to the complete immersion of the spherical surface of the rod. It is seen that with increasing temperature of the rod the value of q increases. Wetting of the surface of the liquid occurred in different experiments at depths of immersion not recurring from experiment to experiment. For $T_w > 420$ K wetting occurred at $h > 170$ μm, when the sharp edge of the transition of the spherical bluntness to the cylinder was below the liquid level in the cuvette.

The values of specific heat flow, measured at $h = 170$ μm, are plotted in Fig. 11.3 in dependence on $\Delta T = T_w - T_l$. When $\Delta T < 73$ K, when wetting occurred at $h < 170$ m, the values of q were linearly extrapolated to $h = 170$ μm. The open points in Fig. 11.3 correspond to the measured values of q

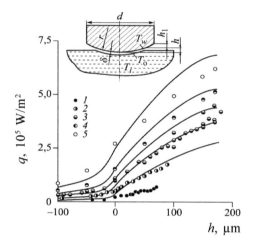

Fig. 11.2. Measurement of specific heat flux in immersion of a rod in a liquid. Solid curves – calculations, points – experiment. Curves *1–5* correspond to T_w = 342; 396; 471; 536; 659 K.

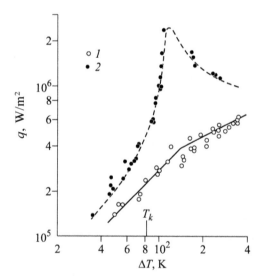

Fig. 11.3. Dependence of the heat flux on ΔT. *1* – in the mode of existence of the vapour layer; *2* – wetted surface. T_k – boiling point.

in the presence of the vapour layer between the rod and the liquid, the dark points – to direct contact with the liquid. It is seen that the upper experimental curve corresponds qualitatively to the usual form of the boiling curve, $\Delta T = 81$ K corresponds to the boiling point of water. The experimental points obtained in the presence of the vapour layer between the liquid and the surface of the rod, show almost the linear dependence of q on ΔT in the range $\Delta T = 50$–160 K and a somewhat weaker dependence at $\Delta T > 150$ K. Measurements of q for $\Delta T < 50$ K were not carried out due to increased errors.

The results of measurements of the heat flux in the presence of the vapour layer on the assumption that the surface temperature of the fluid, bordering with the vapour, is equal to the temperature of the liquid in the volume,

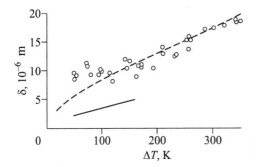

Fig. 11.4. Dependence of the thickness of the vapour layer on ΔT.

were used to calculate values of the vapour layer thickness by the formula $\delta = \lambda'' \Delta T/q$. The coefficient of thermal conductivity of the vapour λ'' was chosen at the average temperature $T = (T_w + T_l)/2$. At $T > 373$ K λ'' of water vapour at a pressure of 10^5 N/m^2 was used, at $T < 373$ K λ'' of air equal to 100% relative humidity was used. The calculation results are represented by the points in Fig. 11.4 as a function of ΔT.

It should be noted that the values of δ, obtained by this procedure, are somewhat too high due to the fact that the actual temperature of the liquid on the surface T_0 with $q \neq 0$ higher than the T_l measured in volume. To assess the accuracy of such experimental data processing, it was assumed that under the same heat fluxes in the case of heat transfer through the vapour layer and in direct contact the surface temperatures of the liquid are equal. This is equivalent to the assumption that the heat in comparable cases is used only to warm the liquid in the cuvette. The surface temperature of the fluid can then be determined from the measurements of T_w for the appropriate values of q in contact of the liquid with the rod. Calculation by the formula $\delta = \lambda''(T_w - T_0)/q$ is shown in Fig. 11.4 as a continuous curve. It is clear that neglecting the part of the heat used for evaporation gives a higher value of T_0 and, therefore, understated value of δ. Thus, the points and the calculated solid line indicate the range in which the actual values of the vapour layer thickness are distributed.

The dashed line in Fig. 11.4 also presents the results of calculations by the formula

$$\delta = 0.5 \left(\frac{\lambda''\left(T_w - T_l\right)d}{l} \right)^{1/2} \left[\left(\frac{2\sigma}{r} + \rho g h \right)\rho''' \right]^{-1/4}, \qquad (11.1)$$

where l is the specific latent heat of vaporization; ρ, ρ'' is the density of liquid and vapour. Formula (11.1) was obtained from the Kutateladze–Borishansky formula [2]:

$$\delta = 1.1 \left(\frac{\lambda''(T_w - T_l)d}{l(1 + c(T - T_l)l} \right)^{1/2} (\rho g \delta' \rho'')^{-1/4}$$

for the spheroidal state of the liquid by replacing the hydrostatic pressure of the spheroid $\rho g \delta'$ by the sum of Laplace pressure drop and the hydrostatic pressure when submerged. The formula (11.1) omits heating of the fluid, reflected by the member $c(T - T_l)/l$. Figure 4.11 shows that the calculation by formula (11.1) agrees with the experimental data at $\Delta T \geq 150$ K. This allows a more detailed analysis of the mechanism of heat exchange between the rod and the liquid.

11.1.3. Processing the measurement results

A model for calculating heat transfer is schematically shown in Fig. 11.2. Between the surface of the rod and the liquid layer there is a vapour layer with thickness

$$\delta = 0.5\left(\frac{2\lambda''(T_w - T_l)\chi_2}{l}\right)^{1/2}\left(\frac{2\sigma}{r}\rho''\right)^{-1/4} \tag{11.2}$$

and the radius χ_2. Here, in comparison with formula (11.1), the hydrostatic pressure of immersion is neglected. Under the assumption that the local heat flux is determined by the thermal conductivity of the vapour gap, the average heat flux at negative h can be calculated by the formula

$$q_{-h}^p = \frac{2\lambda''(T_w - T_l)\left[(r - h)\ln(1 - h_1 / h) - h_1\right]}{h_1(2r - h_1)} \tag{11.3}$$

and at positive h

$$q_h^p = \frac{2\lambda''(T_w - T_l)}{h_1(2r - h_1)}[rh/\delta + (r - h + \delta)\ln(1 + (h_1 - h)/\delta) - (h_1 - h)]. \tag{11.4}$$

In equation (11.4) the first term in brackets reflects the heat flux through the vapour layer, and the rest – through the peripheral portion of the vapour gap. Calculations by formulas (11.3) and (11.4) are plotted as solid curves in Fig. 11.2. It is seen that the divergence of calculations and experiments increases with decreasing temperature of the rod. The most significant difference is observed when $T_w < 150$ K.

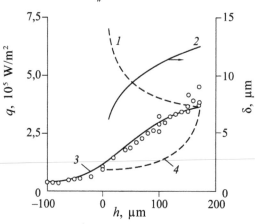

Fig. 11.5. Dependence of the average density of the heat flux, its components and the thickness of the vapour layer on the depth of immersion. $1 -$, formula (11.5); $2 - \delta$, formula (11.2); $3 -$ average heat flux q_{-h}^p, formula (11.3); $4 - q_{-b}^p$, formula (11.6).

Figure 11.5 shows for $T_w < 473$ K the average heat flux (curve *3*) calculated by the formulas (11.2)–(11.4), compared with the experiment.

In these calculations, the formula (11.2) uses the coefficient of 0.11. It also presents the specific heat flux through the vapour layer, calculated as

$$q_\delta^P = \frac{2\lambda''(T_w - T_l)r}{(2r - h)\delta} \tag{11.5}$$

(curve *1*) and heat flux through the peripheral portion of the vapour gap, calculated by the formula

$$q_b^P = \frac{2\lambda''(T_w - T_l)}{h_1(2r - h_1) - h(2r - h)}\left[(r - h + \delta)\ln(1 + (h_1 - h)/\delta) - (h_1 - h)\right] \tag{11.6}$$

(curve *4*). The curve *2* shows that values of δ calculated by the formula (11.2). It is seen that in the whole range of positive h the heat in current q_δ^P is significantly higher than q_b^P. At low positive h the value q_δ^P rapidly increases with decreasing δ. This means that even at small T_w the specific heat flux at the minimum distance between the solid and the fluid has the value sufficient for the formation and maintenance of the vapour layer.

The qualitative agreement between the calculated and experimental dependences of $q(h)$ throughout the entire temperature range can be explained by the fact that defining the parameters, laid down in the estimated model, are geometric. Therefore, despite the quantitative discrepancy between calculations and experiments, the proposed calculation model can be used to analyze the results. The formulas (11.4) and (11.5) and the measured values of q can be used to determine the specific heat flux through the vapour layer for any h as $q_\delta = q(q_\delta^P/q_h^P)$. The result of this processing of the measurements of q at the moment prior to wetting the surface of the rod showed that regardless of the depth of immersion wetting occurs at the average specific heat flux $q_{\delta c}$ of $2.2 \cdot 10^5 \pm 0.9 \cdot 10^5$ W/m². The dependence of the values of $q_{\delta c}$ on ΔT is shown in Fig. 11.6.

11.2. The generality of the phenomena of 'floating droplets', film boiling and coalescence

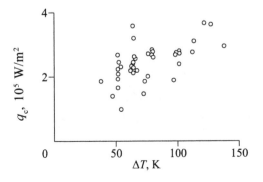

Fig. 11.6. Dependence of the oeat flux through the vapour layer at the moment of wetting the heated surface on ΔT.

Figure 6.11 shows that $q_{\delta c}$ tends to decrease with a decrease in ΔT. The large scatter of points shows that the process of wetting is determined by factors of a random nature. These random factors may include, apparently, the pulsations of the fluid associated with the vibrations of the installation, the surface roughness of the order of several microns, foreign inclusions in the liquid with the size of the same order.

Figure 11.3 shows that when $\Delta T > 142$ K the dependence of the heat flux on ΔT in the case of existence of the vapour layer is close to $q \sim \Delta T^{1/2}$, and at $\Delta T < 150$ K to $q \sim \Delta T$. The shift of the nature of the dependence of q at $\Delta T = 150$ K, can apparently be explained as follows. At high ΔT, the surface temperature of the liquid, facing the surface of a heated body, is close to the saturation temperature; the evacuation of the vapour from the vapour layer is carried out by the convective mechanism. The heat transfer models, based on the convective mechanism of escape of the vapour from the vapour layer, were described in [1, 2]. At $\Delta T < 150$ K, when the temperature of the liquid, which borders the vapour layer, is less than the saturation temperature, the vapour layer is filled with a mixture of air and vapour at a moisture content determined by the process of evaporation and diffusion of vapour out of it. That is, at $\Delta T < 150$ K, the mechanism of escape of vapour from the vapour layer is a diffusion mechanism. The diffusion of vapours of the liquid from the vapour layer into a dry environment creates an excess pressure necessary for the existence of the vapour layer even at temperatures T_w, lower than the boiling point. In our experiments, the vapour layer in immersing the rod into the liquid existed to $\Delta T = 2$ K. In [3] studying the merger of droplets it is shown that the vapour layer can also exist when droplet temperatures are equal.

Thus, in the process of wetting of heated solids by a liquid at any difference of their temperature the fundamental role is played by the vapour layer formed between them, which prevents wetting. All real wetting processes are due to the destruction of the layer due to vibrations or roughness.

In the case when the heated solid with a free surface of the still liquid is replaced by the interaction of the free surface of some amount of the liquid, it is possible to implement the spheroidal state of the liquid at lower temperature

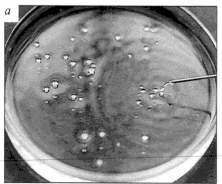
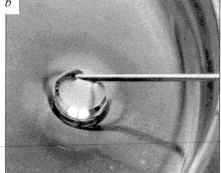

Fig. 11.7. 'Floating droplet' (*a*) and a stationary ethanol spheroid on the free surface of ethanol (*b*).

gradients and in the isothermal case. Figure 11.7 b shows a spheroidal droplet of ethanol with a diameter of $6.3 \cdot 10^{-3}$ m, lying motionless on the surface of ethanol for about 15 seconds. The droplet was obtained using a medical syringe directly on the surface at ambient conditions. Observation of a droplet with a microscope through the transparent bottom of the cuvette in monochromatic light shows the presence of an interference pattern, indicating the existence of the vapour layer with the thickness of the order of a micron between the droplet and the liquid in the cuvette.

In [3] the authors published made an overview of studies of 'floating droplets', started by Makhayan started in 1933. The possibility of prolonged contact of liquid droplets in the air without the merger is also shown and the conditions of their merger are investigated. The role of a viscous air 'cushion' that occurs during the approach of the droplets is determined. Prolonged contact of the droplets without merging is explained by diffusive transport of the vapour of the liquid from the layer.

No data on the existence of 'floating droplets' of the liquid in the medium of its vapour and in non-isothermal cases have been found.

11.3. Coalescence of droplets in non-isothermal conditions

The purpose of this section is to attempt to assess the limit of existence of the vapour layer on the basis of the theory of non-equilibrium evaporation [4]. In [5] the authors demonstrated the applicability of this theory to solve a specific problem – the study of film boiling of He II. In this section we consider the heat and mass transfer between the two droplets of the same liquid having temperature of T_1 and T_2, driven into contact in a saturated atmosphere at a temperature T_1 and pressure p_{s1}. The droplet, which has the temperature $T_2 > T_1$, loses thermodynamic equilibrium with the environment,

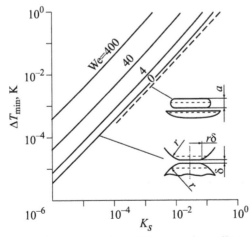

Fig. 11.8. Minimum relative temperature difference depending on the K_s^a complex at various Weber numbers.

and then evaporates. The droplets approach each other with relative velocity u. In the case where the vapour layer of thickness δ remains between them, the droplets are deformed, as shown schematically in Fig. 11.8. We consider the initial moment of deformation of the droplets when $r_\delta \ll r$. In this case, the existence of the vapour layer between them is necessary to ensure that the excess pressure in the layer is consistent with the hypothesis

$$p'' - p \geq 2\sigma/r + \rho u^2 / 2, \quad (11.7)$$

where δ is the surface tension coefficient; ρ is the density of the liquid. If the speed of convergence of the droplets is restricted at the level than the speed of the molecules in the vapour, we can assume that in every moment of the deformation of the droplets in the interlayer the stationary heat and mass transfer takes place between the droplets, described the theory of non-equilibrium processes of evaporation [4]. According to this theory in the vapour environment at distances from the plane of the interface, larger than the thickness of the Knudsen layer, there is the pressure p'', which differs from the saturation pressure and is determined by the specific flux of evaporation and heat flux.

Separately for each drop we can write

$$\frac{p'' - p_{S1}}{p''} = -2\pi \frac{1 - 0.4^2}{2} \tilde{j}_1 - 0.44\tilde{q}_1, \quad (11.8)$$

$$\frac{p'' - p_{S2}}{p''} = -2\pi \frac{1 - 0.4^2}{2} \tilde{j}_2 - 0.44\tilde{q}_2,$$

$$\tilde{j}_i = j_i/\rho'' \sqrt{2RT_i}, \ \tilde{q}_i = q_i / p'' \sqrt{2RT_i}. \quad (11.9)$$

Here, β is the accommodation coefficient of the vapour molecules at the phase boundary; ρ'' is vapour density; R is the gas constant. Signs for $\tilde{j}_i$ and $\tilde{q}_i$ in (11.8) and (11.9) are positive if they are directed deep into the vapour phase. Since in this case $T_2 > T_1$, then j and q are directed from the second drop to the first. In line with this, $\tilde{j}_1$ and $\tilde{q}_1$ are negative $\tilde{j}_2$ and $\tilde{q}_2$ are positive. Provided that the temperatures of the droplets differ only slightly, then it can be accepted $\tilde{j}_2 = -\tilde{j}_1$, $\tilde{q}_2 = \tilde{q}_1$. Then, adding up (11.8) and (11.9) we obtain

$$p'' = \left(p_{s1} + p_{s2} \right)/2, \quad (11.10)$$

i.e. the pressure in the vapour interlayer is equal to the arithmetic mean of the saturation pressure at the droplet temperatures.

11.3.1. Critical temperature drop and critical heat fluxes in coalescence

From (11.10)

$$p'' - p_{s1} = \left(p_{s2} - p_{s1} \right)/2, \quad (11.11)$$

and from (11.11) and (11.7) –

$$p_{s2} - p_{s1} \geq 4\sigma r + \rho u^2. \tag{11.12}$$

From the Clausius–Clapeyron equation in the linear approximation we can obtain

$$p_{s2} - p_{s1} = 2(T_2 - T_1) l \rho'' / (T_1 + T_2),$$

that after substitution in (11.12) gives

$$l \rho'' (T_2 - T_1) / (T_1 + T_2) \geq 2\sigma + \rho u^2 / 2$$

or

$$\Delta \overline{T}_{min} = \frac{T_2 - T_1}{T_1} \geq \frac{4K_s (1 + \text{We}/4)}{1 - 2K_s (1 + \text{We}/4)}. \tag{11.13}$$

Equation (11.13) is a condition of 'non-adhesion' of the droplets as the ratio of dimensionless complexes We $= \rho u^2 r / \sigma$ of the Weber criterion $K_s = \sigma / l \rho'' r$, with the physical meaning of the ratio of the specific volume surface energy and the specific volume latent heat of vaporization.

The ratio of the first term of the right-hand side of (11.8) and (11.9) to the second is of order $p'' / \rho'' l$, which for most liquids at pressures below atmospheric pressure amounts to $\approx 10^{-2}$. With this in mind, from (11.7) and (11.8) we can obtain the condition of non-attachment of the droplets:

$$0.155 K_t \geq 1 + \text{We} / 4, \tag{11.14}$$

where the criterion $K_t = |q| r / \sigma \sqrt{RT_1}$ is the dimensionless complex in the physical sense denoting the ratio of the pressure thermotranspiration (Knudsen effect) and the Laplace pressure.

Equations (11.13) and (11.14) can be used to estimate q_{min} and $\Delta \overline{T}_{min}$ – minimum specific heat flux and the relative temperature difference at which the vapour layer can exist. The minimum thickness of the vapour layer at which the attachment can be defined by (11.13) and (11.14):

$$\frac{\delta_{min}}{r} = 0.62 \frac{K_s}{1 - 2K_s (1 + \text{We}/4)} \frac{\lambda'' \sqrt{T_1}}{\sqrt{R\sigma}}. \tag{11.15}$$

In the case of a spheroid of a liquid having a temperature T_2 and laying motionless on the surface of the liquid at a temperature T_1 in a saturated atmosphere with a pressure p_{s1}, the condition of existence of the vapour layer can be written similarly to (11.12) if we replace the Laplace pressure by the hydrostatic pressure at the bottom of the spheroid:

$$p_{s2} - p_{s1} \geq 2 \sqrt{2\sigma g (\rho - \rho'')}. \tag{11.16}$$

From (11.16) we obtain the expressions

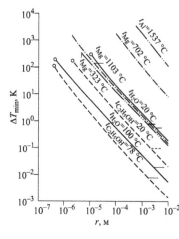

Fig. 11.9. The minimum temperature difference depending on the radius of droplets of various liquids.

$$\Delta \bar{T}^a_{min} = \frac{2\sqrt{2}K^a_s}{1-\sqrt{2}K^a_s}, \quad K^a_s = \sqrt{\sigma(\rho-\rho'')g/l\rho''},$$

$$K^a_s = \frac{|q|}{\sqrt{RT_1}\sqrt{\sigma(\rho-\rho'')g}} \geq 4.54,$$

$$\delta^a_{min} = \delta_{min}/a = 0.44\frac{K^a_s}{1-\sqrt{2}K^a_s}\frac{\lambda''\sqrt{T_1}}{\sigma\sqrt{R}},$$

where $a = \sqrt{2\sigma/(\rho-\rho'')g}$ is the height of the spheroid.

The results of calculations by formula (11.13) are shown in Fig. 11.8 as a dependence of $\Delta\bar{T}_{min}$ on the complex K_s. The dashed line shows the calculation of $\Delta\bar{T}^a_{min}$ in dependence on the complex K^a_s. These relationships can be used to estimate the minimum temperature difference at which the vapour interlayer, preventing coalescence forms between two colliding droplets or between a liquid spheroid and the free surface.

Figure 11.9 presents estimates of $\Delta\bar{T}_{min} = T_2 - T_1$ for the coalescence of droplets of various liquids, water, ethanol, mercury, molten magnesium and aluminium. The calculations for the coalescence of spheroids of these liquids are shown on the labels of the curves.

The calculations show that for ethanol droplets with a radius of $2 \cdot 10^{-3}$ m at room temperature $\Delta T_{min} = 0.12$ K, for the spheroids, respectively, $\Delta T_{min} = 0.1$ K. It is known that the non-isothermal values of the individual droplets in the normal conditions due to evaporation are of the order of several degrees. This fact explains the experiments shown in Fig. 11.7.

The controlling influence of the environment on the process of coalescence of droplets should be noted. We have assumed that the environment is filled with saturated vapour at the temperature of the colder droplet. Expression (11.7) shows that if the environment is filled with vapour at a pressure p_s, higher than $(p_{s1} + p_{s2})/2$, the left-hand side of (11.7) becomes negative and the effect of non-attachment (repulsion of droplets) changes to the effect of attraction of the droplets. When the pressure in the medium is $p_s < p_{s1}$, which means a dry environment for both droplets, the effect of non-attachment of non-isothermal droplets increases. It is easy to see that at $p_s < p_{s1}$ the value of $p'' - p_s = (p_{s1} + p_{s2})/2 - p_s$ is positive even when $T_1 = T_2$; this means that the effect of non-attachment of the droplets in a dry environment also takes place even when their temperatures are equal. In the limiting case of a collision of droplet in absolute vacuum the non-attachment condition can be written as

$$(p_{s1} + p_{s2})/2 \geq 2\sigma/r + \rho u^2/2.$$

11.3.2. Rarefaction effects in the vapour layer

The case where the environment in which the collision of droplets takes place, is filled with a foreign gas in general terms can not be assessed in the present work, since it requires the solution of the diffusion of vapours of the liquid from the vapour layer between the droplets. However, such estimates can be made for the Knudsen flow in the layer. Calculation of δ_{min} by formula (11.15) for water at room temperature gives a value of $5.2 \cdot 10^{-8}$ m. The calculation of the mean free path of the molecules under the same conditions gives a value of $\lambda = 6.3 \cdot 10^{-8}$ m. It is seen that in this case the Knudsen number Kn $= (\delta_{min}/\lambda)^{-1}$ is of the order of unity. Despite the neglect of the effects of rarefaction in the formula (11.15), it is clear that in the conditions realized in vapour layer are close to the free molecule conditions.

It is known that if two cavities filled with gases with molecular masses μ_1 and μ_2, have temperature T_1 and T_2, then when they are connected to the pipeline in the free molecular mode, the pressure formed in them is described by the ratio

$$p_1/p_2 = \sqrt{T_1\mu_1/T_2\mu_2}.$$

If in the event of a collision of two droplets we neglect the discharge of the vapour from the vapour layer between them, the expression takes the form

$$p''/p_s = \sqrt{T''\mu''/T_s\mu_s}, \tag{11.17}$$

where T'', T_s – the temperatures in the layer and the environment, μ'', μ_s – the corresponding molecular weights. From (11.17) and (11.7) we can obtain by the non-attachment condition for free molecular conditions in the layer:

$$p_s\left(\sqrt{T''\mu''/T_s\mu_s} - 1\right) \geq 2\sigma/r + \rho u^2/2. \tag{11.18}$$

For $\mu'' = \mu_s$ the condition (11.18) is qualitatively equivalent to (11.12). For $\mu'' \neq \mu_s$ expression (11.18) reflects qualitatively the role of the baroeffect quality in the non-attachment effect. It is seen that when filling the medium, surrounding the colliding droplets, with a lighter gas compared with the vapour of the liquid the non-attachment effect increases.

11.3.3. The effect of the gas 'cushion'

In addition to the above discussed thermal and baroeffects that play a role in the process of coalescence of droplets, we must specify another reason conducive to the formation of a gas layer between the droplets at temperatures below the saturation in the medium of a foreign gas. It is a gas 'cushion' that forms between the bodies when they approach. An unsteady gas 'cushion' can produce excess pressure in the gap between the solids decreasing with time. However, the maximum value of this pressure differential can not exceed $\rho u^2/2$. This means that even when considering the non-stationary process of collision of the droplets the gas 'cushion' does not fundamentally change the conditions (11.7). For a stationary droplet in the absence of convection on its surface the 'cushion' effect plays no role. In the presence of convection on the surface of the droplet facing the vapour layer, such as thermocapillary convection induced by the Marangoni effect, the effect of the 'cushion' can occur due to the effect of the convection of the liquid surface on the gas dynamics of the vapour interlayer. The influence of the 'cushion' may appear, evidently, also in the case of colder droplets relative to the saturation in the environment. In the process of vapour condensation on droplets near their surface and in the layer between them the foreign non-condensing gas, present in the environment, can build up.

11.3.4. Generalization of phenomena accompanying coalescence

All of the estimated calculations are valid if we assume $\tilde{j}_1 = 0$. This means that the results of the evaluations and all conclusions with the same assumptions are applicable for the qualitative analysis of the processes of interaction of

Table 11.1. Effects on various processes on coalescence processes (* – formed by the Marangoni effect), ** – formed by condensation)

Scheme	Conditions	Consumption effect	Thermal effect	Baro-effect	Gas cushion *	Gas cushion **
Liquid T_1 Liquid T_2 $T, p = p_s + p_h$	$T_1 - T = T - T_2$; $p_s = (p_{s1} + p_{s2})/2$	0	0	0	0	0
	$T_1 = T_2 = T$; $p_s < p_{s1}, p_{s2}$	+	0	+	0	0
	$T_1 = T_2 = T$; $p_s > p_{s1}, p_{s2}$	−	0	−	0	+
	$T_1 = T; T_2 > T_1$; $p_s < (p_{s1} + p_{s2})/2$	+	+	+	0	0
	$T_1 = T; T_2 < T_1$; $p_s > (p_{s1} + p_{s2})/2$	−	−	−	0	+
Solid T_2 Liquid T_1 $T, p = p_s + p_h$	$T_1 - T = T - T_2$; $p_s = (p_{s1} + p_{s2})/2$	0	0	0	0	0
	$T_1 = T_2 = T$; $p_s < p_{s1}, p_{s2}$	+	0	+	0	0
	$T_1 = T_2 = T$; $p_s > p_{s1}, p_{s2}$	−	0	−	0	+
	$T_1 = T; T_2 > T_1$; $p_s < (p_{s1} + p_{s2})/2$	+	+	+	−	0
	$T_1 = T; T_2 < T_1$; $p_s > (p_{s1} + p_{s2})/2$	−	−	0	+	+
Liquid T_2 Solid T_1 $T, p = p_s + p_h$	$T_1 - T = T - T_2$; $p_s = (p_{s1} + p_{s2})/2$	0	0	0	0	0
	$T_1 = T_2 = T$; $p_s < p_{s1}, p_{s2}$	+	0	+	0	0
	$T_1 = T_2 = T$; $p_s > p_{s1}, p_{s2}$	−	0	−	0	+
	$T_1 = T; T_2 > T_1$; $p_s < (p_{s1} + p_{s2})/2$	+	+	+	+	0
	$T_1 = T; T_2 < T_1$; $p_s > (p_{s1} + p_{s2})/2$	−	−	−	−	+

non-isothermal liquid droplets and the solid surface, as well as a spherical rigid body and the free surface of the liquid. Table 11.1 shows the main variants of the conditions of interaction of droplets with droplets, droplets with the solid surface, a solid sphere with the surface of the liquid and the qualitative effects of different processes on the non-adhesion effect. The '+' sign corresponds to the non-attachment effect, the '−' sign − to the attraction of the interacting objects. Symbol '0' indicates no effect of the appropriate process. The consumption effect means evaporation or condensation, corresponding to the first member of the right-hand side of (11.8) and (11.9), the thermal effect corresponds to the second term. The baroeffect is examined on the basis of the ratio (11.12) for $\mu_s < \mu''$. When $\mu_s > \mu''$ the signs in this table column should be changed to opposite.

Several cases of using Table 11.1 will be described. Imagine the surface of a real solid having a roughness with a characteristic radius of the rounded protrusions of 10^{-6} m, heated with respect to the liquid. This case corresponds to the 9th row of the table. The repulsive effect of evaporation processes, thermal and baroeffect is weakened only by the influence of the Marangoni effect. In accordance with Fig. 11.9, the surfaces roughness with this surface curvature in the interaction with the free surface of stationary water at $T_1 =$ 373 K can produce a vapour layer with a temperature difference 75 K. This value correlates with the actual minimum overheating ΔT_{cr2} observed in the processes of film boiling.

Falling of the liquid droplets at low speed on a perfectly smooth surface, having a lower temperature, corresponds to the 14th row of the table. It is seen that all relevant processes cause the repulsion of the droplet from the surface. Figure 11.9 shows that, for example, a droplet of molten magnesium with a radius of 10^{-4} m, has a temperature 1376 K, and can interact with the solid surface without forming a vapour layer at a temperature difference below 18 K. At high temperature drops the attachment of the droplets can be caused only by the influence of the non-ideal form of the surface − roughness and in accordance with Fig. 11.8 a higher rate of incidence. This example is in qualitative agreement with the well-known fact of the reduction of the adhesion with decreasing substrate temperature and decreasing rate of deposition of droplets in the processes of depositing metallic coatings by means of liquid-metal aerosols.

Cooling of the mist corresponds to the third row of Table 11.1. The consumption effect, associated with the condensation of vapour on the mist microdroplets, causes their coalescence, even in the absence of electrical phenomena. Apparently, this is one of the attributes of the mechanism of formation of raindrops. Enlargement of the droplets is prevented only by the secondary effect of condensation − gas 'cushion'. This is in qualitative agreement with the fact that the formation of large rain drops can not be explained only by condensation on them [3].

The qualitative analysis of the processes discussed in this section indicates the possibility and the need for formulating a series of new problems of the interaction of objects related to the processes of heat and mass transfer.

11.4. Impact spray

The vapour and/or gas cushion between two liquid, liquid and solid media can be used to create a fundamentally new technology of dispersing liquids, especially molten metal. If the impact of the jet or melt droplet on the obstacle between the melt and a barrier results in the formation of a vapour or gas layer, then in spreading of the melt on the surface of the obstacle the lack of friction allows a finer liquid film to form. The relatively thin film in spontaneous decay can produce melt droplets, which in solidification will form fine powder particles, even without forced spraying. Estimates show enticing prospects.

The surface energy of 1 m² of the liquid film is equal to $E_\sigma = 2\sigma/\delta$. If the decay of the film is not accompanied by the dissipation of surface energy, droplets of the melt $d = 3\delta$ are formed. If the kinetic energy of the melt $E_{kin} = \rho V^2/2$ is the source of the accumulation of surface energy, the melt drops with the diameter

$$d = 12\sigma/\rho V^2.$$

are formed.

As in chapter 1, the energy efficiency of different processes of dispersion was calculated and the evaluation of the spontaneous decay of the film gives

$$\eta = \frac{6\sigma}{\rho d} \bigg/ \left(L + \frac{6\sigma}{\rho d} \right). \tag{11.19}$$

Calculations by formula (11.19) are shown by the curve *7* in Fig. 11.10.

Figure 11.10 shows that the process of obtaining a powder by spontaneous decay of a liquid film that forms by spreading on the obstacle with ideal vapour or gas lubrication, competes only with the process of refining with the ideal lubricant. It is interesting to test the technical feasibility of such a process.

11.4.1. The experimental setup

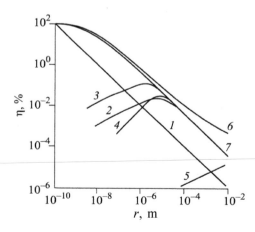

Fig. 11.10. Theoretical efficiency of various processes of production of powders. *1* – the gas-phase method; *2* – spraying with the 'focus' of the jet; *3* – sputtering a thin liquid film; *4* – centrifugal atomization; *5*— shear milling; *6* – milling with perfect lubrication; *7* – the disintegration of the film. The figure is taken from chapter 1 and curve *7* is added.

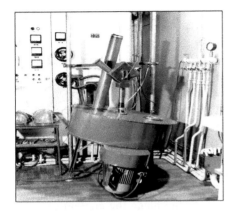

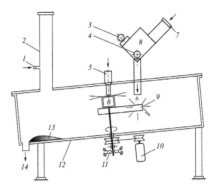

Fig. 11.11. Research facility.   **Fig. 11.12.** The experimental setup.

The setup shown in Fig. 11.11 was designed to implement the above ideas. In this setup, to achieve a sufficiently high kinetic energy of the collision of the melt with the obstacle, the obstacle and not the melt is accelerated. The experimental setup is shown in Fig. 11.12.

The housing *12* contains the rotor with the blade *9*. The rotor is rotated by a motor *10*, rotor speed is measured by sensor *11*. Above the installation there is the source of the melt of the material *8* for conversion into powder. The melting chamber contains the plasma torch *4*. The blank designed to melt, is fed on the opposite side. Window *7* is used for observation of melting, the window of *3* – to illuminate. The melt droplets fall from melting chamber into the body of the installation and come under impact of the vanes *9*. The casing of the setup receives the liquid through the nozzle *5*. The liquid enters the bowl *6*, mounted on the rotor with the blade. As the rotor rotates the liquid is dispersed by the 'bowl'. As a result, the installation is filled with a liquid spray. When the blade is rotated aerosol particles are deposited on its surface and, therefore, its surface is always covered with a thin film of liquid. The high-speed impact of the blade surface, covered with a liquid film, on the melt drop takes place in this way. If the temperature of the melt is much higher than the boiling point of liquid, the contact of the melt with the liquid film on the surface of the blade results in the formation of a vapour–gas layer, drastically reducing the friction of the melt on the surface of the blade, which makes it possible to obtain a thin liquid film, with the melt aerosol formed during its decay. Drops of the melt solidify and are deposited on the walls and bottom of the installation *13*.

The system is provided with pipe *2* with a compensation flexible membrane at the end, designed to dampen pressure pulses that occur inside the body at explosive fragmentation of the melt droplets. A small amount of inert gas *1* is also provided to create an inert atmosphere inside the system.

The body of the system is installed in an inclined position. The vibrations of the body during rotation of the rotor are sufficient to ensure that the resulting powder with the liquid move to the outlet nozzle *14*.

11.4.2. The experimental results

The tests of the installation were carried out in order to establish the possibility of using the method of shock fragmentation of the melt to produce powders of various metals and alloys in different conditions. Powders of copper, brass, zinc, mild steel, steel 45, cast iron and stainless steel were produced. Water was used to produce the aerosol in the system, nitrogen – to create an inert atmosphere. Photos of the powder with the corresponding histograms are shown in Figs. 11.13–11.22.

The experiments revealed the following features of the processes of impact fragmentation of the melt.

1. By increasing the speed of the rotor with the blade the content of fines in the final powder increases.

2. Vapour explosion on contact of the molten droplet with the liquid film on the blade plays an important role in the process of fragmentation of the melt.

3. When using nitrogen to create an inert atmosphere in the installation, the resulting powder particles become more rounded in comparison with the particles obtained by using air. Oxidation of molten droplets by the air causes formation of sharp tails in the droplets.

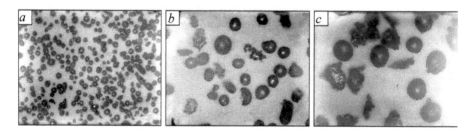

Fig. 11.13. Copper powder obtained by impact; ×100. Fraction, mm: a – <0.05; b – 0.071–0.09; c – 0.09–0.16.

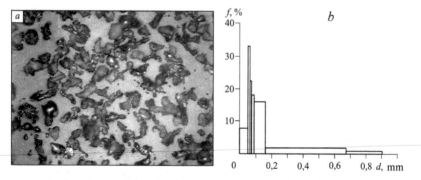

Fig. 11.14. The zinc powder obtained by impact, (6000 rpm); a – 0.05–0.063 mm fraction, × 100; b – size distribution function.

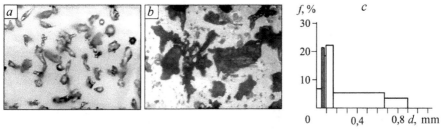

Fig. 11.15. Brass powder obtained by impact, (3000 rpm); a – fraction <0.05 mm, ×200; b – fraction 0.16–0.67 mm, ×50; c – size distribution function.

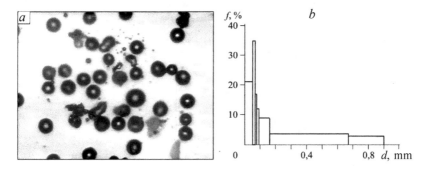

Fig. 11.16. Iron powder obtained by impact, (3000 rpm); a – fraction <0.071–0.09 mm, × 200; b – size distribution function.

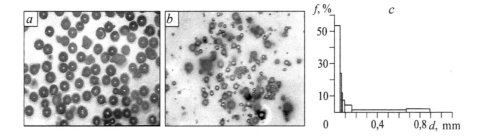

Fig. 11.17. Iron powder obtained by impact, (6000 rpm); a – fraction <0.063–0.071 mm, ×100; b – fraction <0.05 mm, ×500; c – size distribution function.

11.4.3. Disadvantages of the installation revealed in experiments

1. In the histograms in Figs. 11.14–11.22 it is seen that in all cases there are large contents of coarse particles. These particles have an unsuitable form and are formed due to partial misses of the blade when hitting the droplet. Part of the droplet, which is not included in the zone of direct impact of the blade, does not receive sufficient kinetic energy to split into microdroplets.

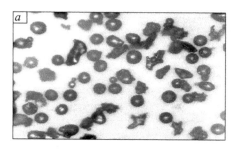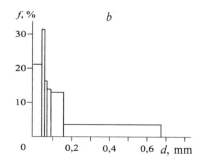

Fig. 11.18. Iron powder prepared in nitrogen atmosphere by impact, (8500 rpm). A total of 376.65 g was produced, the yield (up to 0.67 mm) – 163.21 g – 43.3%. a – fraction <0.05–0.063 mm, ×100; b – size distribution function.

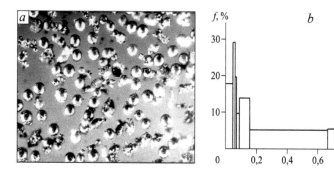

Fig. 11.19. Iron powder obtained with an impact, with a shortage of water (8500 rpm); fraction 0.063–0.071 mm; a total of 508.98 g was produced, the yield (up to 0.67 mm) 128.5 g – 25.2%. a – the dark field, ×100; b – size distribution function.

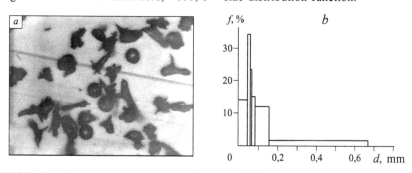

Fig. 11.20. Cast iron powder obtained by impact, (8500 rpm), total amount 186.18 g, the yield (up to 0.67 mm) – 140.11 g – 75.3 %. a – fraction < 0.071–0.09 mm, ×100; b – size distribution function.

This deficiency can be corrected if the system is fitted with synchronization of the rotation of the rotor blade and the fall of the droplets.

2. In the experiments there was a significant heating of the walls of the installation, which changes the operating mode of the installation. The operation of the plant can be stabilized by cooling of the casing walls.

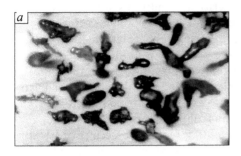 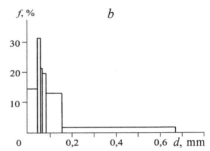

Fig. 11.21. Cast iron powder prepared by impact in nitrogen, (8500 rpm); a total of 214.42 g, the yield (up to 0.67 mm) – 112.6 g – 52.5%. *a* – fraction <0.071–0.09 mm, × 100; *b* – size distribution function.

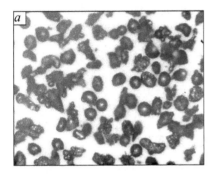 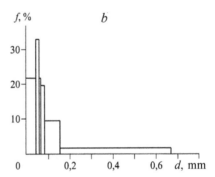

Fig. 11.22. Stainless steel powder prepared in nitrogen, (8500 rpm). A total of 181.68 g; the yield (up to 0.67 mm) – 93.33 g – 51.4%. *a* – fraction <0.063–0.071 mm, ×100; *b* – size distribution function.

3. Melting of the starting material without the use of the crucible makes it impossible to control the temperature of the melt before crushing. Insufficient cooling of the melt affects the shape of the produced particles. To address this shortcoming it is apparently necessary to perform melting in the crucible.

4. When melting without a crucible, the dripping of molten metal from the initial blank and dripping of the melt into the system occur irregularly. This leads to instability of the crushing process and all the settings in the installation, which also makes it difficult to synchronize the dripping and rotation the rotor. To address this shortcoming, in addition to melting the crucible should have controlled dosing of the melt in the system.

Findings

1. It was found that in the process of coalescence of the particles of the dispersed aerosol system an important role is played by the vapour layer formed between the particles or the particle and the surface when they approach each other.

2. The criteria for determining the critical conditions for coalescence – the Weber number, the ratio between the specific volume surface energy and the specific latent heat of vaporization, the relationship between thermotranspiration pressure and Laplace pressure, were determined.

3. The qualitative pattern of the temperature and pressure ranges in which they facilitate or impede the coalescence of four different phenomena: hydro-gas dynamics in the vapour layer, thermotranspiration effect, thermobarodiffusion effect, and the Marangoni effect, was constructed

4. The common features of the film boiling regime, the effect of 'floating droplets' as a single phenomenon of the existence of the vapour layer between two boundaries of the condensed phase were determined. The second boiling crisis is a violation of this phenomenon by real different effects: vibration, roughness.

5. The possibility of using vapour and gas 'cushion' in contact of the liquid droplet and the melt film was shown to create a fundamentally new cost-effective technology for producing metal powders by impact the wet blade on the falling melt droplet.

List of symbols

c	– specific heat capacity of the mass, J/(kg·°C);
d	– diameter, m;
g	– acceleration due to gravity, m/s²;
h	– depth in meters;
h_l	– the depth of immersion of the blunt rod, m;
j	– specific flow rate, kg/(m²·s);
l	– the latent heat of vaporization, J/kg;
p	– pressure, N/m²;
q	– heat flux, W/m²;
R	– gas constant, J/(kg·K);
r	– radius, m;
T	– temperature, K;
u	– velocity, m/s;
We	– Weber number;
δ	– thickness of the vapour layer, m;
λ	– thermal conductivity, W/(m·°C);
μ	– molecular weight, AU;
ρ	– density, kg/m³;
σ	– the surface tension, N/m;
χ	– the radius of the vapour layer, m

Indices

kin	– kinetic;
cr	– critical;
p	– the settlement;

c	– wetting;
b	– a peripheral portion of the steam of the gap;
h	– at positive depth;
$-h$	– at low depths of immersion;
k	– a boil;
l	– liquid;
min	– the minimum;
s	– Wednesday;
w	– a solid surface;
δ	– a steam layer;
0	– on the surface of the liquid;
"	– The vapour phase.

Literature

1. Wachters L.H.J., Bonne H., Nouhuis H.J.V., Chem. Eng. Sci., 1966, V. 21, 929–936.
2. Kutateladze S.S., Heat transfer in condensation and boiling, Moscow, Mashgiz, 1952.
3. Deryagin V.V., Prokhorov P.S., in: New ideas in the study of aerosols, Moscow, Leningrad, 1949, 84–101.
4. Muratova T.M., Labuntsov D.A., Teplofizika vysokikh temperatur, 1969, V. 7, No. 5, 959–967.
5. Labuntsov D.A., Ametistov E.V., Teploenergetika, 1982, No. 3.

Capture of microparticles from gas flows by condensation processes

The need to capture microparticles from gas streams occurs during cleaning of flue gases from industrial installations, the exhaust gases of internal combustion engines, etc. The need to capture is dictated not only by environmental concerns. Often the captured particulate phase is a valuable product. For example, the finest fraction of cement in cement plants is released into the atmosphere, together with the combustion products of fuel used for kilns, while the high content of this fraction in the cement improves its quality. In the production of kaolin by anhydrous technology there is the same situation – the most delicate and most valuable part of the fraction goes away along with the flue gases. Addressing these and similar problems requires the development of effective ways to capture microparticles. There is a huge number of ways but they do not always satisfy the needs of production. At submicron size of the captured particles only electrical and fabric filters are effective.

Tasks of trapping nanosized particles arise in the solution of environmental problems and in some industrial processes associated with the formation of nanoparticles in the gas phase and their subsequent separation from the gas. In these cases the particles can captured only by fabric filters.

Sometimes there is a need for selective capture of nanosized particles from gas streams with fractions of larger microparticles. This problem has not as yet bee solved.

This chapter describes the experience with a simple idea for solving the above types of tasks to capture particles. If we organize a condensation of vapour on submicron and nanosized particles and the condensate grow to micron size, the task of capture becomes much easier, the particles can be captured by many well-known methods. If we can find conditions for selective condensation only on the particles with certain properties, we can solve the problem of decomposition of disperse systems in gases into their constituent fractions. In line with this idea, below we present the results of the modification of the well-known principle of centrifugal trapping of particles, and the less well-known principle of injection. In both cases, these principles are supplemented by the processes of condensation of vapour, additionally mixed into the gas stream with the particles.

The applicability of the above ideas is tested on the task of cleaning the atmosphere of a paint shop to remove aerosols of paint and its fumes formed during spray painting products. The test method of purification is based on the implementation of the following processes:

1) removing paint particles of the micron-size from the aerosol of the sprayed paint from the air of the space;

2) mixing of sprayed paint with vapors of the solvent;

3) the condensation of solvent vapors and other volatile components on the dispersed particles of submicron size;

4) capture of the enlarged particles due to inertial forces in the capture device.

It is assumed that the production line of vehicles engaged in these processes, or the successive stages in a single unit, will solve the problem of complete cleaning of the room air of the painting workshop. The aerosol of the sprayed paint can be removed using exhaust ventilation. The task is to intermediate devices or a set of devices and ventilation, which creates optimal conditions for the implementation of these process steps. The base variant of the trapping apparatus was a centrifugal rotary machine action, successfully used for purification of industrial gases from dust [1, 2].

The main difference between the proposed structures of the devices currently available is the availability of intake of paint solvent vapors in the working chamber of the apparatus. It is assumed that the input of vapour and its mixing with the aerosol will create conditions for joint condensation of the components of the mixture and separation of condensed particles in the volume of the apparatus, as in [3]. The experience gained by the study of the joint condensation of organic compounds and metals, with the formation of microdispersed particles and their enlargement [4, 5] leads us to expect a high efficiency of using this phenomenon for the purification of gases from harmful impurities. Thus, if the sprayed paint condenses with the solvent vapors, and condensed particles will be captured in the volume of the system and separate from the flow of purified air, the task of cleaning will be solved.

12.1. Features of technology of capture of sprayed paint aerosol

The main difficulties in clearing the atmosphere of the paint shop aerosols and vapours of dispersed dyes are as follows:

1. Wide grain size distribution of aerosols of the sprayed paint – a wide range of particle sizes of paint (d_p = 0.001–1000 μm) and a wide range of concentrations of these particles in the atmosphere department (N_p = 1–10²¹ particles/m³).

2. The large concentration of volatile substances, released along with the paint from the gun shop in the atmosphere. These substances may include vapours of paints and organic solvents (acetone, xylene, benzene, ethyl acetate, etc.). For example, the ratio of the amount of acetone needed for diluting the nitrocellulose paint, to the number of colours is 5–20. Since acetone is a low-boiling solvent (t_b = 56.1°C), the pressure of saturated vapour in the

atmosphere of the shop is $p_h \approx 26600$ mm Hg [1], the density of the vapour in the atmosphere of the shop can be very high – $\rho_{ac} = 0.001–1$ kg/m³.

3. A significant difference in the physical properties of each component of the aerosol emitted from the gun, and fast dynamics of changes of these properties in the shop atmosphere due to phase transitions (evaporation and condensation), carried out at different rates for each of the components. For example, the density of matter ρ_p constituting the heavy components of the paint aerosol, equal to $1–11.5\cdot10^3$ kg/m³, and the density of volatile substances (vapors of paint and solvents) vary in the range of $1\cdot10^{-3}–1\cdot10^3$ kg/m³. The properties such as viscosity, heat capacity of the individual components, etc. also vary greatly.

From the analysis of the literature on various aspects of purification of gases from harmful impurities [2–4], we can conclude that the problem of cleaning the atmosphere of the shop to remove simultaneously vapors and aerosols of the dispersed dyes cannot be solved by the existing methods due to the following reasons:

a) dry mechanical dust collectors have constraints on the minimum particle size, giving in 100% capture: dust settling chambers $d_p \geq 20–40$ μm, cyclones $d_p > 3–5$ μm;

b) electric dust collectors do not have such significant constraints on the particle size $d_{p\ min} \geq 0.1–0.01$ mm, but are difficult to use, are expensive and their use for cleaning of spray atomized paint is difficult;

c) wet dust collectors are not satisfactory because of the limited size of captured particles $d_p > 2–10$ μm and high costs of energy and water resources.

Therefore, solutions to this problem must be sought in developing a new method of treatment, which combines the advantages of different methods.

The methods discussed here is based on:

1) condensation to build up the size and mass of aerosol particles and precipitation of volatile components;

2) capture the enlarged particles from the gas stream under the influence of the centrifugal effect in the volume of the trapping device.

Obviously, the novelty of the method is contained in the first point. The Schemes of the feed of the flow of contaminated air into the cleaning unit and return after cleaning to the atmosphere of the room may be different. In this case (Fig. 12.1), the flow of air polluted with the aerosol of the paint from painting items must be absorbed into the cleaning apparatus by vacuum

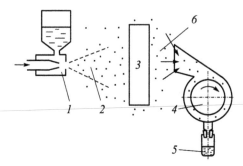

Fig. 12.1. Scheme of painting by spraying with simultaneous cleaning of the atmosphere of the shop. *1* – airbrush; *2* – aerosol spray paint; *3* – painted products; *4* – cleaning apparatus; *5* – bunker for captured paint and solvent; *6* – sprayed aerosol with paint fumes in the room.

created by the apparatus itself, which is the most cost-effective [1, 2], or by vacuum created by the exhaust ventilation plant.

To implement this method of treatment, it is proposed to use the device, basically similar to a rotary centrifugal dust collector [1, 2]. In these devices the efficiency of separation is easily adjustable by selecting the rotor speed. The devices have small dimensions and are easy to use. Figure 12.2 shows the construction of a centrifugal rotary machine TsRA-2 with a multi-disc rotor with the following parameters:

The inner chamber radius, R_1	200	mm
Size of the inlet	100·100	"
The outer radius of the rotor, R_2	160.100.60	"
The inside diameter of the rotor, d_0	32	"
The length of the rotor, L_R	95	"
The width of the blade, B	19	"
Motor type AP-600,	AVP-4,	
The frequency of rotation, n_R	0–10000	rpm

An important feature of the TSRA-2 apparatus is a flow of aerosol through a tangential inlet and outlet of pure gas from the axis of the hollow rotor shaft. The scheme of the apparatus is similar to a centrifugal fan whose hollow shaft which rotates in the direction opposite to the normal direction. There are several features in the design.

1. The inlet valve is equipped with a distributor which supply an additional vapour into the aerosol, and a flap, partially overlapping the front entrance to regulate the inlet speed.

2. The outer diameter of the rotor disks is much smaller than the inner diameter of the housing unit. The annular space between the disks and the

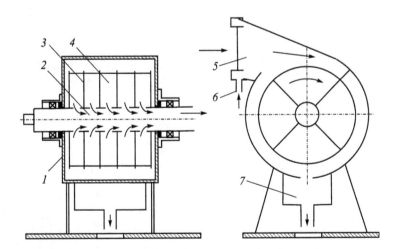

Fig. 12.2. Centrifugal rotary machine (TsRA-2-2) with a multiple disk rotor. *1 –* chamber of the apparatus; *2 –* hollow shaft; *3 –* rotor disk; *4 –* the rotor blades; *5 –* input device for the air contaminated with the aerosol; *6 –* an input device for solvent vapours; *7 –* drain device,

housing is intended for the organization in its vortex flow. In an adiabatic gas flow the vortex cools down causing condensation of vapour on aerosol particles. The vortex flow is an obstacle to the passage of large and aggregated (by condensation) micron particles in the rotor space.

3. At the bottom of the housing there are traps for the capture of particles retained by the vortex flow.

4. On the floor with the side holes of the rotor shaft there are several thin disks with intermediate radial vanes. Several parallel disks separate the gas flow supplied to the rotor into several separate streams, reduce the intensity of secondary flows, thereby reducing the probability of passage of the crude aerosol due to the large-scale fluctuations of turbulent secondary flows.

5. The rotor installed into the housing also has (in addition to bearings) seals that ensure the passage of gases only through the rotor.

It is clear that such a device for sucking contaminated air for cleaning should be provided with an additional fan creating a pressure gradient sufficient to overcome the hydraulic resistance of the device, and provide the calculated flow of purified air.

Figure 12.3 shows a centrifugal rotary machine TsRA-2-2. There are two TsRA-2 systems, connected by a hollow shaft. The first step was works like a TsRA-2 with purified air movement from the periphery through the rotor in the cavity of the rotor shaft, and the second stage as a fan, that is, with the movement of the purified gas stream from the axis to the periphery. The tandem of these two systems allows to combine in a single unit advantages of the cleaner and the fan, and eliminates the need for additional ventilation.

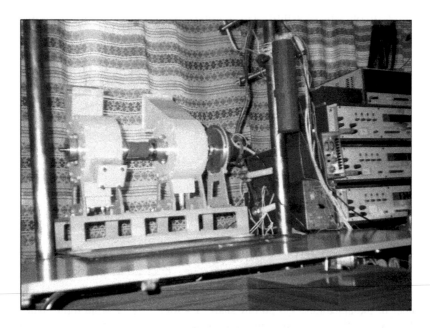

Fig. 12.3. Twin-rotor centrifugal dust collector TsRA-2-2.

This design is advantageous in terms of required energy and makes the device more economical. This assumption is confirmed by the authors [1–3]. By varying the diameters of the rotors of the cleaning and expansion stages, and changing the speed, it is possible to find the optimal conditions for high efficiency of cleaning the contaminated gases and improve the performance of the device, cleaning the spray.

Thus, the cleaning solution was reduced to a search of: a) the optimal operating conditions of each of the devices, and b) factors influencing the improvement of the process of enlargement of the particles due to condensation and further trapping.

12.2. Modelling of gas cleaning processes to remove atomized paint particles

The problem of finding the optimal design and operating parameters of each unit produced is greatly simplified if one can mathematically model the processes occurring in it. In the method chosen in this paper purification of gases from dispersed inclusions by their condensation enlargement and separation from the gas by centrifugal forces, no process can be described completely mathematically. Mathematical modelling of vapour condensation on microparticles is complicated by two features. Firstly, the stochasticity of nucleation of the first nuclei of vapour condensate on aerosol particles requires formulation of the probabilistic multiparameter problems. Secondly, collective participation in the process of polydisperse particles with collective nucleation and condensation, requires setting of dynamic problems of condensation. Problems of this type are currently not resolved.

The modelling of the centrifugal separation of particles from the gas flow is hindered by many side effects that occur in the swirling flow, often associated with the real properties of gases and, especially, aerosols [6–9]. Keeping in mind the applied focus of this paper, this section presents a simplified version of modelling suitable, however, for most estimates of engineering calculations.

12.2.1. The motion of micron-sized particles in the swirling gas flow

Consider the forces acting on a particle moving in a swirling flow of gas. An element of the gas flow moves along a complex trajectory with an angular $\omega = d\varphi/d\tau$, radial $V_r = dr/d\tau$ and axial $V_x = dx/d\tau$ velocities. A particle of mass m_p, carried by a gas moves at velocities $\omega_p = d\varphi_p/d\tau$, $U_r = dr_p/d\tau$, $U_x = dx_p/d\tau$.

The equation of motion of a particle can be written as

$$m_p \frac{d\vec{U}}{d\tau} + (\vec{V} - \vec{U})\frac{dm_p}{d\tau} = \sum_i F_i$$

or in the expansion in the coordinates

$$m_p \frac{dU_r}{d\tau} + (V_r - U_r)\frac{dm_p(r)}{d\tau} = \sum_i F_r,$$

$$m_p \frac{dU_\varphi}{d\tau} + (V_\varphi - U_\varphi)\frac{dm_p(\varphi)}{d\tau} = \sum_i F_\varphi, \qquad (12.1)$$

$$m_p \frac{dU_x}{d\tau} + (V_x - U_x)\frac{dm_p(x)}{d\tau} = \sum_i F_x.$$

A particle moving along the coordinates changes its the mass due to vapour condensation on it in terms of pressure, temperature and vapour concentration. However, for simplicity, consider only the conditions for trapping already aggregated particles using the assumption

$$\frac{dm_p(r)}{d\tau} = \frac{dm_p(\varphi)}{d\tau} = \frac{dm_p(x)}{d\tau} = 0,$$

which means neglecting the change in the mass of the particles in the gas stream in an area where their centrifugal capture takes place.

We consider the forces acting on the particle:
– the centrifugal force, which has only a radial component

$$F_c = m_p U_\varphi^2 / r,$$

– tangential Coriolis force

$$F_C = m_D U_r U_\varphi / r,$$

– the force of aerodynamic resistance (drag) to the motion of a particle in a gas

$$F_a = C_D A_p \rho (U - V)^2 / 2,$$

having components in the coordinates:

$$F_{ar} = C_D A_p \rho (U_r - V_r)^2 / 2,$$

$$F_{ax} = C_D A_p \rho (U_x - V_x)^2 / 2,$$

$$F_{a\varphi} = C_D A_p \rho (U_\varphi - V_\varphi)^2 / 2,$$

where $A_p = \pi d_p^2 / 4$ is the cross-sectional area of the particle; ρ is the density of gas; C_D is the drag coefficient when the particle moves relative to the gas. The diagram of the forces on a particle is shown in Fig. 12.4. In the nozzle at the inlet into the swirling flow at entry speeds of the order meters per second we can take into account only the aerodynamic force F_a. When the particle moves in the swirling flow near the wall bounding the flow, where the velocities are small, the centrifugal force F_c, the reaction of the wall F_w and the force of gravity F_g are added. In the swirling flow the main forces are the

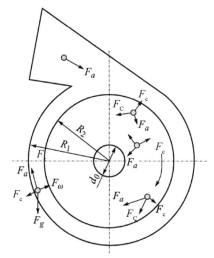

Fig. 12.4. Diagram of the forces on the particle in the swirling flow. F_c – centrifugal force, F_a – resistance force from the air stream, F_C – force associated with tangential Coriolis acceleration.

aerodynamic, centrifugal force and Coriolis force. Substituting the expressions for the forces in (12.1) leads to the equations of motion of the particle

$$m_p \frac{dU_x}{d\tau} = -C_D A_p \rho (U_x - V_x)^2 / 2,$$

$$m_p \frac{dU_r}{d\tau} = m_p \frac{U_\varphi^2}{r} - C_D A_p \rho (U_r - V_r)^2 / 2,$$

$$m_p \frac{dU_\varphi}{d\tau} = m_p \frac{U_r U_\varphi}{r} - C_D A_p \rho (U_\varphi - V_\varphi)^2 / 2. \tag{12.2}$$

The centrifugal particle trapping occurs near the periphery of the rotor. For producing the separating effect of the rotation of the stream, swirled by the rotor, consisting of flat discs and straight blades, it is advisable to introduce the conditions $V_\varphi \gg V_r \gg V_x$, $U_\varphi \gg U_r \gg U_x$.

In addition, we assume that the rotation is carried out at a constant rate and that the Coriolis forces are small. This means equal peripheral speeds and rotational speeds of gas and particles:

$$U_\varphi = V_\varphi, \ \omega = \omega_p = U_\varphi / r = V_\varphi / r = \text{const}. \tag{12.3}$$

Condition (12.3) annihilates the 1st and 3rd equation of (12.2).

For the condition for separation of particles from the gas stream it is logical to take the balance of the particle at any radius:

$$U_r = 0, \frac{dU_r}{d\tau} = 0. \tag{12.4}$$

From (12.2) through (12.3) and (12.4) it follows that

$$m_p U_\varphi^2 / r = C_D A_p \rho (U_r - V_r)^2 / 2. \qquad (12.5)$$

The drag coefficient for the case of small particles $C_D = 24/\mathrm{Re}_p$. If $d_p = 1 \cdot 10^{-4} - 1 \cdot 10^{-9}$ m Reynolds number IS $\mathrm{Re}_p = \rho d_p |U_r - V_r| / \mu \le 1$. Considering the last

$$F_{ar} = 3\pi\mu d_p (U_r - V_r). \qquad (12.6)$$

For particles with the size $d_p < 2$ μm in (12.6) it is necessary to introduce the Cunningham–Milliken correction with takes into account the increase in mobility with decreasing particle size [1.2]:

$$F_{ar} = 3\pi\mu d_p (U_r - V_r) / C_K,$$

where $C_K = 1 + \dfrac{2l}{d_p}(1.257 + 0.4\exp(-1.1 d / 2l))$, the mean free path can be calculated by the formula, $l = \dfrac{\mu}{\rho}\left[\dfrac{\pi M}{(2RT)^{1/2}}\right]$. For example, for air at $T = 293$ K, $p = 1 \cdot 10^5$ Pa, $l = 0.5 \cdot 10^{-8}$ m values for the correction $d_p = 1 \cdot 10^{-6}$; $1 \cdot 10^{-7}$; $3 \cdot 10^{-8}$ m, respectively, $C_K = 1.16$; 24.45; 90. From these data it is seen that for particles larger than 1 μm the correction can be ignored.

From (12.5), (12.6)

$$d_p^2 \rho_p U_\varphi^2 / r = 18\mu V_r. \qquad (12.7)$$

Using equation (12.7), we can construct an algorithm for engineering calculations. If we define the radial velocity through the gas flow rate $V_r = G/(2\pi R_2 L_R \rho)$, then the condition for separation of particles from a stream $V_\varphi \ge (9\mu G / D_p^2 \rho \rho_p L_R \pi))^{1/2}$ or the rotor speed will be as follows:

$$n_R \ge \frac{30}{\pi R_2}\left(\frac{9\mu G}{d_p^2 \rho \rho_p L_R \pi}\right)^{1/2}. \qquad (12.8)$$

12.2.2. Condensation coarsening of the particles

The aerosol fed to centrifugal cleaning may contain particles smaller than the value d_p, appearing in the condition of separations (12.8). These submicron particles can be retained at the periphery of the rotor by centrifugal forces. To keep them, the particles have to be enlarged by condensation of the vapour of a liquid. To carry out the condensation of vapour on the particles the stream of the purified spray has to be supersaturated by the vapours of the liquid. The minimum size of particles that can grow large, must be greater than or equal to the critical nucleus size for the homogeneous nucleation of vapours. According to the Kelvin–Thomson effect

$$d_{pm} = \frac{\sigma\gamma}{k_B T \ln(p_v / p_{vh})}. \qquad (12.9)$$

Here, σ is the surface tension of the liquid whose vapours are used for enlargement of the particles; $\gamma = M/\rho_c$ is the molecular volume of the vapour in the condensate; k_B is the Boltzmann constant, p_v is the partial vapour pressure in the aerosol, p_h is the saturation vapour pressure of the liquid used at the temperature of the aerosol.

Calculation by formula (12.9) is hampered by uncertainty of p_v. It is easier to use consumable parameters, easily determined experimentally. To transfer to the consumption parameters, we can use the solution to the problem of isothermal mixing of two gases.

When mixing two gas streams with the consumptions G_1, G_2, temperatures T_1, T_2, molecular masses M_1 and M_2, heat capacities c_1, c_2 and pressure p_Σ, the mixture parameters are given by:

temperature of the mixture –

$$T_\Sigma = (c_1 T_1 G_1 + c_2 T_2 G_2)/(c_1 G_1 + c_2 G_2), \qquad (12.10)$$

partial pressures of components –

$$p_{1\Sigma} = p_\Sigma G_1 M_2/(G_1 M_2 + G_2 M_1), \quad p_{2\Sigma} = p_\Sigma G_2 M_1/(G_1 M_2 + G_2 M_1). \qquad (12.11)$$

Saturation pressure in the formula (12.9) can be approximated by the formula of the form

$$p_{vh} = K_1 T_\Sigma^{K_2} \exp(K_3/T_\Sigma), \qquad (12.12)$$

where K_1, K_2, K_3 are determined for three values of the table. The value of supersaturation $S_v = p_v (G_1, G_2, p_\Sigma, M_1, M_2)/p_{vh} (T_\Sigma)$ for use in the formula (12.9) can be calculated using the formulas (12.10)–(12.12). It is important that in mixing of vapour to the gas the condition $S_v \geq 1$ is fulfilled. Only in this case the condensation of vapour on the particles is possible. The analysis shows that this condition is satisfied only if the following inequality is fulfilled

$$p_{1\Sigma} \geq \frac{p_\Sigma c_1 c_2 G_1 G_2}{(c_1 G_1 + c_2 G_2)^2} \left(\frac{T_2 - T_1}{T_\Sigma}\right)(1 - K_3 T_\Sigma^{K_2 - 1}). \qquad (12.13)$$

However, calculations show that the inequality (12.13) is not always true. This means that, in general, to obtain the vapour supersaturation in the aerosol flow, after mixing the vapour flow should be cooled.

In the swirling flow at high speeds when the vapour moves in he direction of decreasing rotation radius, extra cooling is due a polytropic process of gas expansion with acceleration taking place there. The following is a simplified algorithm for computing the cooling of the compressed gas in the adiabatic vortex flow of a compressible gas.

12.2.3. Algorithm for calculating the adiabatic vortex flow of a compressible gas

The calculation is carried out within the framework of applied gas dynamics [10], based on the Euler equations, with the following assumptions:

1. Perfect gas, inviscid, non-conductor. The presence of an impurity of the vapour does not affect the gas dynamics.

2. Presence of nano- and microparticles in the gas does not change the gas dynamics.

3. Expendable and thermal effects on the flow of gas, associated with condensation, are not counted.

4. Large-scale turbulence is absent.

5. Resistance to the motion of particles relative to the gas is described by the Stokes formula. This assumption is contrary to paragraph 1, but it is necessary to calculate the behaviour of microparticles in the gas.

Using these assumptions a computer program was created. The calculation is carried out sequentially from the entrance to the housing of a vortex apparatus to the outer radius of the rotor.

The initial parameters are defined:

1. Design: the inner radius of the body of the swirl chamber R_1, the length of the body along the axis H_1, the outer rotor radius R_2.

2. Gas parameters: deceleration temperature T_0, deceleration pressure p_0, gas flow rate G_g, the velocity at the inlet of the vortex apparatus w_{R1}, adiabatic index k, the molecular gas mass M, the viscosity v_g of the gas needed to calculate the resistance force of the particles in the gas.

3. Parameters of the particles in a gas: particle radius r_{pc}, which should be detained in the system.

4. The properties of the liquid and the vapour used enlarging the nanoparticles.

Calculation of gas entry into the body of the vortex system

It is assumed that the purified gas is fed into the vortex unit almost tangentially to the inner radius of the body. The speed ratio at the inlet is [10]:

$$\lambda_{R1} = w_{R1} / \left(\frac{2kRT_0}{k+1} \right)^{0.5},$$
(12.14)

where $R = R_0/M$ is the gas constant. The density, pressure and gas temperature at the inlet are calculated by the formulas

$$\rho_{R1} = \rho_{g0} \left(\frac{1-(k-1)\lambda_{R1}^2}{K+1} \right)^{\frac{1}{k-1}}, \quad p_{R1} = p_{g0} \left(\frac{1-(k-1)\lambda_{R1}^2}{K+1} \right)^{\frac{k}{k-1}},$$

$$T_{R1} = T_{g0} \left(\frac{1-(k-1)\lambda_{R1}^2}{K+1} \right).$$
(12.15)

The cross-sectional area of the slit tangential entrance and the slit width are determined for a given flow rate and the length of the system:

$$F_1 = G_g / (\rho_{R1} w_{R1}), \quad B_1 = F_1 / H_1.$$

The radial velocity at the inlet of the unit is determined from the conditions of conservation of the flow:

$$u_{R1} = G_g / (2\pi R_1 H_1 \rho_{R1}).$$

The circumferential velocity at the inlet is calculated as the projection component of the absolute velocity

$$v_{R1} = (w_{R1}^2 - u_{R1}^2)^{0.5}.$$

The entry angle is calculated from the projections of the absolute velocity

$$\alpha_1 = \text{arctg}(v_{R1} / u_{R1}) \frac{180}{\pi}.$$

Calculation of vortex flow in the space around the rotor

The calculation starts from the inner radius of the housing. The speed ratio, calculated by the formula (12.14), is accepted as the current ratio $\lambda_r = \lambda_{R1}$. For computing at the reduced (still unknown) radius we used the increment of the speed ratio, for example, by the value 0.001:

$$\lambda_r = \lambda_r + 0.001. \qquad (12.16)$$

The absolute velocity at the reduced radius is calculated as

$$w_r = \lambda_r \left(\frac{2kRT_{g0}}{k+1} \right)^{0.5}.$$

Similarly to (12.15), we determine the density, pressure and temperature on the reduced radius:

$$\rho_{gr} = \rho_{g0} \left(\frac{1-(k-1)\lambda_r^2}{k+1} \right)^{\frac{1}{k-1}}, \quad P_{gr} = P_{g0} \left(\frac{1-(k-1)\lambda_r^2}{k+1} \right)^{\frac{k}{k-1}}, \quad T_{gR} = T_{g0} \left(\frac{1-(k-1)\lambda_r^2}{k+1} \right).$$

From the assumption of the conservation of the flow, taking into account the conservation of angular momentum, we determine the current reduced radius:

$$r = \left(\left(\frac{G_g}{2\pi \rho_{gr} H_1} \right)^2 + (v_{R1} R_1)^2 \right)^{0.5} \bigg/ w_r.$$

The radial velocity at the current radius is calculated from the conservation of flow $u_r = G_g / 2\pi r H_1 \rho_{gr}$ and the circumferential velocity – from the conservation of angular momentum $v_r = v_{R1} R_1 / r$. At the current radius we calculate the radius of particles for which the equality of the centrifugal force and the force, applied to the particle by the radial flow of gas:

$$r_p = \left(\frac{9 u_r \upsilon_g r}{2 \rho_p v_r^2} \right)^{0.5}.$$

If $r_p > r_{pc}$ it means that the particles of a given size r_{pc} are not held in a vortex.

In this case, the calculation is repeated starting with the formula (12.16).

Calculation of particle condensation enlargement.

The saturated vapour pressure at a temperature of inhibition is calculated by the formula derived by approximation of the tabulated data:

$$p_{h0} = K_1 T_{g0}^{K_2} \exp(K_3 / T_{g0}).$$

At entry to the rotor the vapour saturation pressure is $p_{h2} = K_1 T_{g2}^{K_2} \exp(K_3 / T_{g2})$. The partial vapour pressure at the inlet of the rotor $p_{v2} = (p_{h0} S_w / 100) p_{g2} / p_{g0}$, where S_w is the relative air humidity at the inlet of the swirl chamber. Supersaturation at the entrance of the rotor is calculated by $S_{h2} = p_{v2}/p_{h2}$. The radius of the enlarged particles is determined with the aid of the Kelvin-Thomson formula

$$r_p^c = 2\sigma\gamma / k_B T_{g2} \log(S_h),$$

in which $\gamma = M/\rho_c$ – the volume of vapour molecules in the vapour condensate.
    We can calculate the vapour consumption in enlargement (consolidation):

$$G_c = N_p (G_g / \rho_{g0})(4\pi r_p^3 / 3)\rho_c$$

assuming all particles in the aerosol are enlarged to a size at which they will be detained at the entrance to the rotor.
    The presented calculation algorithm in the program is continued in the rotary space until the release of the purified gas into the cavity of the shaft. It is envisaged to profile the rotor blades to get the best results in keeping the particles from entering the rotor. After calculation, analysis was carried out to check whether all stages of calculation are consistent. In the case of disagreement the calculation is repeated with modified parameters. The final results are used to design the cleaning apparatus.

12.3. Experimental study of purification of gas from aerosol particles

The kinematic features of devices selected for the purification of gases from aerosol and the principal features of the centrifugal method of trapping microdispersed particles at the maximum possible reduction of energy consumption were studied. In addition, the possibility of using the apparatus for trapping particles of solidified paint was investigated.
    The rotor of TsRA-2 equipment was rigidly fastened to the shaft of the electric motor, the body of equipment was attached to the housing of the motor. The whole structure was mounted on a common footing. The outlet of the device was connected to a hose of the suction fan. When the fan is switched on and the voltage supply to the motor of switched off, the rotor is spun on its own, working as a turbine. At the same time the voltage was measured on the brushes of the collector of the motor. This voltage, generated by an electric motor during rotation of the apparatus by the fan thrust, was used to measure shaft speed using a frequency meter. In addition, the speed

of rotation was measured y interrupting the photoelectric signal. A disk with a hole was attached to the shaft of the motor drive. The luminous flux from the lamp through the hole in the disk was fed to a photodiode. When the disc was rotated, the light beam was interrupted at a frequency $f = \pi n_R/30$, where n_R – the number of revolutions per minute. The frequency was measured with a frequency counter and an oscilloscope.

12.3.1. Separation of solid particles

The results of these preliminary experiments established the following facts:

1. The rotor of the equipment is easily rotated by the energy of the flow, sucked in by the fan.

2. The rotor speed of the equipment was 6000–6400 rpm.

3. The EMF, measured at the motor brushes in the mode of self-rotation, reached 16 V.

4. When feeding the dispersed particles with the gas flow to the input of the equipment, operating in the self-rotation mode, the rotor speed was reduced by an amount of 100–2000 rpm, depending on the concentration of the impurity supplied to the input.

5. Capture efficiency of the sputtered powders depends on their properties.

6. Capture efficiency of the sputtered powders increased significantly, if the input device was fed water vapour together with the powder. The results of points 5 and 6 are shown in Table 12.1.

7. When heated air is applied to the input of the apparatus the emf of the electric motor in the self-rotation mode increased by 0.2–0.4 V.

8. In the operating mode of the device driven by an electric motor with a speed of 5000–6000 rpm the collection efficiency significantly increased. However, this was accompanied by a decrease in the consumption of the gas suspension, passed through the apparatus. The reverse current at the entrance to the gap between the rotor and the housing of equipment increased.

The following conclusions were made on the basis of the results of experimental studies of separation of solid particles:

1. The high efficiency of centrifugal trapping of the solid particles larger than 20 μm without the high energy costs has been shown.

2. The efficiency of separation of smaller particles can be improved in several ways: by increasing the rotor speed or increasing the rotor diameter,

Table 12.1 Efficiency of collection of different dusts

Type of particulate material	Particle size, μm	Dry capture, %	Capture with vapour, %
Quartz sand	20–600	90–96	95–99
Powder of zinc oxide	10–63	85–95	95–99
Chalk	1–20	20–30	50–70

the intensification of the process of enlargement of the particles, the removal of aerosol overflow in the equipment away from the purification zone.

3. It is possible to reduce energy costs for cleaning through the use of the thermal energy of cleaned aerosol by organising energetically favorable thermodynamic conditions inside the rotor system.

12.3.2. Experimental setups for studying trapping of particles of dispersed liquids

To determine the effectiveness of different devices and their modifications for capturing aerosol particles, experiments were conducted to capture the particles of various sputtered liquids, simulating the paint. The experimental setup is shown in Fig. 12.5. In some features it is similar to the setup described in section 12.1. Machine oil or a suspension of powders in water *1* was fed in a spray bottle *2* and dispersed with a stream of compressed air flowing through the regulation valve *6* and the flow meter *5*. The aerosol jet obtained in this way was directed to the input of the trapping unit *4*. The collected aerosol particles settled on the walls of the body of the trap and in the form of liquid flowed into the collecting tank *7*.

The trapping unit had several modifications, as shown in Fig. 12.6:

Figure 12.6 *a* shows a diagram with a single trapping device TsRA-2. In this modification, the rotor is driven by an electric motor *1*. Purified gas stream is fond of the rotor in a twisted movement through openings in the hollow shaft fell into the axial zone of the rotor and through the connecting pipe rolling *4*, *5* fabric filter suction fan *6*.

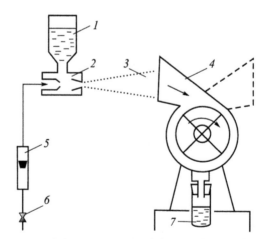

Fig. 12.5. Experimental setup for the study of the separation efficiency of dispersed liquids by various modifications of the trapping system. *1* – sprayed liquid (oil or emulsion); *2* – spray (airbrush); *3* – liquid spray jet; *4* – trapping apparatus; *5* – RS-5 rotameter to measure the flow of air; *6* – air flow control valve; *7* – trapped liquid.

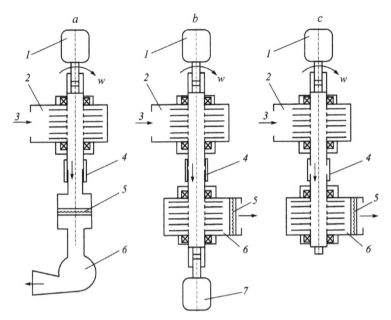

Fig. 12.6. The experimental setup for investigating the efficiency separator tion spraying liquids (top view) *1* – motor; *2* – trapping apparatus; *3* – entry of purified gas; *4* – connecting socket: sliding for *a*, *b*, rigid coupling for *c*; *5* – fabric filter; *6* – fan; *7* – fan motor for *c*.

The scheme of Fig. 12.6 *b* – *a* non-rigid tandem of two devices TsRA-2. In this scheme, each of the devices *2* and *6* had its own drive from the electric motors *1* and *7*; the system *2* was used as a trap, system *6* as a fan. The two devices were connected in the same way as in the scheme *a*, through a sliding coupling sleeve *4*. Fabric filter *5* was mounted on the output of the device *6*. While working on this scheme, each of the devices *2* and *6* could independently rotate with a frequency of 0–12 000 rpm.

The scheme of Fig. 12.6 *c* – rigid tandem of the devices TsRA-2-2. In this scheme the coupling *4*, connecting the hollow shafts, allows you to use to just one electric motor for the drive. As in the circuit *6*, the trapping apparatus *2* carries the main burden of cleaning the gas, the fan *6* is used to create vacuum at the inlet to the cleaning unit. To this end, the diameter of the rotor of the fan *6* was slightly larger than the diameter of the rotor of the trapping apparatus *2*.

When using all circuits, measurements are taken of temperature, the flow of the purified gas and speed of electric motors, and the trapped liquid and the particles not trapped in the devices and settled on the fabric filter are weighed.

12.3.3. Capture of aerosol of sprayed water emulsions

Experiments we carried out using emulsions of chalk powder $CaCO_3$ ($\rho_p = 2 \cdot 10^3$ kg/m^3, $d_p = 1$–10 μm) and zinc oxide powder ($\rho_p = 7.1 \cdot 10^3$ kg/m^3, $d_p =$

10–60 μm). Experiments were conducted according to the scheme shown in Fig. 12.5. The experimental procedure was similar to the method for capturing the powders. In the experiments, measurements were taken of the consumption rate of the sprayed emulsion, the mass and volume of the collected product, the rotor speed in the mode of self-rotation – 3500–5000 rpm.

The following results were obtained in the experiments:

1. The effectiveness of trapping particles of the dispersed emulsion was 85–90% for the emulsion of the $CaCO_3$ powder and 90–95% for the emulsion of the ZnO powder. The losses can be attributed to deposition of the powder particles on the walls of the apparatus and the removal of the particles of the finest fractions.

2. Separation of components of droplets (particles of powder and liquid) in the volume of the equipment. The collected emulsions contained 5–10% less particles in comparison with the original emulsions. This occurred due to deposition of the solid components in the device during short-term experiments.

These shortcomings can be attributed to a number of design flaws of the device, incorrectly selected capture modes, the experimental procedure and the exotic nature of the object – emulsion aerosol.

12.3.4. Capture of aerosol of sprayed oil

Experiments were conducted in systems assembled as shown in Fig. 12.6 a, b, c. Due to a more successful distribution of the incoming flow the level of the radial velocity component was greatly reduced. The experimental procedure is similar to that described in section 12.3.2. By measuring the amount of oil extracted from the aerosol it was possible to determine the degree of recovery of oil – 90–99%. In measuring the amount of oil, deposited on the fabric filter, the experiments showed a high degree of purification of the gas in removing particles of sprayed oil – 99.95%. Cleaning efficiency increased with increasing rotor speed. Along with increasing speed the hydraulic resistance of the device also increased.

One of the identified deficiencies in the course of the experiments is the pulsed release of the aerosol from the inlet nozzle of the system as a result of large-scale turbulence near the entrance. This deficiency was partially corrected using a device for a more gradient input of the purified gas. The gradient of entry of the purified gas is regulated by the degree of restriction of the inlet nozzle.

Of the tested modifications of traps the highest efficiency was shown by the setup in Fig. 12.6 c. The rigid tandem of the rotors of the trap and the fan showed the following advantages.

1. No leakage of the captured sprayed liquid without high demands on seals. This is ensured by the fact vacuum forms inside the housing of the trapping device.

2. The minimal required power of the drive for rotation due to the fact that the thermodynamic processes of expansion and contraction in the trapping apparatus and ventilation in this scheme form a closed thermodynamic cycle,

similar to that used in the turbocompressor internal combustion engine. If a cooler is placed between the trap and the ventilation system, and the input of the capture device receives a heated dusty gas, the similarity of cycles is complete. In this analogy the trapping apparatus is a turbine, the ventilator – a compressor. The source of dusty gas plays the role of the combustion chamber. At sufficiently high temperature of the dusty gas supplied to cleaning, the entire cleaning complex can act as a turbocompressor engine. For the operation of this cleaning complex the electric motor is used only to accelerate the rotor to the required speed.

3. The unique consumption characteristic of operation using this system does not require separate setting of the operating modes of the trap and the fan.

We can assume that the task of capturing the dispersed liquid from the aerosol is fundamentally solved. Only the accommodation of this scheme in the technological direction is required.

12.3.5. Capture of aerosol of sprayed paint

The first samples of capture of the particles of sprayed paint were obtained as shown in Fig. 12.5. The input of the apparatus received NTs-11 dispersed paint. Experiments have shown the fundamental possibility of catching drops of paint. However, a number of shortcomings of this scheme were also found.

1. Paint is deposited on the inner components of the structure of the apparatus, colours them and dries. The end of rotor blades and the inner surface of the housing are coloured.

2. The narrow inlet nozzle leads to subsidence of part pf the the paint on the entrance to the unit.

In line with these experimental experiences and recommendations of calculations performed in section 12.2, the experiments to capture sprayed paint were conducted in the experimental research setting, as shown in Fig. 12.7. The main part of the installation is a tandem with tightly connected rotors as shown in Fig. 12.6 c, driven in rotation by a motor and equipped with a frequency counter. The inlet nozzle of the trap apparatus is equipped with a distributor of the supply of solvent vapour from the evaporator; the required power of the evaporator is measured. The diagram of the inlet of the solvent vapour in the inlet into the unit of cleaned air is shown in Fig. 12.7. The paint sprayer is equipped with a bubbler to saturate the air spraying the paint with solvent vapours. At the output of the fan there is a Venturi tube to measure the flow of air passing through the purification unit. The temperature of the aerosol at the inlet of the unit and the temperature of the exhaust purified air were measured by thermocouples.

An important characteristic of the system is the dependence of the flow of air passing through it on some kinematic parameters. Therefore, special tests were conducted with pumping room air through the system. The flow was measured with a Venturi tube. The results of two series of measurements are shown in Fig. 12.8 as a function of the frequency of rotation of the rotors of the system.

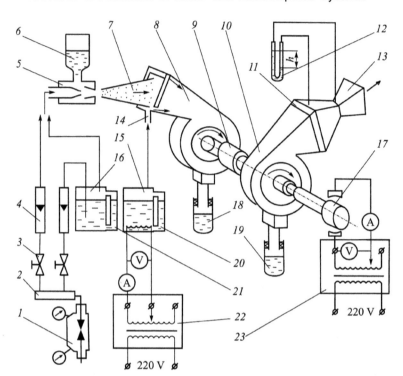

Fig 12.7. Experimental study of the separation of sprayed paint. *1* – high pressure reduction valve; *2* – receiver; *3* – valves; *4* – flow meters RS-5, RS-7; *5* – sprayer; *6* – paint + solvent; *7* – aerosol; *8* – trapping apparatus; *9* – coupling; *10* – fan; *11* – fabric filter; *12* – U-shaped manometer; *13* – Venturi nozzle (for measuring air flow through the capture system); *14* – nozzle for input of vapour of spray solvent; *15* – evaporator of paint solvent; *16* – mixer of air and solvent; *17* – electric motor; *18, 19* – containers for captured aerosol and condensate; *20, 21* – tubes with gauge glasses; *22, 23* – autotransformers.

The experiments revealed that the exhaust air temperature increases with increasing speed of rotation. This indicates the non-adiabaticity of the processes of expansion and compression of air in the unit and allows to analyze the distribution of the required power of the drive the work with the air and mechanical friction losses in the system. The result of this analysis is shown in Fig. 12.9 as the overall dependence of the drive power and the increase of the enthalpy of the air and power losses on the frequency of rotation. The graphs in Fig. 12.9 shows that the losses from friction are always less than half the power consumed, which indicates that the satisfactory quality of the design of the system. Attention is also drawn to the low level of energy consumption for air purification. For example, at a speed of 6000 rpm the pumping of about 7 litres of air per second requires about 200 watts. These results also indicate that the energy consumption can be reduced by thermal insulation of the system.

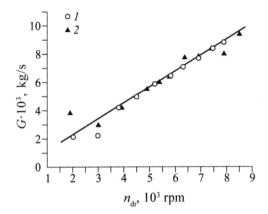

Fig. 12.8. Dependence of the airflow through the purifying unit on the frequency of rotation of the rotor. The points were obtained with a decrease (*1*) and increase (*2*) of the rotor speed

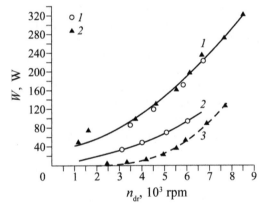

Fig. 12.9. Power consumption of the engine (*1*), the power of the gas (*2*) and the power of losses due to friction (*3*) depending on the rotational speed of the rotor of the purification system.

To prevent drying aerosol droplets of paint in the process of cleaning, prior to cleaning the aerosol should be saturated with the solvent vapors. Supersaturation is required for condensation enlargement of the paint particles. Experimental were carried out to achieve this goal through the evaporation of the solvent with the help of an additional external evaporator and feeding of the vapour in the aerosol at the entrance to the purification unit. This process was simulated with room air without the paint aerosol. Simultaneously with the determination of the flow characteristics of the evaporator the temperature of the mixture of air with solvent vapors at the inlet to the cleaning unit was measured. The measurement results are presented in Table 12.2 which also shows the calculated values of the temperature of the mixture of air with vapors of acetone and the calculated values of saturation. Table 12.2 shows that the experiment is in qualitative agreement with the calculations. However,

Table 12.2 Consumption of acetone and temperature of mixture with the air at the inlet to the unit. Speed 5000 rpm, flow rate $5.6 \cdot 10^{-3}$ kg/s

Heater voltage, V	Acetone consumption, g/cm	Temperature of the mixture, exp./calc.	Saturation, exp./calc.
0	0	299.7	
80	0.25	300.80/301.53	0.066/0.064
90	0.44	301.05/302.78	0.13/0.105
100	0.44	301.42/302.78	0.111/0.105
120	0.74	302.75/304.55	0.173/0.160
130	0.74	303.00/304.54	0.171/0.160
140	1.03	303.42/306.04	0.228/0.205

there is a tendency for the underestimation of the measured temperature of the mixture in comparison with the calculation and the experimental saturation is greater than the calculated value. The latter is obviously due to the cooling of the mixture by the walls of the inlet nozzle of the cleaning unit.

In Table 12.2 attention is drawn to the fact that, despite the large (in comparison with the air) flow rate of the solvent vapour, saturation of air with acetone vapours was not reached. This is due to heating of the mixture by the vapour heat. This disadvantage can be eliminated by artificial cooling of the mixture, which leads to an increase in saturation.

There is another way to saturated the air with the solvent vapours. For this purpose, the installation provides the saturation of the air spraying the paint in the bubbler.

The third way to obtain saturation and supersaturation by vapours of the solvent is the injection of the solvent into the inlet pipe of the cleaner in a spray like paint. Special experiments were conducted to test these methods and the results are presented in Table 12.3. The table shows that the aerosol can be saturated with acetone vapour by saturation of the atomizing air in the bubbler, and by injecting the sprayed acetone. The latter also reduces the temperature of the aerosol before cleaning, which contributes to the condensation enlargement of the aerosol particles. It is evident that some additional acetone is trapped in the capture cascade. The advantage of these saturation methods is that there is no need for additional energy consumption for the evaporation of additional acetone.

12.3.6. Trapping dispersed primer paint

In the experiments, we used a solution of GF-008 primer in acetone. The methodology of the experiments was as follows. A solution of 10 ml of primer in 100 ml of acetone was prepared. The purification unit was brought to the stationary mode with respect to the frequency of rotation. This was followed by injected 50 ml of acetone through the sprayer to bring the system to the desired temperature. The inlet of the system then received the aerosol of the soil solution. After emptying the gun it was filled with acetone in an amount of 50 ml and the final stage of the experiment was carried out – washing out the system. At all stages of the experiment real time measurements were taken of the temperature at the inlet and outlet, the cost of acetone and paint, the amount of trapped acetone and paint. The experimental results for the phase of capture of the paint are presented in Table 12.4. For comparison and obtain information about the parameters not measured in the experiments, the results of calculations under section 12.2 for the corresponding version of the purification unit are also presented. During the experiment, the measured temperature changed over time. The table gives minimum temperature values.

Experimental data on the minimum amount of captured particles were obtained by sectioning and microphotography of the sediment on the fabric filter (Fig. 12.10).

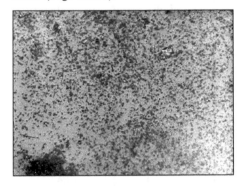

Fig. 12.10. Photo of the cake from a cloth filter, ×1000, Paint GF-4.

Table 12.3. The saturation of air purified by acetone vapour at two different frequencies of rotation

Rotor speed, rpm	5000	6200
Air flow, g/s	5.6	6.95
Consumption of acetone in bubbler, g/s	1.52	0.404
Consumption of acetone in gun, g/s	1.248	1.818
Total consumption of acetone, g/s	2.775	2.22
Capture rate of acetone, g/s	0.3	0.763
Speed of removal of acetone, g/s	2.475	1.457
Inlet temperature, K	273.27	274.4
Outlet temperature, K	305.27	299.25
Saturation in the bubbler	1.237	0.298
Total saturation at the input	2.22	1.452
Saturation after cleaning apparatus	2.02	1.001
Saturation of the exhaust air	0.454	0.307
The entrapment of acetone, %	11.8	34.4

Table 12.4. Example of experiment with capture of paint

Parameter	Experiment	Payment
Maximum diameter of cleaning rotor, m	0.1	0,1
Maximum diameter of the fan rotor, m	0.16	0,16
The height of the rotor of cleaning machine, m	0.1	0,1
Speed of rotation of unit, rpm	5000	5006
Drive power consumption, W	162	–
Flow of purified air, g/s	5.6	5,6
Paint consumption (without acetone), g/s	0.11	–
Consumption of acetone from bubbler, g/s	0.52	–
Consumption of acetone from gun, g/s	0,94	–
Total consumption of acetone, g/s	1.46	1,46
Inlet temperature, K	273.6	300
Exit temperature, K	299.6	–
Captured paint, g	5	–
Captured acetone, g	31	–
Capture rate of acetone, g/s	0.36	–
Paint deposit on the output filter, g	0.08	–
Collection efficiency of paint, %	99.2	–
Collection efficiency of acetone,%	25	–
Minimum radius of trapped particles, μm	0.2–0.4	1.047
The minimum radius of enlarged particles, μm	–	0.00575
Saturation of cleansed aerosol with acetone	–	1.269
Utilization of vapour on enlargement	–	1.10^{-24}
Pressure drop across the rotor, Pa	–	1535.9
Concentration of enlarged particles, m^{-3}	–	1.10^{10}

Table 12.5 shows the data for the other speed of rotation of the cleaning unit.

A few comments should be made on the experimental and calculated data. First, attention is drawn to the high degree of purification of the paint on removing aerosol particles. Secondly, particles smaller than 0.01 μm are captured, which is unattainable by other methods. Third, even solvent vapours are partially captured in the system. Fourth, the calculated and experimental data in the tables show the correlation between them. This indicates the possibility of using the developed method of calculation for the design of such cleaning machines. Fifth, the relatively low return of captured colours is achieved. Most of it remains on the walls of the trapping device. This is due to the short duration of the experiment and can easily be eliminated by

Table 12.5. Example of experiment with capture of paint

Parameter	Experiment	Calculations
Maximum cleaning rotor diameter, m	0.1	0.1
Maximum diameter of the fan rotor, m	0.16	0.16
The height of the rotor of the cleaning machine, m	0.1	0.1
Speed of rotation of the unit, rpm	6200	6199
Drive power consumption, W	235	–
Flow rate of purified air, g/s	6.9	6.9
Paint consumption (without acetone), g/s	0.10	–
Consumption of acetone bubbler, g/s	1.21	–
Consumption of acetone in gun, g/s	1.54	–
Total consumption of acetone, g/s	2.75	2.75
Inlet temperature, K		
Exit temperature, K	277.3	300
Captured paint, g	298.7	–
Captured acetone g	–	–
Capture rate of acetone, g/s	34	–
Paint deposit on the output filter g	0.34	–
Capture efficiency of the paint, %	0.009	–
Capture efficiency to acetone,%	99.9	–
Minimum radius of trapped particles, μm	24	–
Minimum radius of trapped particles, micron	0.2–0.4	0.938
The minimum radius of the particles coarsen, um	–	0.00330
Saturation of acetone cleansed aerosol	–	1.509
The utilization of steam on consolidation	–	$5.8 \cdot 10^{-3}$
Pressure drop across the rotor, N/m^2	–	2357.8
The concentration of particles coarsen, m^{-3}	–	$1 \cdot 10^{10}$

increasing the duration of the experiment. Sixth, the low energy consumption for cleaning should be noted. Seventh, the system, of course, requires further systematic tests.

12.4. Recommendations for industrial use of the technology and apparatus for trapping particles and vapors of sprayed paint

According to the results of this work, calculations and experiments, it can be concluded that the basic theoretical background for an effective technology of hyperfine cleaning of gases from aerosol particles is correct. The results of experimental studies have shown that by using the condensation of vapours of paint and solvent on the surface of aerosol particles it is possible to substantially reduce (from $1 \cdot 10^{-6}$ to $1 \cdot 10^{-8}$ m) the maximum size of particles captured with high efficiency (99.0–99.9%) in the apparatus of the centrifugal type. From the analysis of experiments it follows that the developed apparatus can be used to clean the air of the working premises to remove the aerosol and vapour of sprayed paints and solvents, as well as aerosols of different origins: atomized powders, liquids, mists. The effectiveness of 'dry' cleaning with the help of the developed systems for a fraction $d_p \geq 0.5$ μm at a dust content of $1 \cdot 10^3 – 1 \cdot 10^{10}$ m^{-3} and the purified gas flow rate of 1–20 g/s is 95–99%. The cleaning efficiency with condensation particle coarsening $d_p \geq$ 0.005 μm at the same flow rate of gas and an additional vapour flow rate of 0.3–3 g/s is not worse than 99%. Using these parameters, the cleaning unit has dimensions of 300×390×675 mm and weight in the range of 30 kg. The new laboratory cleaning unit can serve as a basis for the creation of industrial cleaning machines.

However, a number of factors were identified in service that require further investigation of the processes of mixing of vapour with the aerosol, condensation processes in the system, and the thermodynamic cycles. In addition, several deficiencies were found in the design of the unit, which can be removed after the experimental verification of the changes.

The following ways to improve processes in the developed purification units were also identified.

1. Approximation of the polytropic processes in the system to the isentropic ones – to reduce the energy consumption for cleaning.

2. Organization of a more gradient-like entry of the cleaned aerosol to the cleaning system. This will increase the supersaturation achieved in the apparatus and reduce the minimum limit of captured particles.

3. Cooling of the cleaned aerosol at the entrance to the system – to reduce the deadweight loss of vapour used for the enlargement of captured particles. Figure 12.11 presents the results of predictive calculations for the flow of vapour to cleaning with a reduction of temperature at the inlet.

4. Decrease of the degree of turbulence of the flows in the cleaning area – to reduce accidental releases of contaminants to the output of the cleaning unit.

Ways to improve the structures of the system have also been identified.

1. Combination of all processes in the equipment in a single rotor machine. This will simplify the design and reduce friction losses.

2. Profiling blades of the rotors. This will reduce wasteful energy for work with the purified aerosol.

3. Improved composition of the stages of the system, which will reduce the size and weight of the cleaning machine.

In line with the indicated directions for improving the cleaning system, a program was prepared to calculate the vortex flow in the vortex space around the rotor within the framework of the isentropic expansion of the

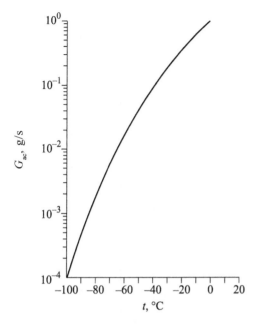

Fig. 12.11. Calculation of removal of acetone vapours from TsRA-2-2 system with decreasing input temperature of the aerosol.

compressed gas and to calculate the profiling of the rotor, which made it possible to harmonize the flows in the space around and inside the rotor. This procedure allowed to significantly increase the calculated speed at the periphery of the rotor without an increase in the large-scale turbulence and to calculate the radial profiles of velocity, temperature and supersaturation. Such a deterministic approach to the design of trapping devices has significantly improved the operating conditions of devices already created and opened prospects for the calculation and design of new devices for a wide range of tasks to capture particles from gas streams.

Tables 12.6 and 12.7 show the results of experiments with the capture of dispersed sulphur and cement. Preliminary calculations have were used to choose the modes of the apparatus.

In both experiments, the system performed capture better than it was in the experiments without calculating the space around the rotor.

12.5. Selective vapour condensation on microparticles

The well-known effect of condensation of supersaturated vapour on the microparticles with a proper understanding of the mechanism of the processes can be used as a basis to create a fundamentally new method of selective capture of ultrafine particles. For this purpose, we must learn how to implement the selective condensation of the vapour of a liquid on the microparticles of any kind for the purpose of enlarging them to further facilitate capture. The importance of the task is not doubted, as the use of this method in principle is not limited to gas purification. It can be used for the selective isolation of the fractions of dust from dusty gases, which are of particular value or a

Table 12.6 Experiment with sulphur capture in TsRA-2-2 equipment

Parameter	Dry capture		With water aerosol
Experiment variant	1	2	3
Maximum cleaning rotor diameter, m	0.1	0.1	0.1
Maximum fan rotor diameter, m	0.16	0.16	0.16
The height of the rotor of the cleaning system, m	0.1	0.1	0.1
Rotor speed, rpm	6000	6000	6000
Power consumption, W	242	242	241.5
Purified air flow, l/s	7.2	7.2	7.2
Inlet temperature, °C	26.85	23.15	11.0
Outlet temperature, °C	33.37	27.75	273.5
Weight of sprayed sulphur*, g	4.5560	3.4871	5.7121
The density of sulphur particles, g/cm	2.06	2.06	2.06
Sulphur deposit in the output filter, g	0.0104	not detected	not detected
The minimum size of captured sulphur particles, μm	1.5	1.5	1.5
Capture efficiency, %	99.77	100	100

*The maximum size of sulphur particles 50 μm was obtained by rubbing through a sieve with 50 μm mesh.

particular environmental hazard. Mathematical modelling of the condensation process shows the reasons for the selectivity of vapour condensation on solid microparticles.

12.5.1. The model of selective condensation of vapour on solid microparticles

In addition to the known dimensional selectivity due to the Kelvin–Thomson effect, here we discuss the selectivity related to the difference in the coefficient of adhesion and the binding energy of the vapour atoms on dissimilar microparticles, and the presence of additional components of vapour or gas. The validity of the model is proved by comparing the model with experimental results.

The basis of the mathematical model of the initial stage of vapour condensation on microparticles is the single-layer Langmuir representation of the processes of adsorption of the vapour atoms on the surface of

Table 12.7 Experiment with capture of cement TsRA-2-2 system

Parameter	Experiment number	
	1	2
Maximum cleaning rotor diameter, m	0.1	0.1
Maximum diameter of the fan rotor, m	0.16	0.16
The height of the rotor of the cleaning system, m	0.1	0.1
Rotor speed, rpm	6000	6000
Power consumption, W	242	242
The density of cement particles, g/cm	3.0	3.0
Flow of purified air, g/s	7.2	7.2
Inlet temperature, °C	26.8	27.0
Outlet temperature, °C	40.0	41.4
Weight of pulverized cement*, g	37.200	43.795
Spraying time, s	120	60
Cement deposit at the output filter, g	0.040	0.064
The minimum size of trapped particles, μm	1.2	1.2
Capture efficiency, %	99.89	99.85

Note. The filter at ×1000 magnification shows individual particles with the size of 1 μm and large formations consisting of micron particles adhering to each other.
* Cement pre-sieved through a sieve with 50 μm mesh.

microparticles, supplemented by taking into account the occupation of vacancies and the presence of condensation centres, traps and sinks of the adsorbate. The following assumptions were made:

1. Adsorbed particles of the vapour, molecules or atoms 'stick' only at the free part of the surface of the microparticle with a constant coefficient pf adhesion. Thermal accommodation of vapour particles on the surface of the microparticles is instantaneous.

2. The atom or molecule of the vapour, adsorbed on the surface of a microparticle, can be desorbed by thermal fluctuations of the oscillations of the surface atoms of a microparticle. The desorbed vapour molecule moves to the vapour environment.

3. The vapour molecule adsorbed on the surface of the microparticle in the process of its surface migration can be captured in a 'trap', it can form more bonds with the atoms of the microparticle and other vapour molecules, or can be absorbed inside the microparticle. The density and the 'capacity'

of 'traps', i.e. sinks for adatoms, are constant. The 'traps' are located on the surface evenly with a characteristic spacing smaller than the mean free diffusion path of adatoms on the surface of the microparticle.

Under these assumptions, the dynamics of the concentration of the vapour adsorbate on the surface of the microparticles can be described by the equation

$$\xi(\tau) = \alpha p(1 - \xi/n) - \xi/\tau + \eta(\xi_h - \xi), \tag{12.17}$$

where α is the coefficient of adhesion; $P = p(2\pi mkT)^{-1/2}$ is the flux of vapour atoms from the vapour phase on the surface of the microparticle; p is the partial pressure of vapour; m is the mass of the atom or molecule of the vapour; T is temperature; k is the Boltzmann constant; n is the density of adsorption vacancies of the vapour atoms on the surface of the microparticle, $\tau = \upsilon^{-1} \exp(\varepsilon/kT)$ is the average re-evaporation time; ε is the binding energy of vapour atoms adsorbed on the microparticle; υ is the frequency of normal vibrations of the adatoms; η i the probability of interaction of the adatoms with the centres of condensation and 'traps' on the surface of microparticles. The value of $\eta\xi$ denotes the flux of the adsorbate into the 'trap', the value of $\eta\xi_h$ is the reverse flux. The value ξ_h corresponds to the surface concentration of adsorbed particles on a microparticle, located in stable equilibrium with the vapour. For our case of condensation of the vapour in the vapour–adsorbate –condensate equilibrium, on the microparticle particle from (12.17) for $\xi_f(\tau) = 0$, $\xi_f = \xi_{hf}$ it follows

$$\xi_{hf} = \alpha_f P_{hf} / (\alpha_f P_{hf} / n_f + 1/\tau_f), \tag{12.18}$$

where

$$P_{hf} = P_\infty \exp(2\gamma\sigma_f / kTr_p)(2\pi m_f kT)^{-1/2},$$

(P_∞ is saturation pressure; σ_f is the surface energy; γ is the molecular volume of the vapour condensate; r_p is the radius of curvature of the microparticle). The index f means that all the parameters correspond to the sorption on the surface of microparticles coated with the condensate. If we assume that the process of establishing equilibrium in the adsorbate is much more rapid in comparison with the change of the partial pressure of the vapour, then under the condition of the vapour–adsorbate–condensation centre equilibrium on the microparticle $\xi(\tau) = 0$, $\xi = \xi_h$, from (12.17) we have

$$\xi_h = \alpha P_h / (\alpha P_h / n + 1/\tau). \tag{12.19}$$

A comparison of (12.18) and (12.19) gives the equation

$$P_h = \frac{P_{hf}\alpha_f / \tau}{\alpha/\tau_f + P_{hf}\alpha\alpha_f[(n_f - n)/nn_f]}, \tag{12.20}$$

meaning that the appearance of the condensate on a microparticle with radius r_p requires the partial vapour pressure different from the equilibrium pressure for the microdroplet of the vapour condensate of the same radius. For simplicity of further analysis we can ignore the difference of the density of adsorption vacancies adsorption on the surface of the microparticle and the condensate $n_f = n$. Then, from (12.20)

$$\frac{P_h}{P_{hf}} = \frac{\alpha_f}{\alpha} \exp\left[(\varepsilon_f - \varepsilon)/kT\right]. \tag{12.21}$$

If the vapour environment contains aggregates of vapour atoms with radius r_p and their stability is considered, then at $\alpha = \alpha_p$, $\varepsilon = \varepsilon_f$ from (2.21) we have $P_h = P_{hf}$.

Equation (12.21) reflects the three types of selectivity of the process of formation of the vapour condensate on the microparticles.

a. Kelvin–Thomson formula $p_h = p_\infty \exp(2\gamma\sigma_f/kTr_p)$ shows that vapour condensation takes place for all aggregates with a radius greater than r_p. This is the dimensional selectivity.

b. For foreign vapours of the microparticles in the general case $\alpha \ne \alpha_f$ $P_h \ne P_{hf}$, which denotes the selectivity of condensation associated with the difference of the coefficient of adhesion of the vapour atoms to the surface of different particles.

c. The difference of the binding energies of the vapour atoms with their own condensate and the surface of microparticles determines the third type of selectivity of condensation.

For the case of the dust-filled two-component vapour or a mixture of the vapour with the non-condensable gas it is also possible to analyze the conditions of selective condensation on the microparticles. Under the same assumptions, the dynamics of the concentration of the adsorbate of the 1st and 2nd components of the vapour on the surface of the microparticles can be written as

$$\xi_1(\tau) = \alpha_1 P_1\left(1 - \frac{\xi_1}{n_1} - \frac{\xi_2}{n_2}\right) - \frac{\xi_1}{\tau_1} + \eta_1\left(\xi_{h1} - \xi_1\right),$$

$$\xi_2(\tau) = \alpha_2 P_2\left(1 - \frac{\xi_1}{n_1} - \frac{\xi_2}{n_2}\right) - \frac{\xi_2}{\tau_2} + \eta_2\left(\xi_{h2} - \xi_2\right). \tag{12.22}$$

Equations (12.22) formally coincide with the equations (4.17) in the fourth chapter, written for the adsorption of the two-component vapour on the aggregates and nuclei with homogeneous condensation and verified by comparison with the experimental data [11]. Analysis of these equations can be found in section 4.5. The trajectories of their decisions can be analyzed from Fig. 4.26.

The trajectories of the solutions 2 in Figure. 4.26 reflect the dynamics of the relative concentrations of the adsorbate on the surface of microparticles, instantly caught in the medium of the vapour mixture. The different trajectories correspond to different initial states of the adsorbate on the microparticle. The concentrations Θ_1^s and Θ_2^s are established on the surface of the microparticle with characteristics time $1/\beta^\pm$. The dashed lines on the phase portrait plot the values of surface concentrations of the adsorbate of the components of the vapour, stable with respect to the vapour phase. Denoted are the extreme cases, corresponding to an infinitely large radius of the microparticle and the minimum radius equal to the radius of an atom or molecule $r_i = r_{ai}$. Each intermediate value of r_i corresponds to the line lying between the readings. The phase portrait in Fig. 4.26 allows the qualitative analysis of the conditions of condensation on the vapour components on the microparticles. If a stable node with coordinates (Θ_1^s, Θ_2^s) is in region I, then perhaps the condensate can form on the microparticles only by the mechanism of capillary condensation in cracks and cavities with the formation of a concave meniscus. Condensation with the consolidation of microparticles becomes impossible. If the node (Θ_1^s, Θ_2^s) is in the region IV, the condensation of both vapour components on microparticles of any size takes place. Areas II and III correspond to the condensation of only 1st or 2nd component on the microparticles of any size. The areas enclosed between the dashed lines correspond to the selective condensation with constraints on the radius of the microparticles.

Such an analysis of the phase portrait can make an important qualitative conclusion: in the case of dust contamination of the vapour mixtures the condensation of individual components on the microparticles of dust occurs at a strong mutual influence of components on each other by a mechanism based on the competition of the components on the adsorption vacancies. Small amounts of non-condensing gas in the vapour on the microparticles require higher supersaturation of the vapour as compared with the condensation of pure vapour. It should also be noted that the features of selective condensation, discussed above, relate to the difference in the coefficients of adhesion and the binding energy of vapour atoms on the surface of the microparticles, are also considered in (12.22).

So, in addition to the known dimensional selectivity of condensation dusty vapour, due to the Kelvin–Thomson effect, we have demonstrated the existence of selectivity associated with the peculiarities of 'sticking' of vapour atoms to the microparticles, the peculiarities of thermal desorption, with the competition of the components of the vapour medium at the adsorption vacancies. These results allow to predict the creation of processes of selective separation of dusty streams: a selective enlargement of dissimilar microparticles and their subsequent separation from the flow, the selective deposition of the vapour components from the vapour mixture on the microparticles. Unfortunately, the mathematical model is difficult to use for the direct forecast calculations due to the lack of specific data on the coefficient of adhesion and the binding energies. However, if the experimental data are available, this model can greatly simplify the prediction of the selectivity of condensation in complex vapour gas dusty systems.

12.5.2. Experimental apparatus and procedure

Investigation of processes of condensation of supersaturated water vapour on the ultrafine particles were carried out on an experimental laboratory setting. The installation has been designed for experimental validation of the theoretical model chosen for the selective vapour condensation on solid microparticles. The general view is shown in Fig. 12.12. The installation

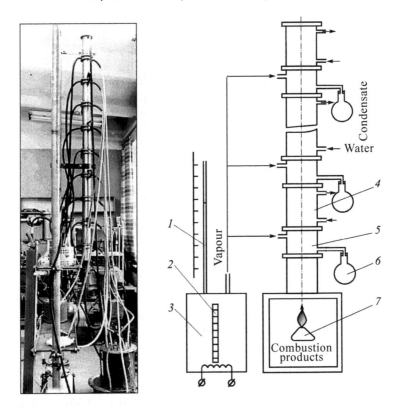

Fig. 12.12. General view and diagram of the installation. *1* – manometer; *2* – level meter; *3* – evaporator; *4* – water-cooled section; *5* – injection section; *6* – flask for the condensate; *7* – burner.

consists of ten identical water-cooled *4* and injector *5*, sections connected in series in a straight flow pipe from the source of ultrafine particles *7* and vapour source *3* with a level gauge *2* and pressure gauge *1*. Injector sections are connected to the flasks for collecting the condensate. The installation of a modular section is shown in Fig. 12.13. It is a water-cooled tube with a flange. Between the sections there are ejector spacer *2* inserted through a seal and equipped with ring injector nozzles *4* and grooves and fittings for draining the condensate *5*. The vapour is fed to the nozzle of the injector through the vapour line and the insulated nozzle *3*.

The vapour source is a sealed stainless steel vessel, equipped with a water gauge tube. On the top there is a sealed glass lid with two holes. The first hole is used to feed the vapour to the injectors, and a glass tube is inserted into the second hole and the rising level of liquid in this tube is used to measure the vapour pressure. Vapour sources are heated on the side surface with a nichrome heater, insulated from the outside.

The source of ultrafine particles was the combustion of kerosene with a lack of air. From the chemical formula for kerosene it follows that the maximum amount of carbon that can be released during incomplete combustion

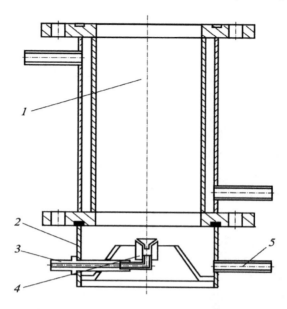

Fig. 12.13. Water-cooled section. *1* – water-cooled housing; *2* – injector spacer, *3* – vapour; *4* – ring nozzle; *5* – condensate drain.

of 1 kg of kerosene is 0.84 kg. In the combustion of kerosene in the air, the flame temperature reaches 1300 K. Partial oxidation of hydrocarbons takes place. The atomic carbon condenses in the flame to soot particles with the size of the order of a hundredth of a micron. The collected ultrafine soot particles of the flame of the kerosene burner are shown in the photograph in Fig. 12.14. Pictures were obtained by in a Neophot-2 optical microscope and a Tesla BS-500 electron transmission microscope. The average particle size was about 0.04 μm. At the vapour pressure $p = 2 \cdot 10^3$ Pa, the aerosol flow rate through the sections is $V = 0.385$ m/s. Knowing the flow rate of kerosene $\approx 1.39 \cdot 10^{-5}$ kg/s, taking the particle size $d_p = 0.04$ μm and the density of soot particles equal to the density of graphite $\rho_p = 2.3 \cdot 10^3$ kg/m^3, we can determine the concentration of soot particles: $n = 1.188 \cdot 10^{17}$ m^{-3}, which corresponds to the specific mass of soot in the flow $m_p = 9.15 \cdot 10^{-3}$ kg/m^3.

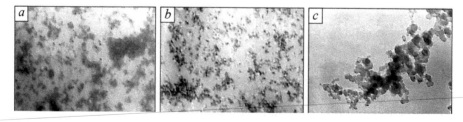

Fig. 12.14. Soot particles of the flame of the kerosene burner. *a* – optical microscope, 2nd section, ×2000; *b* – optical microscope, 5th section, ×2000; *c* – electron microscope, 7th section, ×40 000.

The idea of construction of the device is as follows. Each subsequent section receives a higher vapour flow rate to obtain an increasing supersaturation from the entrance to the pipe to the exit. Thus, each subsequent section should enlarge and capture finer and finer particles from the dusty flow.

Preliminary experiments were conducted at the facility without the injector nozzles. It was found that about 50% water vapour does not condense in the sections, and flies into the air. The analysis results of the vapour flow in the pipe, taking into account the results of preliminary experiments, were used to propose the special design of the vapour injector.

12.5.3. The test results

The described equipment was used for preliminary experiments to determine the ratio of the condensate collected from each section. This experiment allowed to determine the relationship between the amount of supplied vapour and the trapped condensate. The first five sections of the installation, operating in active mode, collect 80% of the condensate after evaporation of 5 kg of water, 12% – in the next five more passive sections, 8% of vapour escapes from the pipe into the atmosphere in the form of mist. The ambiguous effect of 'active' sections on each other was revealed. Then, experiments were conducted to capture the combustion products of kerosene. It is known that at the lack of oxygen the main products of combustion are water, carbon black (soot) particles of submicron size and CO_2.

The soot particles have a low binding energy with respect to the molecules of water vapour, are inert and easily produced. They are the product of combustion of most organic fuels. Experiments with the capture of the combustion products of kerosene showed almost complete capture of soot particles at the vapour pressure of $8 \cdot 10^3$ N/m². In order to reduce the consumption of soot aerosol and increase the concentration of trapped particles in the condensate, as well as to reduce the mutual influence of the sections on each other, the injecting vapour nozzle connected in pairs and in opposite directions.

In order to trap a particle, it is necessary to create a supersaturation at which vapour condensation starts on the particle surface. Ideally, it would be required to vaporize and condense so much water that would cover the total surface of all particles by at least a monatomic layer. In our case, $m_w = 7.165 \cdot 10^{-4}$ kg per 1 m³ of flow of combustion products with the soot particles. In a real experiment, the amount of water condensed on submicron particles is many times greater. This is due to the fact that if condensation starts on the submicron particle, the growth of the condensate layer on it is very fast and stops only when the particle is caught. The process of condensation can be managed by the optimal design solution of the structural parts of the plant in which capture takes place.

The experiments were conducted in the system to trap soot particles in the modes when suitable conditions are created for the dimensional selectivity of vapour condensation on the microparticles. To do this, a positive gradient of supersaturation of water vapour along the tube was formed. The vapour pressure was varied from $2 \cdot 10^3$ N/m^2 in the 1st section to $8 \cdot 10^3$ N/m^2 in the fifth section. The subsequent sections were working in 'passive' mode. Samples of the water condensate from different sections along with the trapped soot particles were dried on an adhesive film and photographed in an optical microscope to determine the size of the particles trapped in each section. It turned out that it is not possible to reveal any quantitative regularity in the size of trapped particles due to association of soot particles in water to form larger aggregates. The soot particles shown in Fig. 12.14 were captured at a vapour pressure of $2 \cdot 10^3$ N/m^2. Qualitatively, it is clear that with increasing numbers of sections, the size of individual particles that make up the aggregates decreases. However, it was observed that at higher vapour pressures the concentration of trapped particles in the condensate water is higher. It was also found that the number of particles trapped in the different sections differs. In the first sections the concentration of particles in the condensate increases, passes through a maximum, and then decreases. The water condensed in the 9th section, did not contain soot particles. This water had a yellowish colour and a pungent smell. Apparently, after the selective in the sections but full trapping of the soot particles, this section trapped the hydrocarbon molecules of unburned fuel. The water, condensed in the 10th section, was quite clear, did not contain soot particles and had no odour, indicating that the 100% capture of combustion products. It can be said that the idea of sequential capture all the finer and finer particles was justified.

In experiments without the combustion of kerosene, when room air was pumped through the partitioned tube, the system was highly sensitive to the room air dust content. A cigarette, lit at a distance of 3 m from the unit, within a few seconds changed the mode of operation of the plant; this was reflected in the pulsed output of water mist from the output section of the pipe.

As a result of tests of the pipe conducted in several modes, the following reliable characteristics were determined:

The inner pipe diameter, mm	100
The minimum diameter near the injectors, mm	80
The number of sections	10
The total length of pipe, cm	220
Specific consumption of water vapour for the full gas cleaning with separation of contaminants to 10 fractions, kg/m^3	0.8
Specific consumption of water vapour for the full purification of gas, without separation of contaminants into fractions, kg/m^3	0.16
Collection efficiency of particles with a size larger than 100 Å,%	100

Due to the high consumption of vapour, this system can be used only in those cases where 100% cleaning the air from dust is required, despite the higher costs, such as surgical operating rooms. However, there is every reason to believe that further improvement of the experimental setup would reduce the vapour consumption and improve collection efficiency.

Findings

1. Isentropic calculations of the space around the rotor, taking into account the compressibility of gas, the calculation of mixing with the addition of vapour to dusty gas, the calculation of vapour condensation on small dust particles and profiling of the rotor blades have raised the well–known method of dust trapping by rotary machines to a new level.

2. Selective condensation of water vapour on the solid microparticles can be used to develop the technology of selective trapping of dust particles.

List of symbols

A_p	– cross-sectional area of the particle, m^2;
B	– width of blade, m;
c	– specific heat, J/(kg·K);
C_D	– drag coefficient;
C_K	– Keninghema-Millikan correction to the calculation of C_D;
d_0	– inner diameter of the rotor, m;
d_p	– the diameter of aerosol particles, m;
F_i	– the power of binding on particle, N;
F_{ts}	– the centrifugal force, N;
F_K	– the Coriolis force, N;
F_a	– the force of aerodynamic drag in the flow particle, N;
f	– frequency, s^{-1};
G	– flow rate, kg/c;
H	– length of the body, m;
$K_1,\ K_2,\ K_3$	– the coefficients in the approximation of saturation;
k_B	– Boltzmann constant;
L_R	– the length of the rotor, m;
l	– the mean free path of gas molecules, m;
$M,\ \mu$	– molecular mass, kg;
m_p	– mass of the particle;
N	– concentration of particles cm^{-3};
n	– the density of vacancies;
n_R	– rotor speed, min^{-1};
n_{dv}	– engine speed, s^{-1};
p_v	– partial vapour pressure in the aerosol;
p	– pressure, Pa;
p_h	– vapour pressure, Pa;
R	– gas constant;
R_1	– the radius of the chamber, m;
R_2	– the outer radius of the rotor, m;
Re	– the Reynolds number;
r	– radius of curvature of the trajectory of a particle or an elementary gas volume, m;
r_{pc}	– the radius of the confined particles, m;
S	– degree of saturation;
S_w	– relative air humidity;
T	– temperature, K;
t	– temperature, °C;
U	– voltage, V;

U_r, U_φ, U_x – radial, circumferential and axial velocity components particle, m/s;

u, v, w – radial, circumferential, and the absolute velocity in the numerical calculations, m/s;

V_r, V_φ, V_x – radial, circumferential and axial velocity componentsn gas flow, m/s;

W – power, W;

x, r, φ – a cylindrical coordinate system, m, rad;

γ – the molecular volume, m³;

μ – dynamic viscosity of gas, Pa·s;

v – oscillation frequency of the adatoms;

v_g – viscosity of the gas;

ρ – density of the carrier medium (gas), kg/m³;

ρ_p – density of the particle;

ρ_c – the density of the condensate;

σ – the surface tension, N/m; surface energy, J/m²;

σ_p – the degree of saturation of the gas vapors;

τ – time, s;

ω – angular frequency of rotation, s⁻¹;

Indices:

au   – a solvent;
c   – the condensate;
f   – sorption;
g   – gas;
h   – the saturation;
m   – the minimum;
p   – a particle;
R, R_1 – rotor, the radius of the chamber;
r   – radius of the current;
v   – partial;
w   – at the entrance to the chamber;
Σ – a parameter in the mixture;
1, 2, 3   – components of the mixture;
0   – parameter of braking;
∞   – over a flat surface.

References

1. Uzhov V.N., Valdberg A., Cleaning of industrial gases from dust, Moscow, Khimiya, 1981.
2. Guide to dust and ash collectionm M. Birger, et al, Moscow, Energoatomizdat, 1983.
3. Author Cert. 325 026 RF, cl. V 01 D 47/00, Installation for flue gas cleaning, N.K. Ovchatov, 1970.
4. Bochkarev A.A., et al., in: Actual problems of physics of aerodisperse systems,

Proc. Reports at the XV All-Union Conference, Odessa, 1989, V. 1, 37.
5.   Bochkarev A.A., et al., *ibid,* 38.
6.   Sow S. Hydrodynamics of multiphase systems, Springer-Verlag, 1971.
7.   Mednikov U.P., Turbulent transport and deposition of aerosols. Moscow, Nauka, 1981.
8.   Green H., Lane V., Aerosols – dust, fumes, mists, Leningrad, Khimiya, 1968.
9.   Kutateladze S.S., et al., Aerodynamics and heat and mass transfer in swirling flows, Novosibirsk, 1988.
10.   Abramovich G.N., Applied gas dynamics, Moscow, Nauka, 1969.
11.   Chukanov V.I., Kuligin A.P., Teplofizika vysokhikh temperatur, 1987, V. 25, No. 1, 70–77.

Conclusion

The material presented in the book can be used to formulate a number of new, previously unknown scientific results relating to the physical processes accompanying the formation of dispersed systems.

1. The methodological analysis of the currently available methods of dispersion of matter shows the high efficiency of the processes of condensation in producing the disperse systems with the characteristic size of 10^{-9}–10^{-7} m in the processes of disintegration of the liquid films in the size range 10^{-7}–10^{-3} m. The analysis results are used to select the main direction of the investigations: the processes of condensation and flow in thin free liquid films.

2. A method is described for formulating the experimental studies of the processes of homogeneous and heterogeneous condensation of metal vapours in vacuum with the known and regulated superheating of the vapour, original method have been developed for measuring the thermal effect of condensation and the efficiency of condensation, and the methods for measuring the size distribution function of the particles formed in the condensation process have been formulated.

3. The experimental and theoretical studies of heterogeneous condensation show that the previously known critical phenomena of 'collapse' of condensation occur as a result of the competition of vapour and impurity gas components at the same adsorption vacancies. Occupying part of the adsorption vacancies, the impurity components prevent adsorption, nucleation, the growth of nuclei of other components and produce dispersed structures.

It has been confirmed that capillary condensation plays the key role in the formation of columnar disperse condensates and during heterogeneous condensation. In the growth of the condensates with the columnar structure the impurity components condense (even if they are not saturated) in the gaps between the columns in the structure.

4. The calculations of the conditions of homogeneous condensation of the metal vapours during the jet expansion in vacuum, including the determination of the critical conditions of the start of condensation, show that a number of studies described in the literature were carried out in the conditions in which the homogeneous nucleation and

condensation cannot take place. Nevertheless, clustered flows of metal vapours were produced in these investigations. The origin of these clusters has not been explained. The experimental studies using the non-destructive methods in the conditions in which the calculations predict homogeneous condensation confirm the expected classic nucleation. Measurements of the size distribution function of the homogeneous condensate particles showed their bimodal nature. In addition to the clusters formed in the homogeneous process, larger particles were found in vapour jets. It has been confirmed that these are clusters of heterogeneous origin formed on the solid walls of the structural components of equipment by the fluctuation mechanism. The theoretical model of their origin has been constructed. The problem of 'unlawful' clusters and the mystery of dust contamination of the vacuum systems have been solved.

Comparative calculations of the rates of homogeneous and heterogeneous nucleation show that the heterogeneous nucleation rate is tens of orders of magnitude higher than the homogeneous condensation rate. It is concluded that to produce the disperse systems, it is more efficient to use heterogeneous nucleation, and the effort of researchers should be directed to determining the conditions of heterogeneous condensation in which the disperse systems form.

5. Experimental studies of microcrystals in the columnar disperse condensates in the heterogeneous process show that their form is similar. The generalized form of all crystals growing in the condensate consists of a surface of the expanding and moving sphere. The similarity of the shapes stimulates the construction of the 'gas-dynamic' theory of the growth of crystals in which the form of the crystals depends on the speed distribution of the molecules of the vapour jet. As a result of developing this theory it has been possible to describe the orientation of the crystals in the columnar condensates, the faceting of their top part and the stresses formed in the condensates.

The formulated homological matrix of the nucleation of the forms of the crystals in the disperse condensates is based on the analysis of the local equilibrium of the surface of the nuclei of the condensed phase with the adsorbed phase and with the vapour medium. The matrix includes all types of structures of disperse condensates described in the literature.

The theory of the local normal growth of the surface of crystals is also described. The numerical calculations carried out using this theory made it possible to investigate the evolution of the shapes of the nuclei of the condensed phase during their growth from the vapour phase. The calculations demonstrate several previously known types of interaction of the adjacent crystals: mutual screening and the associated 'fight for survival', enlargement of the crystals together with the increase of the thickness of the condensate. Previously unknown phenomena were observed: a) increase of the substrate temperature initiates mass exchange between the crystals and this is the reason for the change in the structure of the condensates, observed in the experiments; b) evaporation of the screened areas of the crystals, leading to the evaporation of the base of the crystal, with the latter transferred to the vapour medium. This phenomenon is referred to as the evaporation mechanism of generation of the free clusters in heterogeneous condensation. The phenomenon has been confirmed by the experimental results.

The investigations of the combined heterogeneous condensation of the vapours of different metals and organic compounds show that in these processes there are no restrictions in the formation of completely new materials by joint packing of the components on the molecular level. A number of unique disperse materials have been produced. Suitable conditions have been created for the industrial production of a zinc–butanol nanodispersed composite characterised by useful technological properties.

6. Methods of the efficient transformation of the kinetic energy to surface energy in the cases of mutual collision of the coaxial jets of the liquid, the Taylor bells and the centrifugal system for dispersion of the liquid have been developed. The results are used to construct technologies of producing powders by the melting – disintegration of the melt – solidification of the droplets mechanism.

The coalescence of the droplets of the melt is a harmful phenomenon in the dispersion processes. Investigations of the processes of coalescence of the liquid droplets together and with the solid surfaces showed a number of phenomena supporting or preventing coalescence. The general features of the phenomenon of 'floating droplets' with the Leidenfrost phenomenon and with the film boiling of the liquid are discussed. It is shown that the second boiling crisis is not an independent physical phenomenon and results in the disruption of the vapour layer between the heated solid and the liquid due to random reasons.

The vapour interlayer, formed during contact of the boiling liquid with the superheated surface, is useful for developing a new efficient method of dispersion of the metal melt in order to produce metallic powders.

7. The concept of preliminary condensation enlargement of the nanoparticles in the dusted flows of the gases for trapping by

traditional methods has been tested. An algorithm has been developed for calculating the vortex flow of the mixture of the gas with the condensing vapours of the liquid. Calculations were carried out taking into account compressibility so that it was possible to calculate the relationships governing the supercooling of the mixture and vapour condensation on the nanoparticles. The calculations were used to construct a centrifugal system for cleaning the gases in practice. The tests of the device showed the almost complete trapping of the particles of the paint aerosols, the disperse sulphur particles and the fine fractions of cementite. A device has also been constructed for the cleaning of the products of combustion of kerosene to remove the soot particles and uncombusted organic material.

A shortcoming of this monograph is that it examines only the physical processes in the disperse systems. There are a large number of chemical processes leading to the formation of disperse systems. The authors believe that the chemical processes could provide 'tunnelling' paths for producing the disperse systems without overcoming the energy barriers described in the book which form in the 'pumping' of surface energy into matter. The absence of these barriers in the chemical processes and, more accurately, their hidden nature, may greatly facilitate the process of production of useful disperse materials and reduce the production costs.

Index

A

adsorption
  Langmuir adsorption 283, 297, 315, 316

B

butanol 246, 247, 248, 249, 250, 251, 252, 253, 254, 255, 256, 257, 258, 259, 260, 261, 265, 266, 267, 268, 269, 270, 273, 274, 275, 277, 278, 279, 280, 281

C

chemical power sources (CPS) 198, 215, 242
chemiluminescence 214, 226, 234, 243
clustered flows 114, 117, 120, 124
co-condensation 283, 292
coefficient
  adhesion coefficient 237
  coefficient of adhesion 151, 152, 155, 159
  local coefficients of condensation 171
  polytropic coefficient 126
cold 'wet' galvanizing 277, 279
condensation
  capillary condensation 70, 72, 73, 74, 80, 91, 105, 109, 110
  homogeneous condensation 114, 117, 120, 121, 123, 124, 129, 134, 136, 137, 151, 158
condensation coefficient 25, 68, 69, 77, 80, 82, 108, 109
constant
  capillary constant 328
CPS 198, 199, 200, 202, 204, 209, 211, 215, 217, 218, 231, 232, 233, 234, 242, 243

D

Debye frequency 137, 140
dimethoxyethane 221, 225

E

Milton Keynes UK
Ingram Content Group UK Ltd.
UKHW021903071024
449327UK00021B/1611